Reliability Improvement with Design of Experiments

QUALITY AND RELIABILITY

A Series Edited by

EDWARD G. SCHILLING
Founding Editor
Center for Quality and Applied Statistics
Rochester Institute of Technology
Rochester, New York

ADDITIONAL VOLUMES IN PREPARATION

Reliability Improvement with Design of Experiments

Second Edition, Revised and Expanded

Lloyd W. Condra
The Boeing Company
Seattle, Washington

CRC Press
Taylor & Francis Group
Boca Raton London New York

CRC Press is an imprint of the
Taylor & Francis Group, an **informa** business

CRC Press
Taylor & Francis Group
6000 Broken Sound Parkway NW, Suite 300
Boca Raton, FL 33487-2742

First issued in paperback 2019

© 1991 by Taylor & Francis Group, LLC
CRC Press is an imprint of Taylor & Francis Group, an Informa business

No claim to original U.S. Government works

ISBN-13: 978-0-8247-0527-5 (hbk)
ISBN-13: 978-0-367-39739-5 (pbk)

Visit the Taylor & Francis Web site at
http://www.taylorandfrancis.com

and the CRC Press Web site at
http://www.crcpress.com

To Dr. William L. Larsen, my friend, advisor, and professor of metallurgy at Iowa State University, who taught me not only how to earn a living, but how to live a life.

and

To Dr. John J. Svitak, my mentor, advisor, and friend at Bell Telephone Laboratories, whose valuable advice and youthful enthusiasm have been my guide, example, and inspiration throughout my career.

I hope both of you are as proud to have your names on this page as I am to put them here.

About the Series

The genesis of modern methods of quality and reliability can be found in a sample memo dated May 16, 1924, in which Walter A. Shewhart proposed the control chart for the analysis of inspection data. This led to a broadening of the concept of inspection from emphasis on detection and correction of defective material to control of quality through analysis and prevention of quality problems. Subsequent concern for product performance in the hands of the user stimulated development of the systems and techniques of reliability. Emphasis on the consumer as the ultimate judge of quality serves as the catalyst to bring about the integration of the methodology of quality with that of reliability. Thus, the innovations that came out of the control chart spawned a philosophy of control of quality and reliability that has come to include not only the methodology of the statistical sciences and engineering, but also the use of appropriate management methods together with various motivational procedures in a concerted effort dedicated to quality improvement.

This series is intended to provide a vehicle to foster interaction of the elements of the modern approach to quality, including statistical applications, quality and reliability engineering, management, and motivational aspects. It is a forum in which the subject matter of these various areas can be brought together to allow for effective integration of appropriate techniques. This will promote the true benefit of each, which can be achieved only through their interaction. In this sense, the whole of quality and reliability is greater than the sum of its parts, as each element augments the others.

The contributors to this series have been encouraged to discuss fundamental concepts as well as methodology, technology, and procedures at the leading edge of the discipline. Thus, new concepts are placed in proper perspective in these evolving disciplines. The series is intended for those in manufacturing, engineering, and marketing and management, as well as the consuming public, all of whom have an interest and stake in the products and services that are the lifeblood of the economic system.

The modern approach to quality and reliability concerns excellence: excellence when the product is designed, excellence when the product is made, excellence as the product is used, and excellence throughout its lifetime. But excellence does not result without effort, and products and services of superior quality and reliability require an appropriate combination of statistical, engineering, management, and motivational effort. This effort can be directed for maximum benefit only in light of timely knowledge of approaches and methods that have been developed and are available in these areas of expertise. Within the volumes of this series, the reader will find the means to create, control, correct, and improve quality and reliability in ways that are cost effective, that enhance productivity, and that create a motivational atmosphere that is harmonious and constructive. It is dedicated to that end and to the readers whose study of quality and reliability will lead to greater understanding of their products, their processes, their workplaces, and themselves.

Edward G. Schilling

Preface

It was not an easy decision for me to write a second edition of this book. I had a number of conflicting emotions. On one hand, the subject matter of a book such as this one changes constantly, and I felt the need to update it to reflect the changes. On the other hand, I was not sure that this was the best time to update it. I wanted to correct the inevitable mistakes that seem to show up after the book goes to press, but there were relatively few such mistakes, and they were so obvious that I was pretty sure they wouldn't cause problems to the readers. I was also concerned about the amount of work required to write the second edition. In the end, I decided to do it. (And, yes, it was more work than I anticipated.)

I was gratified by the reception of the first edition, and I probably learned more from it than the readers did. I learned how to organize my thoughts and communicate them to a wide audience. I learned that reviewers can be kind and understanding when they encounter mistakes. I learned that readers, including authors of similar books, can be quite helpful in pointing out areas where improvements can be made. I learned that students who use the book are serious about its contents. And I learned that there can be no final treatment of any subject. On the whole, the total experience was one not to be missed.

A number of things have changed since the first edition was published in 1993. Perhaps the most significant is the demise of the reliability discipline that had been supported by the military specification system. I had hinted at this demise in the first edition, and when it came time to review it for the second edition, I found that there was very little to be said about the system. I consider this a positive development, as do many of my friends in the various defense departments around the world. We all applaud the transition from reliability assurance by compliance to standards to reliability assurance by good design and manufacturing practices.

At the time of publication of the first edition, interest in accelerated testing began to grow significantly. In retrospect, I should have included a reference to accelerated testing in the title, since that's where I perceive that most of the interest was among the users of the book. In the second edition I've expanded the section on accelerated testing and I hope it will be more useful to its readers. Accelerated testing as a discipline is still much misunderstood, and I hope that I've been able to clear up some of the misunderstandings.

The section on design of experiments (DoE) is probably the least changed. Since the first edition, many publications have developed and updated our theoretical understanding of DoE. I have not included those developments here, because the purpose of this work was never to go too deeply into the theory.

The target audience for this book is the same as that for the first edition: engineers and managers who wish to learn enough about DoE, reliability, and accelerated testing that they can begin to apply the principles of the book to immediate improvement of their operations. Also consistent with the first edition are the goals of the book: a basic introduction to DoE, reliability, and accelerated testing for practical use by practicing engineers. As with the first edition, I believe that the best way to learn a new discipline is by example. Therefore, I've left in all the examples from the first edition and added a few more.

Throughout my career as an engineer, which now spans almost 35 years, I've worked with good people. Those people have helped me in a variety of ways, and in some form or other, many of them contributed to this book. Specifically, I'd like to thank my colleagues of the last few years, who have offered advice and support. In particular, I thank David Bond, my colleague at Boeing, who proofread the manuscript for this edition, and helped me improve its content and language in many instances. He helped me avoid a number of mistakes. As for the remaining mistakes, all I can do is quote Hilaire Belloc (in a slightly different context):

When I am gone, I hope it may be said
His sins were scarlet, but his books were read.

Lloyd W. Condra

Contents

PART III.   RELIABILITY ENGINEERING

PART IV.   ACCELERATED TESTING

PART V.   USING DESIGN OF EXPERIMENTS TO INTEGRATE RELIABILITY INTO THE ORGANIZATION

Reliability Improvement with Design of Experiments

Chapter 1

Faster . . . Better . . . Cheaper . . .

1.1 INTRODUCTION

A long time ago, back in the "good old days," engineering was considered the art and science of making trade-offs. In many respects, it still is, but the pressures of the present-day marketplace are so persistent and so relentless, that it is often hard for us to appreciate that we can make trade-offs. It used to be said, with conviction, that of quality, schedule, and price, the customer could have any two, but not all three. Today, however, customers want all three; and if they cannot get them from one source, there is always someone else willing to promise them.

To compete in this marketplace, we need all the tools, knowledge, and resources we can obtain and learn to use effectively. In this book, Design of Experiments (DoE) and reliability engineering are recommended as effective tools for meeting today's market challenges. While each is effective in its own right, they are quite formidable when combined.

Although not part of the title, considerable attention is devoted in this text to the subject of accelerated testing. Accelerated testing grew mainly out of the reliability discipline, and it has become a powerful force in improving reliability. Its combination with both DoE and reliability is explored in some detail in this book.

In this chapter, we lay the groundwork for a journey throughout the worlds of DoE and reliability, get a glimpse of what to expect from this book, and consider some reasons to pursue the disciplines contained in it.

1

1.2 DESIGN OF EXPERIMENTS
AND RELIABILITY

Design of Experiments and reliability are different types of subjects. DoE is a means of obtaining and organizing knowledge, while reliability is a feature of a product. They are alike, however, in that they are applicable across the entire range of the product design, development, and use cycle. They are also alike in that they are most effective when used as tools by professionals in disciplines not usually associated with them, such as design engineering, process engineering, and even marketing and product procurement. Reliability is best assured when it is designed into the product by the design engineer, and built into the product by production personnel, rather than calculated externally by a reliability professional. Likewise, DoE is best applied to a product design by the product engineer, or to the production process by those responsible for production, rather than by a statistician whose career goal is to conduct experiments. These very useful methods are best assimilated if one takes a long, wide, and skeptical view of them. That is, the one who learns best is one who maintains a healthy innocence about a subject, who never forgets what is known from other sources, who never lets go of common sense, and who never forgets basic questions such as [1]:

> What can go wrong?
> How will we know it went wrong?
> What will we do if it does go wrong?
> How can we prevent its going wrong or mitigating its effects?

The goals of this book are modest by the standards of reliability and DoE professionals. They are (1) to impart a working understanding of our two subjects to those who do other things for a living, so that they may incorporate them into their separate disciplines; and (2) to show how they can be used together to improve reliability at all stages of the product development and use cycle.

This introductory chapter is intended to help develop those perspectives about DoE and reliability, and to provide some reasons to learn more about them. Above all, it is about faster, better, and cheaper products.

1.3 FASTER . . .

Figure 1-1 shows a version of the well-established learning curve for manufacturing. Initially, there is a period of less-than-optimum performance, and even monetary loss, as the new product, process, or material is assimilated into factory or field operation (shown as area *A*). As time or manufacturing volume increases, a break-even point is reached, beyond which the product or process is profitable (shown as area *B*). It is assumed that profits generated in this region will exceed losses from the earlier region.

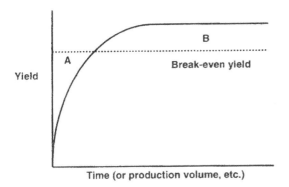

FIGURE 1-1   The traditional learning curve showing profitability after achieving break-even performance.

It is common, however, to experience the situation shown in Figure 1-2. There is not enough time to generate the information or experience to allow profitability. This is especially true if some type of reliability guarantee is required early in the product cycle. The challenge is to operate under the conditions shown in Figure 1-3. The best manufacturers and service providers must find ways to optimize all features of their products at the same time, and to do so quickly at minimum cost.

Table 1-1 shows some examples of recent improvement in times to develop new products, and in times to produce them in manufacturing. It is worth noting

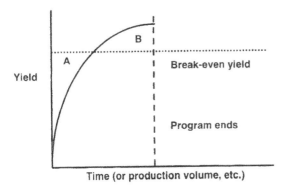

FIGURE 1-2   The traditional learning curve showing reduced time for profitable operation.

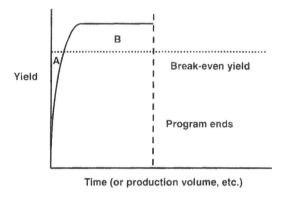

FIGURE 1-3   The new learning curve showing profitability at an earlier point in the product cycle.

that the companies listed are respected leaders in their markets, who use speed as a competitive advantage.

Modern solid-state electronic hardware capability has been governed throughout its entire history (surprisingly, only about 40 years) by Moore's law, which states that the capability of an integrated circuit will double every 18 months. While this law has reduced costs and improved capability significantly,

TABLE 1-1   Time to Develop and Time to Produce Products Rapidly Decreasing

Fast Innovators: Company	Product	Development time	
		Old	New
Honda	Cars	5 years	3 years
AT&T	Phones	2 years	1 year
Navistar	Trucks	5 years	2.5 years
Hewlett Packard	Printers	4.5 years	2 years

Fast Producers: Company	Product	Order-to-ship time	
		Old	New
General Electric	Circuit breakers	3 weeks	3 days
Motorola	Pagers	3 weeks	2 hours
Hewlett Packard	Test equipment	4 weeks	5 days
Brunswick	Fishing reels	3 weeks	1 week

Source: Fortune Magazine.

it has placed enormous pressure on the designers and manufacturers of the products, not only to continue the rate of development, but to ensure the integrity of the products while doing so

1.4 BETTER . . .

Due to constant market pressures, especially in the electronic world, product development is never "finished," but continues throughout the production life of the product [2]. We all have experienced the need to upgrade our computers to incorporate new developments and capabilities. This phenomenon impacts commercial airplanes as well. An FAA report [3] stated that commercial airplanes systems comprised of commercial-off-the-shelf components will "be in a continual state of enhancement because of commercial market pressures levied on vendors to improve product functionality and performance." In all industries, customer expectations for product integrity have never been higher. Minimum standards for quality and reliability exist, below which it is impossible to be in the market. These standards are not often written down anywhere, or even understood quantitatively by the customers, but they do exist, and they must be met. Customer frustrations when they are not met can take many forms, including documentation in technical journals [4].

1.5 CHEAPER . . .

Traditionally, direct labor costs have dominated the thinking of cost-conscious manufacturers. As Figure 1-4 shows [5], this is no longer the case. Knowledge

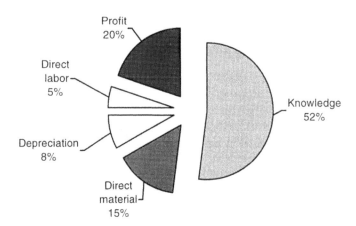

FIGURE 1-4 Knowledge is the largest single factor in the selling price of a medium-volume commercial electronics product. (From Ref. 5.)

TABLE 1-2 Incurred and Committed Costs at Various Stages of the Design-Development-Production Cycle

Process	Percent of costs incurred	Percent of costs committed
Conception	3–5	40–60
Design	5–8	60–60
Testing	8–10	80–90
Process planning	10–15	90–95
Production	15–100	95–100

Source: Ref. 6.

workers, such as product and process designers, manufacturing engineers, production control schedulers, accountants, and managers, account for over half the selling price of an electronics product.

Another way to view the same idea is shown in Table 1-2, which shows that, even though the major costs of a manufacturing program are *incurred* late in its history, most of these costs are *committed* much earlier [6]. By the time a new product reaches the production stage, there is very little potential to improve its cost. Again, the points are made that knowledge workers have the major influence on product cost, and that they must bring their own knowledge to bear quickly and efficiently on the product.

The principle product of the knowledge worker is, of course, knowledge. The challenge is to improve the productivity of this type of worker by obtaining, organizing, and applying knowledge in the most efficient way possible.

1.6 WHAT IS RELIABILITY?

To the typical user of a service or a manufactured product, quality and reliability mean that the product should perform the function for which it was acquired, when it is desired to do so. The reliability discipline has been developed over the years since World War II, with a significant contribution by military personnel and agencies. This effort has resulted in much more reliable products, but it has also resulted in an arcane subject with an esoteric language spoken by a few who are usually out of the mainstream of product design and manufacturing. Furthermore, reliability is often applied off-line from the product design and manufacturing cycle. As reliability takes on new importance, it is becoming apparent that it cannot remain a separate discipline. It must be included in all phases of the product cycle, from product proposal, through design and manufacture, and into use by the customer. It is thus important for those who are not reliability professionals to be able to understand and apply methods, within their own disciplines, which will assure reliability of the product. This is sometimes called "concurrent engineering."

Concurrent engineering is a fashionable term for the simultaneous consideration of more than one discipline in the design and development of products. Most often, it is taken to mean concurrent product design and manufacturing engineering. An illustration of the need for concurrent engineering is shown in Figure 1-5. Figure 1-5a shows a printed circuit board which was laid out by the equipment designer. It is functional, but was found to be difficult and expensive to manufacture. Figure 1-5b shows the same board as laid out by the printed circuit board assembly shop. It accommodates manufacturing and assembly concerns and is easy to build with high yield; however, it does not function reliably due to overheating caused by the concentration of high power devices together on the board. Through this experience, the manufacturer learned a valuable lesson about concurrent engineering, and designed a third board which is both functional and manufacturable.

Concurrent engineering involves more than having the design and manufac-

A   B

FIGURE 1-5   A printed circuit card laid out by A) the equipment designer and B) the same card laid out by the printed circuit card shop. (Photo courtesy of the ELDEC Corporation.)

turing engineers talk to each other. The scope of concurrent engineering must include all areas of product proposal, design, development, manufacturing, and use. It must include the disciplines of marketing, design engineering, manufacturing engineering, production, quality, and reliability. The practitioners of each of these disciplines must be able to obtain and apply enough knowledge about the others to be able to communicate with them and to accommodate their concerns.

1.7 WHAT IS DESIGN OF EXPERIMENTS?

Design of Experiments (DoE) is presented in this book as a means of producing faster, better, and cheaper products. For our purposes, it is defined as follows:

> Design of Experiments (DoE) is a method of systematically obtaining
> and organizing knowledge so that it can be used to improve operations
> in the most efficient manner possible.

By this definition, DoE can be applied to any area in which it is important to obtain and apply knowledge quickly and efficiently; which is to say, any field of human endeavor. With this definition, DoE should be seen as an opportunity for almost limitless benefits to the user. If the user concentrates on the knowledge acquired by DoE, and on what can be learned from the organization and application of that knowledge, instead of on the statistical process of obtaining and manipulating it, the benefits will be truly dramatic.

The most important fact for the DoE practitioner to remember is that the ultimate capability and responsibility for improving the process, design, or other operation lies with the person whose job assignment requires it. DoE and the statistical methods associated with it can generate no knowledge; they can only help extract and organize the knowledge already there. It is unrealistic to expect a statistician, whose main skill is in manipulation of data, to be able to draw accurate conclusions and make proper recommendations about an operation with which he or she has no hands-on experience. This book is about equipping the knowledge professional with the tools to obtain and organize the knowledge necessary for improvement without "farming out" a major portion of that task to a statistician.

DoE is usually seen as primarily applicable in the technical disciplines in manufacturing or research. It is earnestly hoped that users of this text will not so restrict their opportunities. If, as stated earlier, DoE can be used anywhere knowledge is important, then it can be applied across a broad range of disciplines, such as marketing, accounting, management, procurement, and reliability.

1.8 THE GOAL OF THIS TEXT

After this somewhat lengthy introduction, the goal of this text is stated now in three parts:

1. To present DoE as a means of obtaining and organizing information, which would otherwise not be available, so that those responsible for each discipline may understand the impacts of their decisions on others, and the output of the entire enterprise may be optimized
2. To present Reliability as an integral part of the product
3. To equip the reader to meet today's challenges by using DoE to improve reliability at all stages of the product cycle.

REFERENCES

1. R. A. Evans, Reliability Engineering, Ancient and Modern, IEEE Transactions on Reliability, 47(3):209, September 1998.
2. T. Dellin, Impact of Future Components Technology on Their Use in Military and Space Applications, Commercialization of Military and Space Electronics Workshop, February 1999.
3. Federal Aviation Administration, Report of the Challenge 2000 Subcommittee of the Federal Aviation Administration Research, Engineering, and Development Advisory Committee, p. lxv, March 6, 1996.
4. A. Katz, The Small Appliance and Computer Industry: Do they Forget Reliability Lessons? IEEE Transactions on Reliability, 48(1), March 1999, pp. 4–5.
5. Jerry Carter, IC Management, Inc., private communication, 1990.
6. D. S. Woodruff and S. Phillips, A Smarter Way to Manufacture, Business Week, April 30, 1990, pp. 110–117.

Chapter 2

Quality and Reliability

2.1 DEFINITIONS AND WHY THEY ARE IMPORTANT

Published definitions of reliability vary somewhat from industry to industry, and from user to user. Most of them involve some form of statistical reference such as [1]:

> The probability that an item will perform its intended function for a specified time interval under stated conditions;

or the even more formal [2]:

> Reliability is (1) the conditional probability, at a given (2) confidence level, that the equipment will (3) perform their intended functions satisfactorily or without failure, and within specified performance limits, at a given (4) age, for a specified length of time, or (5) mission time, when used in the manner and for the purpose intended while operating under the specified (6) application and operation environment stress.

The Reliability Society of the Institute of Electrical and Electronic Engineers (IEEE) takes a broader view [3]:

Reliability is a design engineering discipline which applies scientific knowledge to assure a product will perform its intended function for the required duration within a given environment. This includes designing in the ability to maintain, test, and support the product throughout its total life cycle. Reliability is best described as product performance over time. This is accomplished concurrently with other design disciplines by contributing to the selection of the system architecture, materials, processes, and components—both software and hardware; followed by verifying the selections made by thorough analysis and test.

All of the above are excellent definitions, but for this text, we will use a simpler definition, based on the viewpoint of the user of the product. To the typical user, there is no great distinction between quality and reliability, and the following definition captures both ideas:

A reliable product is one that does what the user wants it to do, when the user wants it to do so.

Although it may not seem so, there is a profound distinction between the formal, or statistical, definitions of reliability, and the one proposed here. The former are based on the point of view of the producer of the product or service, and the latter is based on the point of view of the customer. The distinction matters. In the former case, the emphasis is on the performance of certain analyses, calculations or tasks. In the latter, the emphasis is on making the customer happy, which is the only way to thrive, or even to survive.

Throughout this book and, more importantly, in applying its lessons to our work, we must never lose sight of this distinction. It is unsatisfactory to view quality or reliability as just another box to be checked in the design review, or as a necessary sign-off to ship the product. It is insufficient just to collect data, conduct tests and analyses, and issue reports. Although those tasks are important, and are dealt with at length in this book, they are not enough. All the while we are doing them, we must ask ourselves constantly, "Will this make the product do what the customer wants, when he or she wants it to do so?"

2.2 THE DIFFICULTY OF RELIABILITY AS A DISCIPLINE

To measure quality, we make a judgment about a product today. To measure reliability, we make judgments about what the product will be like in the future. To guarantee reliability effectively, we must do something today that will result in a product that will operate successfully in the future. We must be able to predict and, to some degree control, the future.

In theory, quality is associated primarily with production, and reliability is associated mostly with design. The separation is never complete, though, and concurrent engineering must include both quality and reliability. In addition to the requirements for concurrent design and manufacturing to produce a cost-effective, high-quality product, we must include reliability to produce a product that is cost-effective throughout its life.

It is difficult to predict and guarantee the future; it is not surprising, therefore, that manufacturers and users of some products have gotten together to define a series of actions which, when properly performed, will be viewed as resulting in a product which can be defined as reliable. These actions make up the disciplines known collectively as "reliability."

2.3   RELIABILITY CATEGORIES: MARKET

There are many ways to categorize reliability assurance methods. One of them is by the market for which the products are intended. We look at some of them here, and at examples of each.

2.3.1   The Consumer Market

The first way to categorize reliability assurance disciplines is by market. The *consumer* market, for instance, has a large volume of an identical product spread over a wide customer base. In this market, the cost of a single failure is relatively small, since it may be noticed by an individual user who is a very small part of the total user base. An example of this market is consumer electronics. Manufacturers for consumer markets can, and usually do, operate on the axiom that quality equals reliability, i.e., that the product that is best as it comes off the production line will be the most reliable in the future. Furthermore, since the cost of a single failure is relatively small, field failures can be allowed to occur at an acceptable rate, and the feedback can be used to improve reliability of future products. Producers of reliable products in this market usually have very effective statistical quality control systems, and most operate free of any outside standard except customer satisfaction.

2.3.2   The Industrial Market

A second market category is the *industrial* market. As defined here, the industrial market may have large or intermediate production volumes, but fewer and more sophisticated customers. These customers exert some form of control over their suppliers in the form of industrial standards, such as SAE, JEDEC, ASTM, UL, etc. In this market, the supplier and the customer usually negotiate some form of reliability contract. The typical form of this contract is that, if a sample of the product passes a defined set of tests, it is *qualified* and the manufacturer is allowed to build and ship the product. Although a field failure in this market is more costly than in the consumer market, some level of field failures is usually considered acceptable, and often is specified in the contract. Since production volumes are often large in this market, there is allowance for initial reliability to be low, but to improve over time.

2.3.3 Military and Similar Markets

A third category is the *military* and similar markets, where there is often only a single customer, such as the government, or a prime contractor to the government. The commercial aerospace market also fits this category. A field failure in this market is the most costly, and may result in bad publicity or even legal action. Furthermore, since production volumes are almost always low, there is no chance for improvement over time. Traditionally, reliability methods in this market have been defined by the customer in the form of government standards, such as the United States Military Standards (MIL-STDs).* These standards define not only the required reliability of a product, but how the reliability must be demonstrated, and even how the product must be manufactured, tested, and screened. There is little room for deviation. The reliability function in military and similar markets is a formal one, and is usually audited by the government.

2.4 RELIABILITY CATEGORIES: INTENT

Another way to categorize reliability methods is by intent. There are three categories of intent:

  Methods to measure and predict failures
  Methods to accommodate failures, and
  Methods to prevent failures.

2.4.1 Methods to Measure and Predict Failures

When people think of the reliability discipline today, they usually think of that part of it which measures and predicts failures. The basis of these activities is the description of failures in time with mathematical equations of a statistical nature. These equations are discussed in more detail in later chapters.

  Because of the relative difficulty or unfamiliarity of the mathematics, probability measurement usually is practiced by statisticians and others who are not directly involved in the design and production of the product. It is therefore common to emphasize the statistics; little connection is made to the actual structural changes which take place in the product at the time of failure, and little distinction is made among types of failures. This is called *probabilistic* reliability. (At the other end of the spectrum is *deterministic* reliability, which is discussed later in this chapter.)

---

* This situation is changing rapidly. See Chapter 15 for a more detailed discussion.

Failure rate distributions usually are selected, and their parameters calculated, on the basis of data obtained from actual operation of the product in service. These data then are used to predict the likelihood of failure for future identical or similar products before they are placed in service. This approach is particularly well-developed in the electronics industry, where several formal prediction methodologies are in use today. They provide equations to calculate failure rates for electronic parts based on field failure history. This type of data is difficult to obtain for several reasons: (1) There are many different users, and their ability and willingness to collect and record failure data varies greatly. (2) Operating conditions vary widely from one application to the next. (3) The amount and type of maintenance provided varies greatly. (4) It is often difficult to tell if the items actually failed, or were caused to fail by some external conditions. (5) The definition of failure varies widely; for example, some users may replace an unfailed item in the process of repairing a system which failed for other causes. (6) There is seldom time or funding available to analyze and categorize failures.

For obvious reasons, reliability prediction is an inexact science. Nevertheless, it has been applied intensively since the end of World War II, especially in the military, aerospace, and electronics industries, and is at least partially responsible for the spectacular reliability improvements experienced in those industries.

2.4.2 Methods to Accommodate Failures

Measuring and predicting failures do little to reduce their effects; failure accommodation is one way to do so. In this approach, it is assumed that failures will occur, and that pro-active measures can be taken to minimize their effects. One of the most straightforward failure accommodation methods is redundancy. If an item is likely to fail in service, the system in which it operates is designed to have two or more of the items available, so that the system will operate even if one of them fails. An example of redundancy is to have two light bulbs operating in parallel, when only one is needed.

Sometimes, a system can be designed so that, even if a failure occurs, the system can still operate with its capabilities only slightly impaired. Two disciplines used in this type of analysis are failure mode effects analysis (FMEA), and failure mode effects and criticality analysis (FMECA).

Another way to mitigate the effects of a failure is to provide a warranty. If a user knows that a failure will not cause disaster, and that the failure will be corrected at no cost to the customer, this can be an acceptable failure accommodation method. Although still widely used, warranties are losing their attractiveness as customers recognize the true costs of failures, such as inconvenience, time, and irritation, which are not addressed by warranties.

An interesting method of failure accommodation which is widely used in military and other high reliability products is *failure allocation*. In a system consisting of many items which might fail, a system *failure budget* is developed by the program manager, and pieces of this budget are allocated to the various subsystems or components. Some interesting discussions are likely to occur as those responsible for various parts of the system negotiate their individual portions of this budget.

Maintenance is another failure accommodation method. It can be either preventive, i.e., replace items according to some schedule before they fail (we hope), or reactive, i.e., replace them as they fail.

Their are numerous other failure accommodation methods in use today, but they are not discussed here. It should be apparent that a better technical understanding of the system, its operation, and its components is required here than for the purely statistical tools required for measurement and prediction. We are thus moving farther along the spectrum from probabilistic toward deterministic reliability methods. The preponderance of current technical expertise in reliability, however, is more at the system level than at the component or structural level.

2.4.3   Methods to Prevent Failures

The most desirable way to deal with failures is to design and build products that do not fail. Many studies have demonstrated the financial benefits of failure prevention, and much progress has been made in almost all industries.

Failure prevention requires a thorough understanding of the product and its components; their behavior in service (both operating and environmental); the likely mechanisms of failure at the structural level; a physical understanding of these mechanisms; and the ability to specify or control the product, its operation, and its operating environment. This expertise must be applied by product designers, manufacturing personnel, and those who use the product.

Failure prevention usually requires testing or analysis to determine the capabilities of the product and its components in a wide range of actual or simulated operating and environmental conditions. Testing almost always involves causing failures to occur in known conditions, and failure analysis is a critical required skill. The results must then be interpreted using the best and most current scientific information to describe potential failures at the structural level. This is sometimes called *physics-of-failure* analysis. The results of this analysis are then used to produce products with minimum likelihood of failure.

Obviously, failure prevention is at the *deterministic* end of the spectrum of reliability methods. It assumes that every failure has a cause, and that by understanding these causes, and the conditions that produce them, we can take steps to reduce or eliminate product failure.

2.5   SUMMARY

In the past, practitioners in the three categories described above worked in relative isolation. Statisticians developed new and more elaborate equations for reliability measurement and probabilistic failure prediction; physicists, chemists, and materials scientists conducted laboratory experiments to describe structural behavior of products; and product designers and manufacturers concentrated on shipping product. Although there have been significant improvements in the reliability of almost all products in recent years, there is room for much more improvement.

Reliability has risen to the forefront of customer concerns. Numerous studies have shown that this is the product feature most desired by customers at the beginning of the twenty-first century. Much progress has been made in quality improvement in the recent past, and a similar improvement must be made in reliability in the near future. Products must meet customer expectations, not just when they are placed in service, but for the expected life of the product. It is a requirement for everyone involved in the product to understand the need for reliability, and the impact of his or her contribution to reliability.

Clearly, the best reliability methods are those which prevent failures, and deterministic methods are required to do this. This is in no way a disparagement of probabilistic methods, however, and the most effective system is one which takes advantages of both methods. It has been said that, if the only available tool is a hammer, then every problem will be treated as a nail. Much progress has already been made in reliability with the tools available, and more tools are becoming available all the time. One of the most effective of these is Design of Experiments.

REFERENCES

1. N. B. Fuqua, Reliability Engineering for Electronic Design, Marcel Dekker, New York, 1987.
2. D. B. Kececioglu, Proceedings for the 15th Annual Reliability Testing Institute, Tucson, AZ, 1989.
3. http://www.ewh.ieee.org/soc/rs/mem.htm, The IEEE Reliability Society Home Page, 2000.

Chapter 3

Design of Experiments

3.1 THE LASER WELD EXPERIMENT

An electronics manufacturer contracted to deliver 2400 units of a small custom device at the rate of 30 per month for 80 months. A further contractual requirement was to deliver 20 prototypes by a certain date, in order to receive a nonrecurring engineering (NRE) payment of $500,000. The key manufacturing step was a laser welding process. Past experience with this process showed that, with new products, the initial yields were in the 5% range, and that as production volumes grew, the yield gradually increased to the 60–80% range. Three weeks before the prototype delivery due date, 20 samples were made, and only one was found to be acceptable.

Using recently acquired knowledge in Design of Experiments (DoE), an experiment was conducted on 36 samples at a total cost of less than $2000, and an optimum set of process parameters was determined. These settings were significantly different from the previously used "best practice" settings, which had resulted in the low yield of the first attempt. Using the new settings, 21 samples were made, and 20 were found to be acceptable. The prototype shipment was made on time, and the NRE payment was received. This payment, plus the estimated scrap savings compared to the previous settings, resulted in a financial benefit of over $750,000 over the life of the program.

Additional savings, which are not quantifiable but nevertheless real, re-

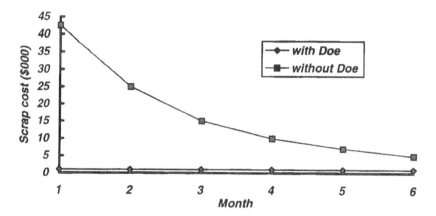

FIGURE 3-1   Cost savings with DoE for the laser weld experiment.

sulted from reduced manufacturing engineering time to work on a low-yielding process, no missed shipments, fewer production schedule disruptions, improved inventory stability, and better customer relations.

This example is dramatic, but it is not unusual. In fact, almost every company or individual who routinely uses DoE can point to a similar story in its own experience. It represents almost all the major benefits of DoE: reduced costs, rapid generation and organization of quantitative information, optimization of a process, and minimum project cost. The curves in Figure 3-1, which show comparisons of costs incurred using DoE with those which would have been incurred using traditional methods, are very similar to the "learning curves" of Figures 1-1 through 1-3.

3.2   WHAT IS A DESIGNED EXPERIMENT?

The term "Design of Experiments" sounds technically sophisticated. In a sense, the designation is unfortunate, for it conjures up images of statistical tables and arcane formulae, which are frightening even to those skilled in other technical disciplines. It connotes esoteric exercises conducted by mathematical wizards in remote laboratories. In reality, DoE is a dynamic force which can be used over a wide range of functions to transform an organization, and it is accessible to everyone.

We shall define a designed experiment as:

A technique to obtain and organize the maximum amount of *conclusive* information from the minimum amount of work, time, energy, money, or other limited resource.

There are several systems of Design of Experiments in use today. The two main types are classical and Taguchi arrays. Both methods have been used successfully for many years. For a variety of reasons which are explained later, Taguchi DoE methods are used primarily in this text. A Taguchi example is used here to illustrate the approach to problem solving with DoE. The purpose of this example is not to teach DoE skills, but to show how it can be used, and to familiarize the reader with the terms and structure of a designed experiment.

A manufacturer of electronics assemblies had a process sequence which produced highly variable results. Because of this variability, a large amount of scrap was produced. Analysis of the problem indicated that the variability was due almost entirely to a manual process step which was strongly dependent on the skills of the operators, and that one of the operators was clearly better at producing good products than the others.

Based on traditional thinking, this is a straightforward problem: the cause of the variability has been identified, and now the task is to remove that cause. Two methods are available to remove the cause of variation: (1) allow the best operator to perform all of the work, or (2) to purchase automated equipment for the process step.

Neither of the above solutions appealed to the manufacturer, however, since it is impractical and very risky to have only one operator for a key process step, and automated equipment is very expensive. Instead, the manufacturer asked, "What can I do to reduce the variability of the process step using *existing* methods and *regardless* of who performs the work?"

The manufacturer conducted an experiment in which the following six factors were evaluated: (1) raw material type, (2) raw material manufacturer, (3) process method, (4) process temperature, (5) type of material preparation, and (6) the type of manual tool used. Samples were made by several different operators, not with the goal of determining who was the best (or the worst), but to select the levels of the factors being evaluated which yielded the best results from all the operators taken as a group.

Two levels (values) for each factor were selected for experimentation, and combined in an array requiring eight separate sets of experimental samples to be built. This array is called an L_8, and is shown in Table 3-1.

The generation and use of this array are explained in a later chapter; the focus here is on the benefits available from its proper application. The data from these eight sets of experimental samples were used to select a set of factor levels which reduced variability, allowing the manufacturer to avoid the expense of automated equipment, and to continue to assign several different operators to the process. Since one of the materials selected was less expensive than the material that had been in use, the cost of the process was actually reduced, in addition to the reduction of costly scrap.

Although it would have been possible on a theoretical basis to implement

TABLE **3-1**   A Taguchi L_8 (2^7) Array

Run no.	1	2	3	4	5	6	7
1	1	1	1	1	1	1	1
2	1	1	1	2	2	2	2
3	1	2	2	1	1	2	2
4	1	2	2	2	2	1	1
5	2	1	2	1	2	1	2
6	2	1	2	2	1	2	1
7	2	2	1	1	2	2	1
8	2	2	1	2	1	1	2

Source: This, and all other Taguchi arrays, are taken from G. Taguchi and S. Konishi, *Orthogonal Arrays and Linear Graphs.* (© 1987 American Supplier Institute, used by permission.)

either of the two solutions which were suggested originally, it was not possible to do so because of cost in one case, and of unacceptable risk in the other. In this respect, then, the use of automation and of operator selection are *uncontrollable* factors. The *controllable* factors are those which the experimenter can, or wants to, control. In this example, six of the controllable factors were those investigated in the experiment. The experiment then became one of determining which levels of those controllable factors would minimize the variability of product, given that the uncontrollable factors would continue to vary as they had been doing.

3.3   THE BENEFITS OF A DESIGNED EXPERIMENT

The above example illustrates many of the benefits of a designed experiment, which are listed here.

3.3.1   Simultaneous Optimization of Several Factors

Six factors (which traditionally have been called independent variables) were considered and optimized in one experiment. The traditional one-at-a-time procedure, which most engineers learn in engineering school laboratories, is neither efficient nor effective. DoE allows the experimenter to collect, organize, and analyze data efficiently, and to use them along with good engineering judgment to draw effective conclusions.

3.3.2   Simultaneous Cost Reduction and Quality Improvement

If the proper data are collected and organized, it is often possible to select the combination of parameters which will not only improve the operation, but reduce cost. Even a result which shows that a given factor has no influence on quality is useful, because it allows the investigator to be confident in choosing the option which costs less. In this sense, the traditional conflict between cost and quality often is not a conflict at all; it arises only from lack of knowledge. The engineer should never approach a problem with the attitude that quality or reliability can be achieved only at higher cost.

3.3.3   Elimination of the Effect of the Cause Without Eliminating the Cause

In this example, the main causes of the product variation were the skill of the operator and the manual process. Those factors were not changed, but their deleterious effects were removed by changing other factors. This is a key benefit in the application of DoE to reliability, and it is dealt with in some detail in a later chapter.

3.3.4   Use of a Fractional Factorial Design to Reduce Size and Cost

In this example, six factors at two levels each were evaluated. It required only eight sets of samples for each repetition. To test all combinations would have required 2^6, or 32 sets of samples for each repetition. (In fact, the array used was capable of evaluating seven factors, but one column was left empty.)

3.3.5   Rapid Data Collection and Decision Making

Data from all factors were available after the first experiment, and they could all be considered in optimizing the process. In the one-at-a-time approach, the first factor is optimized, then the process is run for a while to see if there is sufficient improvement. There is usually not enough improvement, so a second factor is optimized in a later experiment, and so on. In practice, this type of experiment usually is repeated indefinitely, with very little improvement in the process.

3.3.6   Considering Noise in the Experiment

With any manufacturing process or product design, there usually are many factors which are not evaluated specifically, but nevertheless contribute to the outcome. These factors are called noise. In one-factor-at-a-time experiments, the experimenters usually eliminate the noise on purpose (often at significant cost), thereby sterilizing the experiment. Then, when an attempt to apply the results to production is made, the outcome is different; and a question such as, "Why does it always work when the engineers do it, but never in production?" is often asked. By contrast, in a designed experiment, the noise is purposely left in, and the process is optimized in spite of it. This is called *robust* design.

3.4 USING DESIGNED EXPERIMENTS

DoE is one of the tools used to improve quality and productivity, along with other tools such as statistical process control (SPC), and Pareto diagrams, quality function deployment (QFD). In Japan, where great improvements in quality have been made over the last 40 years, the progression has been from inspection to SPC to DoE [1]. In 1950, there was almost total dependence on inspection to ensure that good product was shipped. Product was made using the processes and materials available, and at the end, the good parts were separated from the bad. The good parts were shipped, and the bad parts were scrapped or reworked.

Statistical process control was introduced about 1950, and by 1970 had become the major method of ensuring quality. In SPC, the important controllable characteristics of a process are determined and their variation is monitored and controlled using simple statistical methods. This results in good product with only minimal inspection. It is significantly less expensive than inspection.

At the present time, Design of Experiments is the major means of achieving quality. DoE is used to quantify the optimum combinations of process materials and parameters to minimize rejects and product cost. DoE does not compete with other quality methods, but should be used in conjunction with them. A company which finds itself in the position of complete reliance on inspection to achieve quality must plan to go through a similar progression, but must do so in significantly less than 40 years.

Different companies and different industries will implement the quality tools differently. Table 3-2 shows a comparison of SPC and DoE. From this table, it may be seen that SPC is more easily implemented in a highly structured organization with large production volumes of a relatively small number of product designs. DoE, on the other hand, fits better in a less-structured company with a strong engineering organization producing small volumes of custom designs. The choice, however, is not one of *either-or*, but one of *emphasis*.

DoE is often called an *off-line* quality method; that is, it is usually done on product which is not intended for shipment.* Figure 3-2 is a diagram showing

---

* In the author's experience, many DoE projects are begun as off-line experiments, but in fact result in mostly shippable product.

TABLE 3-2 Comparison of Some Features of SPC and DoE

SPC	DoE
Most successful in a production environment	Most successful in an engineering environment
Technically easy	Technically difficult
Culturally difficult	Culturally easy
Requires long-term coordination of many functions	Can be implemented by individuals on a project basis
Does not measure financial impact of results	Can be used to measure financial impact

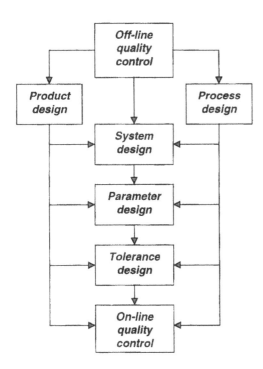

FIGURE 3-2 Diagram showing where Design of Experiments can be used for various functions in product and process design.

how DoE is used as an off-line technique in both process and product design prior to implementation in production, where *on-line* quality methods, such as SPC, are more dominant. This figure also makes the point that concurrent engineering involves reliability assessment at every step.

Figure 3–2 also shows three types of design, called system, parameter, and tolerance design. *System design* is the selection of the major design architecture or process method. Examples would be to use an internal combustion engine over a steam turbine as a source of power, or to use digital rather than analog signal-processing methods, or to use vapor phase soldering rather than wave soldering to assemble a circuit card. Often, it is considered the area where the most innovation can be applied. *Parameter design* is the selection of the optimum conditions, after the system design choices have been made.[†] Examples are the temperature of the wave solder, the diameter of a shaft, or the value of a resistor. *Tolerance design* is the specification of the allowable limits of the parameters.

Historically, American engineers have spent the overwhelming majority of their time in system design, and almost no time in parameter or tolerance design. This should not be surprising, since the primary focus of U.S. college engineering programs is on system design, and since most U.S. industries give greater recognition to design engineers than to manufacturing, quality, or reliability engineers. Although system design is in many ways the most fun, the creative application of parameter design can be quite rewarding. DoE is quite effective in producing good parameter design, but it is also applicable to system and tolerance design.

Design of Experiments is the prototype of a discipline that is best learned by doing. There are many excellent texts and articles which describe DoE methods, but the best way to learn DoE is to conduct DoE projects, beginning with simple ones and progressing to more difficult ones. The student should not be concerned with learning a great deal before beginning, and should not be uncomfortable with many unanswered questions in the beginning. DoE is as much a mind-set as a technical discipline, and this mind-set must be developed as the ability to use the technical tools grows.

3.5 PRACTICAL APPLICATION OF DoE

DoE can be used in at least six different areas in a manufacturing organization. Each is discussed briefly here, with examples where appropriate.

[†] It is tempting to think of system, parameter, and tolerance design as occurring sequentially, that is, parameter design is not begun until after the system is defined, etc. In reality, the best designs are produced when there is considerable overlap. This is another example of concurrent engineering.

3.5.1 Process Optimization and Improvement

The manual manufacturing process experiment discussed earlier in this chapter is an example of process optimization and improvement; and others are discussed later. In most manufacturing operations there are so many variables, both controllable and uncontrollable, that the selection of the optimum conditions using unstructured intuition and judgment is impossible. As an example of this, listed below are 32 factors considered important for laminating a printed wiring board [2]. Clearly, there is no way to identify and optimize them without a structured approach.

1. Analysis frequency
2. Oxide uniformity
3. Oxide reducer time
4. Oxide lamination hold time
5. Before lamination break
6. Oxide agitation
7. After reduced oxide rinsing
8. Whole process rinsing amount
9. Lamination pressure
10. Hot DI rinse temperature
11. Lamination kiss cycle
12. Oxide reducer temperature
13. Lamination heat rinse
14. Oxide temperature
15. Oxide thickness
16. Oxide reducer cuposit, %
17. Oxide panels
18. Oxide reducer age
19. Oxide reducer concentration
20. Oxide concentration
21. After oxide reducer rinse type
22. Microtech rate
23. First rinse after oxide temperature
24. Lamination prepreg resin type
25. Conditioner cuposit, %
26. Chemical clean chromate removal
27. Microtech time
28. Conditioning time
29. Predip concentration
30. Lot number
31. Predip time
32. Stripper break point

3.5.2   Product Design and Improvement

In the same way that process design parameters can be optimized, DoE can be used to optimize product designs. System level design decisions can be evaluated at the same time as design parameters and tolerances are evaluated. Not only can optimum values for design parameters be determined, but the effects of variation of these values can be determined in various operating conditions. If computer simulation or finite element analysis is used in the design process, DoE is an excellent way to reduce this effort and to organize the data from it.

3.5.3   Process Development

The laser welding process discussed at the beginning of this chapter is an example of process development with DoE. Sometimes a manufacturer must commit to deliver a product for which a process does not exist, and DoE is the only way to develop the process on schedule without unacceptable costs.

3.5.4   Capital Equipment Authorization, Specification, Acceptance, and Setup

One supplier in the automotive industry has a rule that no capital equipment purchases to improve capability will be approved unless DoE results are available which show that the existing equipment is incapable.

Often, capital equipment is purchased subject to approval in the manufacturer's factory prior to shipment. The author has used DoE to test the equipment over its entire range of operating parameters in a few days prior to acceptance. This also can be done after the equipment is received to familiarize operators and technicians with its operation and to set it up in the factory.

3.5.5   Quick Response to RFPs

Sometimes, a request for proposal (RFP) involves an operating or environmental requirement which has not been evaluated by the manufacturer. DoE can be used to produce data within the schedule requirements of an RFP, and the proposal can be more accurate and realistic.

3.5.6   Communication

An electronics company wanted to use batteries as a power source on a new product. Since the company was unfamiliar with batteries, they approached a battery manufacturer, who was equally unfamiliar with the product design requirements, to develop a test program to ensure the safety and reliability of the design. Since both companies were familiar with DoE, this common language allowed them to outline an entire development program in one day. This would have been a long and expensive process without DoE.

Another area where DoE is invaluable is in communication between management and technical personnel within the same organization. Without some form of common language, these conversations are burdensome and frustrating, and often are avoided. Technical and nontechnical personnel can use DoE terms as a kind of shorthand to discuss factors, levels, and responses being evaluated as part of a design, process, or problem-solving project.

3.6 USING DoE IN CONCURRENT DESIGN, PROCESS, AND RELIABILITY ENGINEERING

The controllable factors in a product are the design and processing parameters used to produce it. In DoE language, these parameters are called *inner array* factors. The uncontrollable factors in a product include those within the design and manufacturing process which cannot be controlled, as well as the operating and environmental conditions to which the product will be subjected after it is shipped to the customer. By evaluating these factors in an *outer array*, and by applying some reliability and accelerated testing knowledge, some predictions can be made about the long-term reliability of the product during the design and production stages. More importantly, changes can be made in these stages to improve the reliability. We are not yet ready to consider the technical details of this process, but it is outlined here and discussed in greater detail in later chapters.

3.6.1 Define Realistic System Requirements

Usually, this is done at least in part by the customer, although the supplier must be sure that every system requirement which impacts performance, cost, quality, and reliability is defined. Every product is required to operate over a range of environmental conditions for a specified length of time. This includes operating profile requirements, number of missions or required operating life, and input and output variation.

3.6.2 Define the Design Usage Environment

This includes definition of all ambient operating and storage conditions, including stresses generated by electrical and mechanical loading of the product. Examples are temperature, temperature changes, humidity, operable duty cycle, vibration, applied voltage. Both here and in step 1, the supplier and the customer must interact to assess the trade-offs available for a given product in a given application.

3.6.3   Identify Potential Sites and Mechanisms

Critical product components and component details, and potential failure modes and mechanisms, must be identified early in the design. Potential product and stress interactions must be defined and assessed. Appropriate measures must be implemented to reduce, eliminate, or accommodate expected failures. Properly designed experiments can facilitate this step.

3.6.4   Purchase Reliable Materials and Components

All purchased materials and components must be characterized, and their key characteristics must be controlled. Key characteristics may include types and levels of defects, as well as expected variation of properties, dimensions, etc. These defects and variations can have significant impact on product performance. The control system for purchased materials and components must be documented and auditable. DoE can be used to obtain and exchange information between supplier and customer.

3.6.5   Design Reliable Products, Within the Capabilities of the Materials and Manufacturing Processes

The design must be evaluated and optimized for reliability in the expected use conditions, both operational and environmental, before production begins, to identify and expose design flaws. Modeling and computer simulations can be used here, and DoE can streamline the process. Accelerated testing can be used to assess reliability.

3.6.6   Qualify the Manufacturing and Assembly Processess

All manufacturing and assembly processes must be optimized and capable of producing the product. Key characteristics of the processes must be identified and measured. DoE is major feature of this activity.

3.6.7   Control the Manufacturing and Assembly Processes

All manufacturing processes must be operated under control; that is, they must operate continuously within their control limits. Key process characteristics must be identified, and some form of control charts must be used to verify that they are within control limits. Environmental stress screening can be used to eliminate early failures and improve reliability. DoE is a key continuous improvement tool.

3.6.8 Manage the Life Cycle Usage of the Product

Closed-loop procedures must be used to collect data from tests performed in design, manufacturing process development, procurement, production, and field operation to assess the quality, reliability, and cost-effectiveness of the product on a continuous basis. The data must be used to improve the product.

In view of these tasks, it is apparent that a new relationship among the design, manufacturing, and reliability disciplines must be developed. The use of designed experiments as an educational and communications tool can facilitate this development.

3.7 SUMMARY

In this chapter, we have seen a broad range of areas in which Design of Experiments can be used to improve products, and can used by those who design and improve them to improve their processes, and the ways in which they do their jobs. The subject of concurrent engineering is mentioned in a variety of situations. This book is not about concurrent engineering, but it is about people working together to improve their efficiency, effectiveness, and profitability. It is hope that the reader will grasp the significance of DoE as a tool to do this. In that sense, it can be viewed as a unifying theory, or as a unifying discipline for an organization.

Each of the functions of the DoE organization must find ways to apply DoE uniquely to its unique problems. Fortunately, DoE is flexible, and can be applied in many different ways. In addition to being used within a discipline or a function, DoE can be used across functions, to solve problems related to more than one discipline, and to improve communication. It is in this sense that DoE is perhaps most valuable. The DoE organization must not be shy about using DoE in almost any situation, or to solve almost any problem. Of course, there are situations in which some other approach is more useful, and the DoE organization must be able to understand its limiations as well.

REFERENCES

1. K. Ishikawa, Quality Standardization: Progress for Economic Success, Quality Progress 1, 1984.
2. G. McQuarrie, DoE Discipline Leads to TQC Rewards, PC Fab, pp. 50–53, December 1992.

Chapter 4

Introduction to Design of Experiments

4.1 HISTORICAL BACKGROUND

The basic concepts of Design of Experiments were developed by Sir Ronald A. Fisher in England during the 1920s. He applied them to agricultural experiments, and published his classic book, *Design of Experiments*, in 1935. Other statisticians used his methods, and further developed them in the next few decades. One publication which deserves mention is "The Design of Optimum Multifactorial Experiments" [1], which provides the basis for a class of experiments, known as screening experiments, used to identify the variables which most strongly affect the results of an experiment. Further information on the history of DoE, some interesting anecdotes, and an extensive bibliography, can be found in Ref. 2.

American and European statisticians have continued to develop experimental design and theory since the early part of the twentieth century but, until recently, the only effective application of their methods was in agriculture. The fact that American farmers have the capability to feed almost the entire world is a testament to the success of this application. (The story of the development of hybrid seed corn is a fascinating and instructive one, even for manufacturing and design engineers.) It is appropriate to consider this for a moment, since agriculture represents almost an ideal application of DoE.

Agricultural experiments are expensive. Entire fields must be planted and tended under a variety of conditions in order to obtain data.

Agricultural experiments are time-consuming. It often requires an entire growing season, or lifetime of an animal, to obtain a single set of data points.

Agricultural experiments are subject to many uncontrolled variables. Weather, parasites, genetic conditions, and interactions among them are examples of the many uncontrollable factors which agricultural experimenters must deal with.

In consideration of these facts, it is easy to see that some form of systematic design of experiments is not only convenient, but necessary, to produce meaningful results in a finite length of time at a finite cost. It also is obvious that, at the close of the twentieth century, the above points can be made about other disciplines, such as manufacturing and the service industries. The best-known pioneer in applying DoE to manufacturing has been Dr. Genichi Taguchi.

Dr. Taguchi joined the Electrical Communication Laboratory of Japan (similar to Bell Telephone Laboratories) in 1949, and began modifying western DoE techniques to fit Japanese needs. He did this in response to the urgent need to accomplish monumental tasks with very little available time and money, a situation not unlike that which faces many businesses in today's marketplace. His goal was to develop methods that are easily learned and applied by a broad range of practicing engineers, rather than those which could by applied by a few skilled theoreticians to a few problems. A good introduction to his methods is contained in reference 3.

The success of Dr. Taguchi's approach is legendary. Experimental design is considered an essential part of a Japanese engineer's training, and thousands of practicing engineers continue to learn his methods. More important, however, is the widespread application of DoE to Japanese manufacturing. In 1976, a Toyota subsidiary reported the use of orthogonal arrays in 2700 experiments. In some companies, the number of completed experiments is a performance metric for engineers and managers.

Some controversy surrounded Taguchi's methods when they were introduced into the United States, but it has receded as statisticians have published theoretical bases for his work, and especially because the original methods continue to produce dramatic improvements in all applications.

Design of Experiments can take many forms, but its two major distinguishing elements are (1) simultaneous variation and evaluation of several factors, and (2) systematic elimination of some of the possible test combinations to reduce experimental time and cost.

4.2   THE LANGUAGE OF DoE

The language used in DoE is somewhat different from that used in traditional experiments. A complete list of terms is contained in the glossary, but some of the more important ones are listed here.

Factors are the independent variables in an experiment, sometimes called the input variables. These are the variables which are intentionally changed according to a predetermined plan. There are rules governing the assignment of factors to an array, but there may be as many factors in an experiment as the investigator is willing to pay for.

Levels are the values at which the factors are set in an experiment. These can be either parametric, i.e., *6 lb, 70 degrees*, or nonparametric, i.e., *fixture A, supplier X.*

A *combination treatment* is a unique combination of levels of the various factors in an array. It is also referred to as a *run.*

An *array* is the set of all combinations of levels of all factors evaluated in an experiment. It may be a *full factorial*, in which all possible combinations are evaluated, or a *fractional factorial*, in which some combinations are eliminated according to statistical rules.

Effects are the dependent variables in an experiment, sometimes called the output variables. They are the results of an experiment. The same set of samples may be evaluated for several effects, such as color, weight, voltage, etc. In some cases, the conditions which produce a desirable result for one effect may produce an undesirable result for a different effect.

A *response* is another name for an effect.

A *response table* is a table showing the responses from an experiment, organized according to combination treatment or levels of various factors.

An *interaction* is the influence of the variation of one factor on the results obtained by varying another factor.

Main effects are the effects of the factors in an experiment, as opposed to their interactions. In some experiments, only main effects are evaluated. If these experiments evaluate a large number of factors at only a few levels, they are sometimes called *screening experiments.*

Controllable factors are the factors which the experimenter can, or wishes to, control in an experiment. In theory, almost any factor can be controlled if the experimenter is willing to bear the cost; in practice, many factors are not controllable. Controllable factors are systematically varied, and thereby evaluated, in an experiment.

Uncontrollable factors are the factors which are not controlled. They are factors which the experimenter does not consider important, or which are too expensive to control, or which are unknown.*

---

* Sometimes, uncontrollable factors are introduced systematically into an experimental array, in order to determine the sensitivity of the controllable factors to them. This is explained in greater detail in later chapters.

Noise is the effect of all the uncontrolled factors in an experiment. In some cases, all the noise factors are known, but in most cases only some of them are known.

4.3 DoE AND SPC

It was pointed out in Chapter 3 that most companies experience a progression from statistical process control (SPC) to DoE, which might seem to imply that the two are in competition with each other. This is not the case: SPC and DoE both should be viewed as tools to improve quality. Usually, manufacturing companies learn SPC first for two major reasons: (1) The mathematics of SPC are easier, and (2) it is necessary to learn about the process and eliminate special causes of variation, that is, to get the processes under control, before the structure of DoE can be applied.

The most successful users of both methods develop their own criteria as to where each can be applied, and to how they can complement each other. A typical application is to use SPC data to determine whether or not a process needs to be improved, to use DoE to improve the process, and to measure the long term results of DoE with SPC.

Some people respond to their first exposure to DoE with a statement such as, "I don't have time to do this along with my regular job." That is an accurate statement. DoE is only effective if its methods are used *instead of* the old, inefficient methods used by most engineers prior to learning it. DoE must become a way of life for designers, manufacturing engineers, and others who wish to operate efficiently. Often, DoE tells us to do something quite different from the way we "used to do it," or from the way that seems intuitive; and it is difficult for engineers to trust the results of a designed experiment. The biggest barriers to the success of DoE are usually cultural and emotional, not technical. That is why many companies are not willing to trust the results of DoE until they get into extreme distress using traditional methods.

4.4 THREE TYPES OF EXPERIMENTS

There are, in general, three ways to obtain experimental data: one factor at a time, full factorial, and fractional factorial. Strictly speaking, all three may fit the definition of a designed experiment. For our purposes, however, we shall limit our definition of DoE to fractional factorials, since they are the most efficient. All three types are discussed here briefly to illustrate the benefits of fractional factorial designs.

4.4.1 One Factor at a Time

Most of us who spent much time in college engineering and physics labs are very familiar with one-factor-at-a-time experiments. We spent many Tuesday and Thursday afternoons with a piece of test or analytical equipment and a set of instructions which went something like this:

1. Set all knobs at setting 1. Record the results.
2. With all other knobs at setting 1, move knob A to setting 2. Record the results.
3. Move knob A to setting 3. Record the results.
4. Etc.

We did this until all combinations of settings for all knobs were observed and the results recorded. This always took 3 hours. Then we went home and spent countless hours plotting results, drawing conclusions, and writing reports. After having obtained this skill at great expense, we made sure to apply it in the laboratories and factories where we went to work after graduation.

A one-factor-at-a-time experiment is illustrated in Table 4-1. It is designed to evaluate the effects of seven variables: A, B, C, D, E, F, and G. The operational sequence is:

TABLE 4-1 A One-Factor-at-a-Time Experiment to Evaluate the Effects of Seven Variables

Run no.	A	B	C	D	E	F	G	No. of samples
1	1	1	1	1	1	1	1	3
2	2	1	1	1	1	1	1	3
3	1	1	1	1	1	1	1	3
4	1	2	1	1	1	1	1	3
5	1	2	1	1	1	1	1	3
6	1	2	2	1	1	1	1	3
7	1	2	2	1	1	1	1	3
8	1	2	2	2	1	1	1	3
9	1	2	2	2	1	1	1	3
10	1	2	2	2	2	1	1	3
11	1	2	2	2	2	1	1	3
12	1	2	2	2	2	2	1	3
13	1	2	2	2	2	1	1	3
14	1	2	2	2	2	1	2	3
Total								42

Conditions selected: $A_1B_2C_2D_2E_2F_1G_2$

1. Hold B, C, D, E, F, and G constant, and collect results with A at levels 1 and 2. This result is shown as run numbers 1 and 2 in Table 4-1. The results show that level 1 (designated A_1) is preferred.

2. Hold A at its preferred level (1) and C, D, E, F, and G at their former levels, and collect results with B at levels 1 and 2. This is shown as run numbers 3 and 4 in Table 4-1. The results show that level B_2 is preferred.

3. Hold A and B at their preferred levels, and D, E, F, and G at their former levels, and collect results with C at levels 1 and 2, as shown in run numbers 5 and 6. Level C_2 is preferred.

4. Repeat the process until all variables have been evaluated. This will require 14 runs. The optimum settings chosen for our example are A_1, B_2, C_2, D_2, E_2, F_1, and G_2.

If three samples are evaluated for each run, a total of 42 samples are required to complete the experiment. It can be pointed out that run numbers 4 and 5 are identical, as are 6 and 7, 8 and 9, 10 and 11, and 12 and 13. Thus the size of the experiment can be reduced to 27 samples. While this is true, it is rarely taken advantage of in industrial settings because this type of experiment is usually run over a long time period. Due to production schedule and cost pressures, only runs 1 and 2 are conducted in the first campaign. Since the engineers usually have good insight, this results in temporarily improved results, and further work is suspended. As time passes, results deteriorate, and the second variable is evaluated in runs 3 and 4. The process is repeated for the remainder of the runs as needed. It is not unusual for a one-factor-at-a-time experiment of this size to take many months, and still not have obtained meaningful results because of the variation of other factors. In fact, the same set of variables are evaluated in this manner for years in some manufacturing processes, with no measurable improvement.

Another serious drawback of this type of experiment is that all factors are evaluated while the other factors are at a single setting. Level 1 was chosen for A while all the other variables were set at level 1. Subsequent experimentation showed that level 2 was preferable for five of the six remaining variables. Would level 1 still be preferable for A if the other variables were set at their preferred levels? We have no way of knowing from this experiment, as it cannot evaluate interactions among factors.

4.4.2 Full Factorial

One way to get around the interaction problem is to evaluate all possible combinations of factors in a single experiment. This is called, obviously, a full factorial experiment. Figure 4-1 is a graphical representation of a full factorial experiment for our same seven factors, A through G, evaluated in the one-factor-at-a-time

				A₁				A₂			
				B_1		B_2		B_1		B_2	
				C_1	C_2	C_1	C_2	C_1	C_2	C_1	C_2
D_1	E_1	F_1	G_1								
			G_2								
		F_2	G_1								
			G_2								
	E_2	F_1	G_1								
			G_2								
		F_2	G_1								
			G_2								
D_2	E_1	F_1	G_1								
			G_2								
		F_2	G_1								
			G_2								
	E_2	F_1	G_1								
			G_2								
		F_2	G_1								
			G_2								

No. of cells = 8 x 16 =128

FIGURE 4-1 A full factorial experimental design to evaluate seven factors. Data must be collected for all cells.

experiment. Each box represents a unique combination of the levels of all seven factors.

The benefit of a full factorial experiment is that every possible data point is collected. The choice of optimum conditions is thus relatively easy. The catch is that it is very expensive and time consuming. The experiment in Figure 4-1 contains 128 cells, which means that 128 runs will have to be made. At three repetitions, or three samples per run, the experiment will require 384 samples. For all but the most wealthy organizations with no time pressures for completion, the time and money costs are prohibitive.

Another problem for the experimenter is that full factorial experiments are boring. The conclusion usually is apparent well before the experimental data are all collected, and the rest of the work adds little to the excitement of discovery.

Many experimenters have wished that they had not committed to such a task early on in their work.

4.4.3 Fractional Factorial

It should be apparent to the reader that we have saved the best for last. The third type of experimental design, the fractional factorial, allows the experimenter to obtain information about all main effects and interactions while keeping the size of the experiment manageable, and also conducting it in a single, systematic effort.

In a fractional factorial experiment, only a fraction of the possible combinations are evaluated. Figure 4-2 shows the same array of cells for the seven factors, A through G, used in the full factorial experiment. This time, however, we have selected only eight of the possible 128 combinations, which are designated with

| | | | | A_1 | | | | A_2 | | | |
| | | | | B_1 | | B_2 | | B_1 | | B_2 | |
				C_1	C_2	C_1	C_2	C_1	C_2	C_1	C_2
D_1	E_1	F_1	G_1	R_1							
			G_2								
		F_2	G_1								
			G_2					R_3			
	E_2	F_1	G_1								
			G_2						R_5		
		F_2	G_1							R_7	
			G_2							R_8	
D_2	E_1	F_1	G_1						R_6		
			G_2								
		F_2	G_1				R_4				
			G_2								
	E_2	F_1	G_1								
			G_2	R_2							
		F_2	G_1								
			G_2								

No. of cells = 8

FIGURE 4-2 A fractional factorial experimental design to evaluate seven factors. Data must be collected only for cells labeled R_1 through R_8.

TABLE **4-2** A Taguchi L_8 Array

Run no.	1	2	3	4	5	6	7
			L_8 (2^7)				
R_1	1	1	1	1	1	1	1
R_2	1	1	1	2	2	2	2
R_3	1	2	2	1	1	2	2
R_4	1	2	2	2	2	1	1
R_5	2	1	2	1	2	1	2
R_6	2	1	2	2	1	2	1
R_7	2	2	1	1	2	2	1
R_8	2	2	1	2	1	1	2
	A	B	C	D	E	F	G

R's in Figure 4-2. (The selection of these combinations, and the generation and manipulation of arrays, is discussed in more detail in later chapters.) The eight combinations are shown in a Taguchi L_8 array in Table 4-2. With this array, only eight runs are required, and with three repetitions, the experiment can be completed with only 24 total samples. The savings in cost and time are obvious, but another benefit is that the data analysis is also very manageable.

Table 4-3 shows a comparison of sample sizes required for full and frac-

TABLE **4-3** Comparison of Sample Sizes for Full and Fractional Factorial Designed Experiments

No. of factors	No. of levels	Full factorial Total runs	Taguchi fractional factorial Array name	Taguchi fractional factorial Total runs
3	2	$2^3 = 8$	L_4	4
7	2	$2^7 = 128$	L_8	8
11	2	$2^{11} = 2,048$	L_{12}	12
15	2	$2^{15} = 32,768$	L_{16}	16
4	3	$3^4 = 81$	L_9	9
5	4	$4^5 = 1,024$	L_{16}	16
1	2	$2^1\, 3^7 = 4,378$	L_{18}	18
7	3			

Runs shown for fractional factorials are for main effects only. If interactions are evaluated, fewer main effects can be evaluated in same size array.

TABLE **4-4** Comparison of Sample Size, Costs, and Time Required for a Fractional Factorial Experiment Conducted by the Author; and the Costs Which Would Have Been Incurred with a Full Factorial*

	Full factorial	Taguchi L_{18}
Number of samples	$(2 \times 3 \times 3 \times 3 \times 3 \times 3 \times 3 \times 3)$ x 3 $= 13,122$	18 x 3 = 54
Cost of samples at $50	$656,100	$2700
Time required 20 min	4374 hr = 2.29 yr	18 hr

* The costs of the full factorial are prohibitive.

tional factorial experiments of various sizes. It is obvious that, as more factors are evaluated, the advantages of fractional factorial designs become enormous. (This table is applicable for main effects only. If evaluation of interactions is desired, more space is required in the fractional factorial array.)

Table 4-4 shows the sample size, cost, and time required for an experiment conducted by the author. One factor was evaluated at two levels, seven factors were evaluated at three levels, and three samples were made for each combination treatment. The results of this experiment were used to set up a new manufacturing process in a matter of weeks. It is safe to say that the experiment would not have been run if a full factorial were the only alternative.

4.5 COMPARING EXPERIMENTAL DESIGN APPROACHES: THE AUTOMOTIVE SUSPENSION EXPERIMENT

There are good reasons to believe that, at least on a theoretical basis, fractional factorial experiments can be more effective and efficient than other methods of making technical decisions. Fortunately, the published literature contains proof that this is true in practice as well. As an example, we consider an exercise conducted by an automobile manufacturer to optimize the design of a strut suspension for a small, front-wheel-drive car, as reported by the experimenters [4], with further analysis to illustrate the points made above [5].

The purpose of the design exercise was to minimize the peak force on the strut while driving on rough surfaces by finding the best combination of seven design features.† To begin, a cross-functional team of engineers selected the val-

---

† It is premature to discuss the details of planning, conducting, and analyzing the results of the experiment. The focus here is to observe the results, and to see the benefits of the fractional factorial

TABLE 4-5   Factors and Levels for the Automotive Suspension Design Experiment.

Factor	Initial value*	DoE level 1[†]	DoE level 2[†]
Jounce bumper length (JBL)	Base	+30 mm	−30 mm
Jounce bumper length entry stiffness (JBES)	100%	300%	33%
Compression damping (CD)	719	719	719
Extension damping ED)	400	1600	100
Spring rate (SR)	100%	120%	80%
Jounce travel (JT)	100%	117%	83%
Rebound travel RT)	100%	122%	78%

* Values selected by engineering judgment.
[†] Values selected for evaluation in experimental arrays.
Source: Adapted from Refs. 4 and 5.

ues for each of the seven factors which, in their best engineering judgment using traditional design methods, would result in the minimum peak load. Those factors are listed in Table 4-5,[§] with the values of the factors selected by engineering judgment listed in the column headed "Initial value."

The peak load was calculated for this combination of factors using computer simulation, and the result was assigned a normalized value of 1.0. Results obtained from other methods were then compared to this value, with a peak force ratio less than 1.0 indicating improvement, and a peak force ratio greater than 1.0 indicating that the method yielded a poorer result.

The DoE levels shown in Table 4-6 are those chosen by the designers for evaluation in one-factor-at-a-time, full factorial, and fractional factorial analysis. Because the time required to complete each analysis using computer simulation was short (approximately 10 min.), the engineers decided to conduct a full factorial experiment. As may be noted from the discussion earlier in this chapter, the number of runs required to evaluate seven factors at two levels each is ($2^7 = $) 128. For this full factorial experiment, then, a total time of 1280 min, or a little over 21 hours, of computer time was required. The engineers considered this to be acceptable since the the cost was relatively low, and the cost of getting a wrong answer was high. Because a full factorial experiment evaluates all possible combinations of factors, it was a simple matter to select the treatment combina-

---

approach. If, after assimilating the information in Chapters 4 through 9 of this text, the reader wishes further information, a review of the bibliographical references is encouraged.

[§] The factors are described in the unique language of the automobile suspension designer; and they have little meaning for the layman. It is fortunate that, for our purposes, a complete understanding of the automotive suspension design process is not necessary.

TABLE 4-6 Optimum Levels of the Factors for the Automotive Suspension Design Experiment, as Chosen by Engineering Judgment, One-Factor-at-a Time, Full Factorial, and Fractional Factorial Methods.

Factor	Engineering judgment	One factor at a time	Full factorial (run 68)	Full factorial (run 76)	Fractional factorial (Taguchi L_8)
JBL	Base	−30	−30	−30	+30
JBES	100%	300%	33%	300%	300%
CD	719	719	719	719	719
ED	400	1600	100	100	100
SR	100%	120%	80%	80%	80%
JT	100%	83%	117%	117%	117%
RT	100%	78%	122%	122%	122%
No. of runs	—	14	128	128	8
Peak force ratio	1.00	1.485	0.447	0.447	0.540

Source: Adapted from Refs. 4 and 5.

tion that yielded the best result (the lowest peak force ratio). In this experiment, there was a tie between runs 68 and 76, which both had a peak force ratio of 0.447. The factor levels for runs 68 and 76 are shown in Table 4-6; they are considerably better than those obtained by engineering judgment.

Because of the large number of data available reported for the full factorial experiment [4], it is possible to construct both one-factor-at-a-time and fractional factorial experiments, and to compare their effectiveness and efficiency in solving this problem [5]. The results from all five different approaches are shown in Table 4-6.

Table 4-6 shows that, although the 128-run full factorial experiment yielded the lowest peak force ratio, it required 16 times as many runs (or samples) as the Taguchi L_8 experiment. The additional cost was justifiable in this instance, but it may not be in others. For the practical application of DoE, *cost should always be an important consideration in planning and conducting experiments*.

It is noteworthy that only one of the factor levels chosen by the fractional factorial was different from those chosen by run 76 of the full factorial, and only two were different from run 68. The effects of these factors apparently are not too signficant, since the peak force ratios are only marginally different. Another important consideration for the practical application of DoE is that *the focus of DoE is on what you do as a result of the experiment, rather than on detailed statistical descriptions of the data and results*.

It is also worth noting that the one-factor-at-a-time approach (which is still the most commonly used problem-solving technique in most industries) resulted

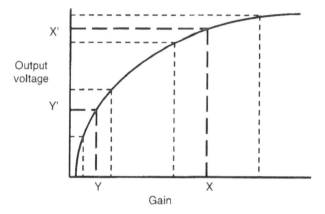

FIGURE 4-3 Output voltage vs. transistor gain for two different transistors. Even though the tolerance range of the gain of transistor X is wider than that for transistor Y, the output range of X is narrower. (Copyright 1986 American Supplier Institute, used by permission.)

in a peak force ratio almost 50% *higher* than that obtained using engineering judgment. In this case, the one-factor-at-a-time approach was not only inefficient, but it produced results that were significantly *worse* than those obtained without any formal analysis or experimentation.

4.6 MEAN AND VARIATION

The two most important features of any distribution are its location and its variation. There are many good statistical methods to describe and analyze these features, but the practicing engineer is more concerned with controlling them than with describing them. Since the time of Shewhart, we have known that it is much more important to be able to control the variation of an output than to control its location. This is illustrated by the following example.[1] Suppose a designer of an electronic circuit wishes to produce an output voltage Y′ using a transistor. The curve of transistor gain vs. output voltage is shown in Figure 4-3. The first choice of the designer is to select a transistor whose gain produces the desired output. From the curve in Figure 4-3, this value would be Y. All parameters of all products have some variation, however, and in this case, the resulting output voltage variation is wider than desired. The usual response to this problem is to specify a tighter tolerance on the transistor gain, or to test a large sample of

---

[1] This example is taken from Ref. 3: G. Taguchi, *Introduction to Quality Engineering*, American Supplier Institute, 1986. It is used by permission of the publisher.

transistors and select those which have acceptable values. These options are expensive, time-consuming, and in some cases impossible to implement. Another choice is to select a transistor with a different nominal gain, shown in Figure 4-3 as X. Since the curve of output voltage vs. gain is not linear, this results in a much narrower output variation than our previous choice. It is then a relatively simple matter to change the series resistance to shift the voltage mean from X′ to Y′. It also may be noted that the gain variation about X is actually greater than the gain variation about Y. Since a looser tolerance is thus allowed, it may be possible to obtain the transistors at a lower cost.

Several types of statistics are available to evaluate the results of designed experiments. It is almost always preferable to place a higher value on those which consider the variation of the responses, rather than just their mean.

4.7 CLASSICAL AND TAGUCHI EXPERIMENTS

There are two major types of Design of Experiments: classical and Taguchi. Both types have their proponents and opponents, and sometimes the discussions between the two have been less than uplifting. Fortunately, industrial and scientific institutions have come to realize that it is more productive to use DoE than to argue about how to use it, and much of the recent theoretical work has been directed toward accommodation of the two methods and developing the best features of both. A good understanding of this discussion can be found in Refs. 6 and 7.

Table 4-7 shows examples of eight-run classical and Taguchi arrays. The most obvious difference is that "+" and "−" are used to designate levels in classical arrays, and "1" and "2" are used for the same purpose in Taguchi arrays. Except for that difference in nomenclature, the two arrays are identical. Analyses of results are similar and, from a practical point of view, there is little difference between them. In the author's experience, the differences between the two methods are mostly differences among those who use them. Some points of comparison are discussed here.

4.7.1 Mathematical Rigor

Classical DoE analyses usually are conducted with more mathematical and statistical rigor than Taguchi analyses. Most texts on classical methods spend a great deal of time and effort on confounding and interaction of factors. Taguchi texts emphasize the evaluation of a large number of main effects, rather than interactions. Classical methods include more hypothesis testing and statistical inference methods than do Taguchi methods. In classical methods, considerable emphasis is placed on understanding the distribution of the data and using calculation methods appropriate to that distribution. Taguchi practitioners operate on the premise

TABLE 4-7   Comparison of Eight-Run Classical and Taguchi Arrays

Run no.	1	2	3	4	5	6	7
Classical array							
1	−	−	+	−	+	+	−
2	−	−	+	+	−	−	+
3	−	+	−	−	+	−	+
4	−	+	−	+	−	+	−
5	+	−	−	−	−	+	+
6	+	−	−	+	+	−	−
7	+	+	+	−	−	−	−
8	+	+	+	+	+	+	+
Taguchi array							
1	1	1	2	1	2	2	1
2	1	1	2	2	1	1	2
3	1	2	1	1	2	1	2
4	1	2	1	2	1	2	1
5	2	1	1	1	1	2	2
6	2	1	1	2	2	1	1
7	2	2	2	1	1	1	1
8	2	2	2	2	2	2	2

that no data fit any distribution exactly, and they take a much more casual approach to this consideration.

4.7.2   Choice of Optimum Results

Classical methods emphasize the choice of optimum results by calculation of response surfaces from a single, large experiment. In Taguchi methods, the emphasis is on collecting data quickly and efficiently, and iterating the experiment several times if necessary. This is sometimes called "Pick the Winner" statistics. As mentioned earlier in this chapter, the practical application of DoE emphasizes what the user does with the results over exact mathematical descriptions of the results. Often, extensive mathematical analyses result in the same course of action that would have been chosen with much less rigor.

4.7.3   Types of Experimenters

Not surprisingly, statisticians are much more prominent in the design and analysis of classical experiments than in Taguchi experiments. In Taguchi experiments, the emphasis is on the work being done by those most knowledgeable of the process or design, and who are most responsible for implementing the results.

4.7.4 Cost

Cost is an ever-present consideration in Taguchi experiments. In fact, it is not unusual for the levels of some factors to be selected purely on the basis of cost. Although classical methods are capable of evaluating costs, this consideration is not usually as prominent in reporting of results.

4.7.5 Management of Uncontrollable Factors

The contribution of uncontrollable factors is of considerable concern in Taguchi experiments. In fact, the consideration of these factors in outer arrays (explained in Chapter 10) is a major part of Taguchi methods. By contrast, classical methods are more concerned with controllable factors.

Based on the above points, it may be obvious to the reader that the two types of DoE have different spheres of application, albeit with significant overlap. In general, classical DoE is preferred in applications where the cost of the experiment is high; or where the time required is long and options for iteration are limited; or where a precise and rigorous result is required; or where uncontrollable factors can be limited; or where the emphasis is on the results obtained, rather than on the process knowledge to be gained by conducting several iterations. Examples of such applications are space shuttle experiments, some medical experiments, and some basic research projects.

Taguchi methods are more applicable where there are many uncontrollable factors; where it is important for the experimenter to obtain results quickly; and where it is possible to iterate the experiment several times.

It is the author's conclusion, based on experience with both types of DoE, that Taguchi methods are more easily learned and applied in manufacturing and product design engineering environments. Therefore, Taguchi experiments have been chosen for presentation in the remainder of this book. More comprehensive information can be obtained from Refs. 8–16.

REFERENCES

1. R. L. Plackett and J. P. Burman. The Design of Optimum Multifactorial Experiments, Biometrika, 33:305-325, 1946.
2. S. Bisgard, Industrial Use of Statistically Designed Experiments: Case Study References and Some Historical Anecdotes, Quality Engineering, 4:(4)547–562, 1992.
3. G. Taguchi, Introduction to Quality Engineering, American Supplier Institute, Dearborn, MI, 1986.
4. M. F. Lamps and E. C. Ekert, Improving the Suspension Design Process by Integrating Multibody System Analysis and Design of Experiments, SAE Technical Paper No. 930264, Vehicle Suspensions and Steering Systems (SP-952), SAE, Warrendale, PA, 1993.

5. L. W. Condra, Value-Added Management with Design of Experiments, Chapman and Hall, London, 1995.
6. Quality and Reliability Engineering International, vol. 4, no. 2 April–June 1988. This is a special issue devoted to Design of Experiments.
7. Journal of Quality Technology, vol. 17, no. 4, October 1985. The entire issue is dedicated to a discussion of Taguchi methods.
8. Y. Wu, and W. H. Moore, Quality Engineering: Product and Process Design Optimization, American Supplier Institute, Dearborn, MI, 1985.
9. P. J. Ross, Taguchi Techniques for Quality Engineering, McGraw-Hill, New York, 1988.
10. M. S. Phadke, Quality Engineering Using Robust Design, Prentice-Hall, Englewood Cliffs, NJ, 1989.
11. T. Mori, The New Experimental Design: Taguchi's Approach to Quality Engineering, ASI Press, Dearborn, MI, 1990.
12. S. R. Schmidt, and R. G. Launsby, Understanding Industrial Designed Experiments, (2nd edition), CQG Ltd. Printing, Longmont, CO, 1988.
13. G. Taguchi, System of Experimental Design, ASI Press and Kraus International Publications, 1987.
14. R. L. Mason, R. F. Gunst, and J. L. Hess, Statistical Design of Experiments: with Applications to Engineering and Science, Wiley-Interscience, New York, 1990.
15. R. H. Lochner, and J. E. Matar, Designing for Quality: an Introduction to the Best of Taguchi and Western Methods of Statistical Experimental Design, Chapman and Hall, London, 1990.
16. G. Taguchi, and D. Clausing, Robust Quality, Harvard Business Review, 90(1):65–75, January–February, 1990. Reprint no. 90114.

Chapter 5

Conducting a Main Effects Experiment

5.1 INTRODUCTION

This chapter begins with an explanation of the use of DoE to set up and optimize the operating parameters in developing a new hybrid microcircuit assembly process. There are many possible factors to be considered, and they may interact with each other; but since the process is new, we want to obtain information about a wide range of parameters. Therefore, we shall use a Taguchi L_{18} array to evaluate as many factors as possible, without considering interactions among them. This is sometimes called a *screening* experiment, or a *main effects* experiment.

The components to be attached to the substrate are a MOSFET integrated circuit and two resistors. They are placed on a ceramic substrate with a metallized conductor pattern, which has previously had solder paste screened onto the component attachment sites. The substrate is then baked in an oven to dry it, and placed on a conveyor belt which passes through a furnace. The furnace melts the solder, reflows it, and attaches the components mechanically and electrically to the metallized pattern on the substrate. The atmosphere inside the furnace is controlled to prevent oxidation of the substrate and solder during reflow.

The setup of a main effects experiment is uncomplicated. The design and setup of experiments with interacting factors is dealt with in later chapters.

5.2 SELECTING FACTORS AND LEVELS

Six factors were chosen for evaluation in this experiment. Their levels are shown in Table 5-1, and they are described briefly here:

   A. *Screen (two levels)*. This is the screen through which the solder paste is forced in applying it to the component attachment sites on the substrate. Two different combinations of mesh, wire diameter, and emulsion thickness were evaluated as levels 1 and 2. Note that this is a nonparametric factor; that is, no single numerical value can be assigned to a given level.

   B. *Profile (three levels)*. This is the time-temperature profile of the assemblies as they pass through the furnace. The furnace has eight zones, each of which can be controlled independently, and the profile is the result of the settings of all eight zones. The three levels evaluated in this experiment were differentiated only by changing the settings of the fifth and sixth zones, with the settings of the other zones held constant. Note that, although the *assembly temperature* is the parameter affecting the results, we are using the *furnace setting* as the factor level. This is because the furnace setting is the parameter under the control of the operator. Obviously, the furnace must be kept calibrated to assure that relationship between the temperature and the furnace setting is known.

   C. *Atmosphere (three levels)*. The atmosphere inside the furnace can be

TABLE **5-1**   Description of Factors and Levels for Hybrid Component
Attachment Experiment

Factor	Level 1	Level 2	Level 3
A: Screen	80 mesh, .0037 in. wire diameter, 0.003 in. emulsion	80 mesh, 0.0020 in. wire diameter, 0.002 in. emulsion	
B: Profile, °C	150–200–250–300–350–350–290–240	150–200–250–300–370–370–290–240	150–200–250–300–390–390–290–240
C: Atmosphere	100% nitrogen	50% nitrogen–50% hydrogen	100% hydrogen
D: Gas flow rate	40 cfh	60 cfh	80 cfh
F: Belt speed	4.5 in./min	5.5 in./min	6.0 in./min
G: Drying temperature	Room temp.	50°C	90°C

controlled to reduce oxidation of the assemblies by changing the relative amounts of nitrogen and hydrogen.

D. *Gas flow rate (three levels)*. Another type of atmosphere control is the rate at which the gases flow into the furnace.

E. *Empty*. Although the L_{18} array allows the evaluation of eight total factors, only six were evaluated here. This is because, for logistical problems not related to the design of the experiment, two of the factors originally chosen could not be evaluated. This means that two columns of the array will be left empty, and those two columns were designated E and H. There is a benefit to doing this because the two empty columns can be used to evaluate the overall variation, or noise, in the data. This will help to tell us if we have considered all relevant factors. This idea is dealt with later in this chapter.

F. *Belt speed (three levels)*. The speed at which the belt conveys the samples through the furnace was evaluated at three levels. This is a good example of a parametric factor.

G. *Drying temperature (three levels)*. The temperature at which the samples were dried prior to reflow was also evaluated at three levels.

H. *Empty*.

5.3 ASSIGNING FACTORS TO THE ARRAY

The Taguchi L_{18} array selected for this experiment is shown in Table 5-2, with the factors assigned to the columns. It may be noted that columns E and H are empty. Since this is a main effects array, the interactions among the factors are spread more or less equally among the columns, and the factors may be assigned arbitrarily to the columns. This is not generally the case with other arrays, however, and rules must be followed in assigning factors. This subject is discussed in detail in Chapters 6 and 7.

The array in Table 5-2 is called an L_{18} array because it has 18 runs, or combination treatments. As a general rule, the number of runs in a Taguchi array is denoted by the subscript. This array is capable of evaluating one factor at two levels, and seven factors at three levels each, as indicated by the notation $(2^1 3^7)$. Other Taguchi arrays are shown in Appendix A, and they are also published separately [1]. Each row in the array represents a run, or treatment combination. The first run was conducted with all factors at level 1, or $A_1B_1C_1D_1E_1F_1G_1H_1$. The second was conducted with factors A and B at level 1 and the remainder at level 2, and is designated $A_1B_1C_2D_2E_2F_2G_2H_2$. For columns E and H, the designation is meaningless, since nothing was actually changed in going from level 1 to level 2, or from level 2 to level 3.

TABLE 5-2   A Taguchi Array with Factors Assigned to Columns for the Hybrid Component Attachment Experiment

				L_{18} (2^{1}3^7)				
Run	A Screen	B Profile	C Atmosphere	D Gas flow rate	E Empty	F Belt speed	G Drying temp.	H Empty
1	1	1	1	1	1	1	1	1
2	1	1	2	2	2	2	2	2
3	1	1	3	3	3	3	3	3
4	1	2	1	1	2	2	3	3
5	1	2	2	2	3	3	1	1
6	1	2	3	3	1	1	2	2
7	1	3	1	2	1	3	2	3
8	1	3	2	3	2	1	3	1
9	1	3	3	1	3	2	1	2
10	2	1	1	3	3	2	2	1
11	2	1	2	1	1	3	3	2
12	2	1	3	2	2	1	1	3
13	2	2	1	2	3	1	3	2
14	2	2	2	3	1	2	1	3
15	2	2	3	1	2	3	2	1
16	2	3	1	3	2	3	1	2
17	2	3	2	1	3	1	2	3
18	2	3	3	2	1	2	3	1

Note that columns E and H are empty.

5.4 SELECTING EFFECTS AND ANALYSIS CRITERIA

The actual experiment involved the evaluation of four effects for each of the three components. For instructional purposes, however, this discussion will be limited to the evaluation of two effects on one of the resistors. They were (1) percent voids in the solder, and (2) shear strength of the solder joints. Since three repetitions of the experiment were made for each run, the total number of data points is

(18 runs) $\times$ (3 repetitions) $\times$ (1 component) $\times$ (2 effects) = 108

There were three repetitions, or three sample substrates, for each run. Thus the entire experiment required 54 substrates. (It is noted, however, that a significantly larger number of data points from these same substrates was actually collected.)

It is also important to note that the two effects are, for analytical purposes, independent of each other. The results for percent voids, for example, may be evaluated independently of those for shear strength. Thus, two separate and independent analyses of the data from this experiment were conducted, and the effects of the various factors were assessed via two different criteria.

The first assessment criterion, percent voids, was measured nondestructively with a technique called scanning laser acoustic microscopy. The best possible joint would have 0% voids, and the worst possible one would have 100% voids (in fact, there would be no attachment at all). Shear strength was tested destructively, and was measured in kilograms.

5.5   THREE TYPES OF ANALYSIS STATISTICS

For the first effect, per cent voids, the smallest numbers are associated with the best results. This effect was evaluated using *smaller-is-better*, (S-type) statistics. Other examples of smaller-is-better effects are wear rate for tires, surface roughness of a machined part, warpage of a laminate, and flaws in a diamond.

For the second effect, shear strength, the best results are associated with the highest numbers. It was evaluated with *bigger-is-better* (B-type) statistics. Other examples of bigger-is-better effects are lifetime of component, hardness of a drill bit, and strength of an I beam.

A third category of statistics is *nominal-is-best* (N-type) statistics, which is not used in this example. Examples are diameter of a shaft, resistance of a resistor, and output voltage of a power supply.

5.6   CONDUCTING THE EXPERIMENT

The 18 runs of the experiment were completed, but not necessarily in the order shown. Proponents of classical experimental design strongly encourage the use of random run orders. From a statistician's viewpoint, this is a good thing to do. From a practical point of view, however, it can be quite inefficient and expensive. Often, it is less expensive to conduct runs in the order which requires the least setup time, which should always be considered by cost-conscious engineers. In this experiment, for example, a furnace profile change (factor B) took about four hours. Therefore, all the runs in which factor B was at level 1 (runs 1, 2, 3, 10, 11, and 12) were completed first, then those for level 2 (runs 4, 5, 6, 13, 14, and 15), and finally those for level 3 (runs 7, 8, 9, 16, 17, and 18).

There is no clear-cut rule for making the trade-off between the potential costs and benefits of randomization. Statisticians usually prefer to randomize the runs and incur the higher costs, and engineers usually tend toward the opposite direction. Perhaps the best advice here is to understand the trade-off as well as possible, and make a responsible decision with the best information available.

5.7   RESULTS FOR PERCENT VOIDS

The results for percent voids are shown in Table 5-3. The columns labeled "Run no." and A through H are a repeat of the array of Table 5-2. They show the combinations of factor levels used for each run. To the right are columns labeled Y_1, Y_2, Y_3, $\bar{Y}$, and S/N.

In columns Y_1, Y_2, and Y_3, the results for each of the three samples for a given run are listed. For the three samples of run 1, for example, the percent voids were 0, 100, and 80. The average of these, $\bar{Y}$, is 60.0. Similarly, the readings for run 2 are 95, 50, and 80 for a $\bar{Y}$ of 75.0, etc.

Looking down the $\bar{Y}$ column, we see a wide range of results, ranging from 5.0 in runs 6, 12, 13, and 14, to 96.7 in run 15. (Remember, smaller is better.) For our purposes, this is good, because it tells us we have selected factors and levels which strongly affect the results.

5.8   THE SIGNAL-TO-NOISE RATIO

The abbreviation S/N stands for *signal-to-noise ratio*. This effect is similar to the signal-to-noise ratio used by electrical engineers, and is measured in decibels. It gives us more information than $\bar{Y}$, since it is a measure of both the location and the dispersion of the measured effects. It is calculated using the equation

$$S/N = -10 \log \text{MSD} \tag{5-1}$$

MSD stands for *mean square deviation*, which is also a measure of the dispersion of the data. It is calculated differently for B-type, N-type, and S-type effects:

For B-type (bigger-is-better):

$$\text{MSD} = \frac{1/Y_1^2 + 1/Y_2^2 + \cdots + 1/Y_n^2}{n} \tag{5-2}$$

For N-type (nominal-is-best):

$$\text{MSD} = \frac{(Y_1 - Y_0)^2 + (Y_2 - Y_0)^2 + \cdots + (Y_n - Y_0)^2}{n} \tag{5-3}$$

For S type (smaller-is-better):

$$\text{MSD} = \frac{Y_1^2 + Y_2^2 + \cdots + Y_n^2}{n} \tag{5-4}$$

It may be noted that these equations are devised such that, in every case, the larger the signal-to-noise ratio, the better the result.

The mean square deviation is a measure of variation, and it may be under-

TABLE 5-3 Results for Resistor #1 Percent Voids

Run no.	A Screen	B Profile	C Atmosphere	D Gas flow rate	E Empty	F Belt speed	G Drying temp.	H Empty	Y_1	Y_2	Y_3	ΣY	$\bar{Y}$	S/N
1	1	1	1	1	1	1	1	1	0	100	80	180	60.0	−37.7
2	1	1	2	2	2	2	2	2	95	50	80	225	75.0	−37.7
3	1	1	3	3	3	3	3	3	70	90	80	240	80.0	−38.1
4	1	2	1	1	2	2	3	3	90	95	90	275	91.7	−39.2
5	1	2	2	2	3	3	1	1	95	95	60	250	83.3	−38.6
6	1	2	3	3	1	1	2	2	5	5	5	15	5.0	−14.0
7	1	3	1	2	1	3	2	3	80	90	90	260	86.7	−38.8
8	1	3	2	3	2	1	3	1	10	0	5	15	5.0	−16.2
9	1	3	3	1	3	2	1	2	5	5	10	20	6.7	−17.0
10	2	1	1	3	3	2	2	1	70	90	100	260	86.7	−38.8
11	2	1	2	1	1	3	3	2	95	85	80	260	86.7	−38.8
12	2	1	3	2	2	1	1	3	5	5	5	15	5.0	−14.0
13	2	2	1	2	3	1	3	2	0	5	10	15	5.0	−16.2
14	2	2	2	3	1	2	1	3	0	5	10	15	5.0	−16.2
15	2	2	3	1	2	3	2	1	95	95	100	290	96.7	−39.7
16	2	3	1	3	2	3	1	2	75	60	80	215	71.7	−37.2
17	2	3	2	1	3	1	2	3	50	80	70	200	66.7	−36.6
18	2	3	3	2	1	2	3	1	95	90	80	265	88.3	−38.9

stood by comparing it to the more familiar concept of statistical variance. The equation for variance of a population is

$$\sigma^2 = \frac{\sum\limits_{i=1}^{n} (Y_i - \mu)^2}{n} \tag{5-5}$$

and the equation for nominal-is-best mean square deviation is

$$\text{MSD} = \frac{\sum\limits_{i=1}^{n} (Y_i - Y_0)^2}{n} \tag{5-6}$$

These equations tell us that the variance is a measure of the variation about the *mean*, while the mean square deviation is a measure of the variation about the *target value*. This is illustrated by the plots of Figure 5-1. The MSD illustrated in Figure 5-1 is for nominal-is-best, in which the target value is the value desired by the experimenter. With a little thought, it will be apparent that, for smaller-is-better, the target value is 0, and for bigger-is-better, the target value is infinity.

The signal-to-noise ratios shown in Table 5-3 were calculated using S-type statistics [Eq. (4-1)]. In this case, the term n is equal to 3, since three samples were measured for each run. The best signal-to-noise ratio is the highest, which in this case is the least negative. For percent voids, the best S/N ratio was -14.0, obtained in runs 6 and 12; and the worst was -39.7, from run 15. (The reader should calculate a few of these as an exercise.)

It may also be noted from the $\overline{Y}$ and S/N columns of Table 5-3 that, although there is general agreement between $\overline{Y}$ and S/N, this agreement is not exact. This is because S/N includes dispersion, while $\overline{Y}$ does not. Note, for instance that runs

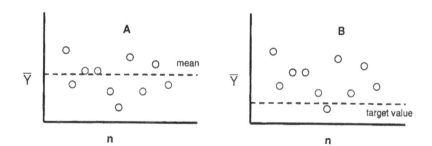

FIGURE 5-1   An illustration of the variation of a sample (A), and the nominal-is-best mean square deviation of the same sample (B). The variation is a measure of the variation about the mean, and the mean square deviation is a measure of the variation about a target value.

8 and 12 both have $\bar{Y}$'s of 5.0, but the S/N for run 8 is -16.2, while S/N for run 12 is -14.0. In this case run 12 is considered better since its data are more tightly grouped. Typically, more weight is usually given to S/N than to $\bar{Y}$.

One way to draw conclusions from this experiment at this point would be to select the conditions of the run which produced the highest S/N ratio. This would be premature, however, since only 18 of a possible 4374 treatment combinations were considered, and it is highly unlikely that one of them would be the best combination. We can, however, use the results of these 18 treatment combinations to point to the best combination. To do so, we will construct a *response table*, as shown in Table 5-4. This table shows $\bar{\bar{Y}}$ and $\overline{S/N}$ for each factor.

For the two-level factor A, the screen type, there were nine runs (1–9) at level 1, and nine runs (10–18) at level 2. $\bar{Y}$ for A_1 is then the average of all the $\bar{Y}$'s of the runs in which factor A was at level 1, or

$$\frac{60.0 + 75.0 + 80.0 + 91.7 + 83.3 + 5.0 + 86.7 + 5.0 + 6.7)}{9} = 54.8$$

(Some of the math may be slightly in error due to rounding.) $\bar{\bar{Y}}$ for A_2 is the average of all the $\bar{Y}$'s of the runs in which factor A was at level 2, or 56.9.

Similarly, the $\overline{S/N}$ for A_1 is -30.8, and $\overline{S/N}$ for A_2 is -30.7.

$\bar{Y}$ and $\overline{S/N}$ are calculated in a similar manner for factors B through H. They are three level factors, however, and n is equal to 6 in each calculation. $\bar{Y}$ and $\overline{S/N}$ for B_1 are calculated data from runs 1–3 and 10–12; for B_2 using data from runs 4–6 and 13–15; and for B_3 using data from runs 7–9 and 16–18. For D_1, runs 1, 4, 9, 11, 15, and 17 are used; for D_2, runs 2, 5, 7, 12 ,13, and 18; and for D_3, runs 3, 6, 8, 10, 14, and 16. (The reader should calculate the remainder as an exercise.)

It is also important to note that the sums of all the $\bar{\bar{Y}}$'s for all factors with the same number of levels are the same. This is also true for $\overline{S/N}$. This is shown in Table 5-4: for factors B through H, the sum of $\bar{Y}$ is always 167.5, and the sum of $\overline{S/N}$ is always -92.2 (with rounding errors eliminated). Aside from being a convenient check on the math, this tells us an important fact: the total variation in the experiment is being considered in the evaluation of each factor. Let's consider this further by going back to the original array.

In the evaluation of factor A, screen design, results from runs 1–9 are compared with those from runs 10–18. Factor A is *constant* for runs 1–9, and it is also constant, but at a *different level*, for runs 10–18. However, all other factors vary during this analysis; that is, they are considered *noise* for the evaluation of factor A. Because of this, we can be confident that the better level of factor A has been selected, in spite of the noise contribution of all other factors. This provides an enormous advantage over one-factor-at-a-time experiments, and results in a *robust* design.

TABLE 5-4 Response Table for Resistor 1 Percent Voids

Factor	Factor level	$\sum Y$	n	$\overline{\overline{Y}}$	$\overline{S/N}$
A: Screen	A_1	1480	27	54.8	-30.8
	A_2	1535	27	56.9	-30.7
	Total	3015	54	111.7	-61.5
B: Profile	B_1	1180	18	65.6	-34.1
	B_2	860	18	47.8	-27.3
	B_3	975	18	54.2	-30.8
	Total	3015	54	167.5	-92.2
C: Atmosphere	C_1	1205	18	66.9	-34.6
	C_2	965	18	53.6	-30.7
	C_3	845	18	46.9	-30.8
	Total	3015	54	167.5	-92.2
D: Gas flow rate	D_1	1225	18	68.1	-34.8
	D_2	1030	18	57.2	-30.7
	D_3	760	18	42.2	-26.7
	Total	3015	54	167.5	-92.2
E: Empty	E_1	995	18	55.3	-30.7
	E_2	1035	18	57.5	-30.7
	E_3	985	18	54.7	-30.9
	Total	3015	54	167.5	-92.2
F: Belt speed	F_1	440	18	24.4	-22.4
	F_2	1060	18	58.9	-31.3
	F_3	1515	18	84.2	-38.5
	Total	3015	54	167.5	-92.2
G: Drying temp.	G_1	695	18	38.6	-26.7
	G_2	1250	18	69.4	-34.3
	G_3	1070	18	59.4	-31.2
	Total	3015	54	167.5	-92.2
H: Empty	H_1	1260	18	70.0	-34.9
	H_2	750	18	41.7	-26.8
	H_3	1005	18	58.8	-30.5
	Total	3015	54	167.5	-92.2

For percent voids, the best levels of all factors are chosen by selecting the level of each factor that gave the best response.

There was little difference between A_1 and A_2, which means that this factor had little effect, and there would be little risk in selecting either level. Because of this, judgment on factor A may be reserved until after the analysis of results from the second effect, shear strength.

For factor B, level 2 produced a lower mean percent void and a higher $\overline{S/N}$. Thus it should be selected on the basis of mean percent void results. Similarly, C_3, D_3, F_1, and G_1 are selected.

Responses from the two empty columns, E and H, are interesting. In column E, $\overline{Y}$ and $\overline{S/N}$ are essentially equal for all levels. This means that, as expected, no factor had an effect here, and the results represent only the underlying variation of the overall experiment. In column H, though, there is a difference among the levels. This means that some factor which was not considered had an effect that showed up here. This illustrates the very important point that all factors will have their effects, whether we recognize them and include them in our experimental plan or not. If we are lucky, their effects will be evenly distributed among all other columns, or they will show up in an empty column. If we are unlucky, they will appear as variation associated with one of the factors assigned to a column, and we will draw inaccurate conclusions. This is the main reason to put considerable effort into including all known factors in an experiment.

In this experiment, it must be decided whether or not to put further effort into tracking down and evaluating the factor in column H. This is an engineering decision, not a statistical one. Are the results obtained from this experiment good enough to implement a satisfactory process, or must we do further work? Before answering that question, results from the other effect, shear strength, will be considered.

5.9   RESULTS FOR SHEAR STRENGTH

The results and response table for shear strength are shown in Tables 5-5 and 5-6, respectively. The analysis for shear strength is identical to that for percent voids. The major difference to be noted is that the B-type (bigger-is-better) equation is used to calculate mean square deviation and thus signal-to-noise ratio. There are also some positive values for S/N ratio. The reader should run through a few of these calculation to get a feel for them.

The selection of optimum values is done in the same way that it was done for percent voids. In this case, obviously, the larger $\overline{\overline{Y}}$ represents the best level.

For the screen, factor A, level 1 is superior to level 2, at least by the $\overline{S/N}$ criterion. As a rule of thumb (call it a *heuristic* rule if you want to impress the boss, or confuse him), a separation of 3 dB indicates a significant difference in $\overline{S/N}$; thus A_1 is significantly better than A_2.

TABLE 5-5 Results for Resistor 1 Shear Strength

Run no.	A Screen	B Profile	C Atmosphere	D Gas flow rate	E Empty	F Belt speed	G Drying temp.	H Empty	Y_1	Y_2	Y_3	ΣY	$\bar{Y}$	S/N
1	1	1	1	1	1	1	1	1	1.65	0.01	0.45	2.11	0.70	−35.2
2	1	1	2	2	2	2	2	2	0.25	0.65	0.75	1.65	0.55	−8.3
3	1	1	3	3	3	3	3	3	0.01	0.25	0.25	0.51	0.17	−35.2
4	1	2	1	1	2	2	3	3	0.20	0.25	0.30	0.75	0.25	−12.4
5	1	2	2	2	3	3	1	1	0.60	0.35	0.40	1.35	0.45	−7.6
6	1	2	3	3	1	1	2	2	2.55	2.90	2.90	8.35	2.78	8.8
7	1	3	1	2	1	3	2	3	1.40	0.65	0.40	2.45	0.82	−4.8
8	1	3	2	3	2	1	3	1	2.95	2.85	3.45	9.25	3.08	9.7
9	1	3	3	1	3	2	1	2	1.95	1.95	1.30	5.20	1.73	4.3
10	2	1	1	3	3	2	2	1	0.40	0.50	0.01	0.91	0.30	−35.2
11	2	1	2	1	1	3	3	2	0.01	0.01	0.25	0.27	0.09	−38.2
12	2	1	3	2	2	1	1	3	0.85	1.05	1.60	3.50	1.17	0.5
13	2	2	1	3	3	1	3	2	2.50	3.15	2.75	8.40	2.80	8.8
14	2	2	2	1	1	2	1	3	2.70	2.80	1.25	6.75	2.25	5.2
15	2	2	3	1	2	3	2	1	0.01	0.20	0.01	0.22	0.07	−38.2
16	2	3	1	3	2	3	1	2	0.55	0.55	0.50	1.60	0.53	−5.5
17	2	3	2	1	3	1	2	3	1.35	0.70	0.50	2.55	0.85	−3.4
18	2	3	3	2	1	2	3	1	0.85	1.20	0.60	2.65	0.86	−2.1

TABLE 5-6   Response Table for Resistor 1 Shear Strength

Factor	Factor level	ΣY	n	$\overline{\overline{Y}}$	$\overline{S/N}$
A: Screen	A_1	31.62	27	1.17	−9.0
	A_2	26.85	27	0.99	−12.0
	Total	58.47	54	2.16	−21.0
B: Profile	B_1	8.95	18	0.50	−25.3
	B_2	25.85	18	1.43	−5.9
	B_3	23.70	18	1.32	−0.3
	Total	58.47	54	3.25	−31.5
C: Atmosphere	C_1	16.22	18	0.90	−14.1
	C_2	21.82	18	1.21	−7.1
	C_3	20.43	18	1.14	−10.3
	Total	58.47	54	3.25	−31.5
D: Gas flow rate	D_1	11.10	18	0.62	−20.5
	D_2	20.00	18	1.11	−2.2
	D_3	27.37	18	1.52	−8.7
	Total	58.47	54	3.25	−31.5
E: Empty	E_1	22.58	18	1.25	−11.1
	E_2	16.97	18	0.94	−9.0
	E_3	18.92	18	1.05	−11.4
	Total	58.47	54	3.25	−31.5
F: Belt speed	F_1	34.16	18	1.90	−1.8
	F_2	17.91	18	1.00	−8.1
	F_3	6.40	18	0.36	−21.6
	Total	58.47	54	3.25	−31.5
G: Drying temp.	G_1	20.51	18	1.14	−6.4
	G_2	16.13	18	0.90	−13.5
	G_3	21.83	18	1.21	−11.8
	Total	58.47	54	3.25	−31.5
H: Empty	H_1	16.49	18	0.92	−18.1
	H_2	25.47	18	1.42	−5.0
	H_3	16.51	18	0.92	−8.4
	Total	58.47	54	3.25	−31.5

For furnace profile, B_2 produced a slightly better $\overline{\overline{Y}}$, but B_3 produced a significantly better $\overline{S/N}$. B_3 is chosen here.

For atmosphere, C_2 was best by both criteria.

For gas flow rate, D_2 was better for $\overline{S/N}$, but D_3 produced a higher mean.

For belt speed and drying temperature, F_1 and G_1 were chosen.

The empty columns E and H produced the same type of result that we saw for percent voids. Clearly, another factor is operating in the process.

5.10 COMPARISON OF RESULTS FOR TWO DIFFERENT EFFECTS

The preferred levels of all six factors with regard to two different effects have been chosen:

Percent voids	A_1 or A_2	B_2	C_3	D_3	F_1	G_1
Shear strength	A_1	B_2	C_2	D_2 or D_3	F_1	G_1

Here the two effects produce slightly different results. Again, an engineering judgment, not a statistical one, must be made. (Additional statistical criteria, resulting form analysis of variance, are presented in Chapter 8.) This is a good illustration of the principle that the experimental design and analysis should be kept within the control of those responsible for implementing the results. Several other factors may be considered, such as the expense associated with operating at each level, or the magnitudes of the differences between the responses, or which level is preferred by the operator. It is also legitimate to reiterate the experiment with new levels chosen on the basis of these results. In the actual experiment used in this example, other effects were evaluated, but for brevity are not included in this chapter. Using those results, and the best engineering judgment, the following levels were chosen:

$$A_1\ B_{2\text{-}3}\ C_3\ D_3\ F_1\ G_1$$

For factor B, a value midway between level 2 and level 3 was chosen.

5.11 CONFIRMATION RUN

The last task in the experiment is to conduct a confirmation run, in which all factors are set at their chosen levels, and a final run is made to confirm that we can indeed produce an optimized product. Five samples, labeled Y_1 through Y_5, were run, with the following results:

Sample number	Y_1	Y_2	Y_3	Y_4	Y_5	$\bar{Y}$	S/N
Percent voids	5	5	5	5	5	5	−14.0
Shear strength	3.4	3.5	3.8	3.4	3.6	3.5	10.9

It may be seen that the percent void results are as good as those of any of the experimental runs, while the shear strength results are significantly better.

While the confirmation run results are good, two questions must be asked: (1) Is there room for further improvement? (2) Is the process good enough? The answer to the first question is obviously yes. The levels were originally selected on the basis of the best engineering judgment, but they are probably not the best possible levels. At least part of the experiment could be repeated with new levels selected on the basis of the knowledge gained. Also, interactions among the factors were not considered, and could be considered in further work. Finally, remember the empty column H, which showed an effect of an unidentified factor. Results could be improved by finding and optimizing that factor.

Further work should only be considered if the answer to the second question is no. On a philosophical basis, we are never good enough, and should always strive to improve. Practically, however, we must consider whether or not we have a process which produces acceptable product, and whether greater opportunities or greater needs exist in other processes or products. We know, however, that we have produced a good result, and we know what to do to produce further improvement.

REFERENCE

1. G. Taguchi and S. Konishi, Orthogonal Arrays and Linear Graphs, American Supplier Institute, Dearborn, MI, 1987.

Chapter 6

Dealing with Interactions

6.1 INTRODUCTION

In Chapter 5, we discussed the setup and analysis of a main effects experiment. This type of experiment is extremely important, since it is often the first experiment conducted on a new process or design. Its purpose is to obtain as much information as possible over a wide range of potential factors without too much concern about the influence of any individual factor, which is why it is also called a screening experiment.

A common admonition from Taguchi experimenters is to "dig wide, not deep," since it is more costly to miss a factor entirely than to gauge its effects incorrectly. Although it is tempting to try to keep costs down by running small experiments initially, this is often counterproductive. Every factor in a process will have its effect in the experiment. If the effect is planned and accounted for in the array, it will be seen and evaluated. If it is not put into the array by the experimenter, it will appear as noise, or it will influence the effect of a factor that has been put into the experiment. In either case, its effect will not be observed uniquely, and it could mislead the experimenter.

The purpose of the screening experiment to identify the important factors, and to "screen out" the unimportant ones. In some cases, as in the example of the preceding chapter, that is all that is necessary. The results produce a satisfactory design or process, and we can go on to other problems. Sometimes, however, we want to learn more about the process, with the goal of improving it further. We also need to be aware that, in addition to the main effects of the factors, the factors may interact, and the interactions may also be important. In this chapter, we consider interactions in an experiment. This consideration also reveals information as to how the arrays are generated.

6.2 INTERACTIONS BETWEEN FACTORS

Figure 6-1 shows a plot of cure time vs. temperature for an epoxy, from an experiment that includes the two-level factors curing temperature (T_1 and T_2) and epoxy lot (A and B). The effect is the time, in minutes, required for the epoxy to cure. Lot A cures in 8 min at T_1, and in 12 minutes at T_2. The effect of the temperature difference is $12 - 8 = 4$ min. Lot B cures in 14 min at T_1, and in 18 min at T_2, and the effect is also 4 min. Since the effect of temperature is the same for both lots A and B, at least over the range of temperature investigated, these two factors can be considered independent of each other; that is, they do not interact.

Figure 6-2 shows a similar plot from the same experiment, but this time the two factors are curing temperature (still at T_1 and T_2) and relative humidity (RH 1 and RH 2). The cure time for RH 1 is 4 min at T_1, and 8 min at T_2, for an effect of 4 min. For RH 2, however, the effect of temperature is 10 min. Since the effect of temperature is different for different levels of relative humidity, these two factors are said to *interact*.

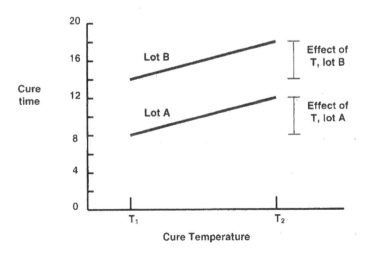

FIGURE 6-1 Factors that do not interact.

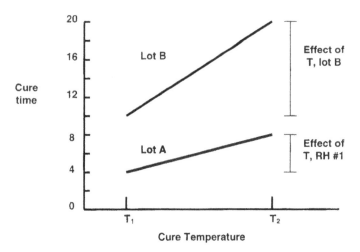

FIGURE 6-2 Factors that interact.

The degree of interaction between two factors is indicated by plots, as shown in Figures 6–1 and 6–2. The more nearly parallel the lines, the less the interaction. The larger the deviation from parallel, the stronger the interaction. The degree of interaction cannot be quantified from such plots, but it can be calculated from analysis of variance, which is described in Chapter 8. It is important to consider and include all relevant interactions among factors when designing an experiment since, like main effects, interaction will show up in the effects, whether they are accounted for in the array or not.

Interactions may occur among two, three, four, or higher numbers of factors, but interactions among three or more are usually considered less likely than those among two factors. This possibility should not be ignored, however, and it is up to the experimenter, not the statistician, to make the judgment as to whether or not to include them in the array.

6.3 INTERACTIONS IN THE ARRAY

For all experimental design arrays except special case main effects arrays like the one illustrated in Chapter 5, possible interactions occupy some of the columns. To illustrate this, consider the Taguchi L_4, or 2^3 array shown in Figure 6-3. This array has four rows and three columns (for arrays with only two-level factors, the number of columns is always 1 less than the number of rows). The numbering scheme for the levels is apparent: in column 1, to which factor A is assigned, the factor is first set at level 1, while factor B (assigned to column 2) is varied

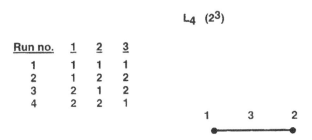

FIGURE 6-3   A Taguchi L_4 array with linear graph.

between its two levels. This accounts for the first two rows, which are also called runs, or treatment combinations. Factor A is then changed to level 2, and the process is repeated in runs 3 and 4. This is a full factorial array for factors A and B, since it includes all possible combinations of them. From columns 1 and 2, however, we cannot tell if there is an interaction between A and B. This must be obtained from column 3.

Column 3 in Figure 6-3 is obtained by "multiplying" the numbers in columns 1 and 2. If the numbers in columns 1 and 2 are the same, the result in column 3 is a "1." If they are not the same, the result in column 3 is a "2." Column 3 is thus where the interaction between columns 1 and 2 is shown. (Another way of saying this is that columns 1 and 2 are orthogonal with respect to each other, but column 3 is orthogonal with respect to neither.) The effects of this interaction, which is designated by A × B, are analyzed in exactly the same way as the main effects. A small effect in column 3 indicates little or no interaction; a large effect indicates a significant interaction.

The device to the right of the array in Figure 6-3 is called a *linear graph.* It shows two points labeled 1 and 2, indicating that the main effects are assigned to columns 1 and 2. The line connecting them is labeled 3, indicating that the interaction between columns 1 and 2 is present in column 3. For the L_4 array, the linear graph is quite simple. For higher-level arrays, however, they can be quite complex, and there are many possible linear graphs for each array. The generation of these graphs is beyond the scope of this text, but considerable attention will be given to their use. Several of the more common arrays and their accompanying linear graphs are shown in Appendix A.

6.4   CONFOUNDING OF MAIN EFFECTS AND INTERACTIONS

If the two main effects A and B are assigned to columns 1 and 2, and A × B is assigned to column 3, all possible main effects and interactions can be evaluated in the L_4 array. Sometimes, however, it is known by the experimenter that A × B is impossible, or at least highly unlikely. In this case, a third factor C may be assigned to column 3. (This is a choice which is entirely at the discretion of the experimenter; the statistician can provide no help in making the decision.) If the experimenter is correct, the effect of column 3 will be the effect of factor C. If the experimenter is incorrect, and A × B does in fact exist, the effect in column 3 will be that of both factor C and A × B. In DoE language, they will be *confounded*.

The L_4 array is not large enough to evaluate the effects of A, B, and C, as well as A × B. The experimenter must decide whether to take the risk of confounding, or to incur the expense of conducting a larger experiment in which confounding is not possible. To further illustrate this point, consider a slightly larger array, the L_8.

6.5   INTERACTIONS AND CONFOUNDING IN AN L_8 ARRAY

Figure 6-4 shows the array and two possible linear graphs for a Taguchi L_8 array. Suppose the experimenter wants to evaluate the following main effects and interactions:

A, B, C, D, B × C, and B × D

To assign these main effects and interactions to the correct columns, we use one of the two linear graphs. Both linear graphs are appropriate to the L_8, but they have somewhat different applications, depending on the combination of main effects and interactions to be evaluated. This will be discussed in more detail later, but for this experiment, the graph labeled *i* will be used.

We begin assigning the main effects to the points, which designate columns 1, 2, 4, and 7. (These are also the orthogonal columns of the array.) Since the only factor not involved in an interaction is A, we will assign it to the unconnected point 7. B, C, and D can then be assigned to points 1, 2, and 4 in any order, and we will choose 1 for B, 2 for C, and 4 for D. When this is done, the two interactions are automatically assigned to the appropriate connecting lines: 3 for B × C and 5 for B × D. This assignment is shown in Figure 6-5.

It may also be noted from Figure 6-5 that column 6 is empty in the array and, correspondingly, line 6 is unassigned in the linear graph. This column would evaluate C × D, if it existed. The experimenter has decided that C × D does not exist, but if the assumption is wrong, an effect will be seen in column 6. If the assumption is correct, then column 6 will show no effect, and will be a measure of the overall noise of the experiment.

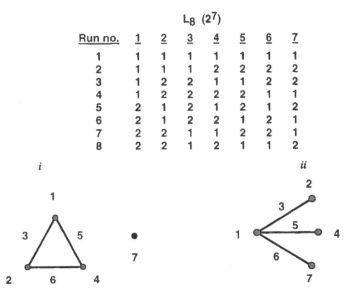

L8 (2⁷)

$L_8 (2^7)$

Run no.	1	2	3	4	5	6	7
1	1	1	1	1	1	1	1
2	1	1	1	2	2	2	2
3	1	2	2	1	1	2	2
4	1	2	2	2	2	1	1
5	2	1	2	1	2	1	2
6	2	1	2	2	1	2	1
7	2	2	1	1	2	2	1
8	2	2	1	2	1	1	2

i

ii

FIGURE 6-4 A Taguchi L₈ array with linear graphs.

$L_8 (2^7)$

Run no.	1	2	3	4	5	6	7
1	1	1	1	1	1	1	1
2	1	1	1	2	2	2	2
3	1	2	2	1	1	2	2
4	1	2	2	2	2	1	1
5	2	1	2	1	2	1	2
6	2	1	2	2	1	2	1
7	2	2	1	1	2	2	1
8	2	2	1	2	1	1	2
	B	C	B×D	D	B×C	B	A

i

FIGURE 6-5 A linear graph used to determine assignment of interactions to a Taguchi L₈ array.

6.6   HOW TO USE LINEAR GRAPHS

Several different types of linear graph are possible for each array. Figure 6-6 shows an L_{16} array and five possible linear graphs associated with it. (A student once remarked that they look like astrological tables; but, while they are powerful, there is nothing supernatural about them.) The linear graph is chosen on the basis of the expected interaction pattern of the factors of interest. If interactions are expected to be equally distributed among all factors, for instance, the "equi-axed" graph *i* would be chosen. Graph *iii* is chosen if it is expected that a single factor

$$L_{16}\ (2^{15})$$

Run	1	2	3	4	5	6	7	8	9	10	11	12	13	14	15
1	1	1	1	1	1	1	1	1	1	1	1	1	1	1	1
2	1	1	1	1	1	1	1	2	2	2	2	2	2	2	2
3	1	1	1	2	2	2	2	1	1	1	1	2	2	2	2
4	1	1	1	2	2	2	2	2	2	2	2	1	1	1	1
5	1	2	2	1	1	2	2	1	1	2	2	1	1	2	2
6	1	2	2	1	1	2	2	2	2	1	1	2	2	1	1
7	1	2	2	2	2	1	1	1	1	2	2	2	2	1	1
8	1	2	2	2	2	1	1	2	2	1	1	1	1	2	2
9	2	1	2	1	2	1	2	1	2	1	2	1	2	1	2
10	2	1	2	1	2	1	2	2	1	2	1	2	1	2	1
11	2	1	2	2	1	2	1	1	2	1	2	2	1	2	1
12	2	1	2	2	1	2	1	2	1	2	1	1	2	1	2
13	2	2	1	1	2	2	1	1	2	2	1	1	2	2	1
14	2	2	1	1	2	2	1	2	1	1	2	2	1	1	2
15	2	2	1	2	1	1	2	1	2	2	1	2	1	1	2
16	2	2	1	2	1	1	2	2	1	1	2	1	2	2	1

FIGURE 6-6a   A Taguchi L_{16} array. Five different linear graphs for this array are shown in Figures 6–6b through 6-6f.

i

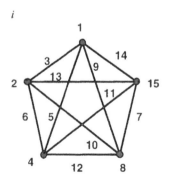

FIGURE 6-6b

ii

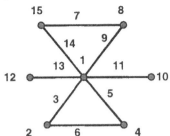

FIGURE **6-6c**

iii

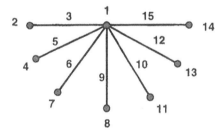

FIGURE **6-6d**

iv

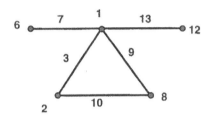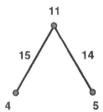

FIGURE **6-6e**

v

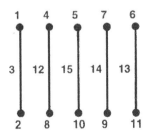

FIGURE **6-6f**

would interact with several other factors; and graph v would be chosen if several interaction pairs are expected.

As an exercise, consider the assignment of the following two-level factors to the L_{16} array of Figure 6-6:

$$A, B, D, C, D, E, F, G, H, A \times D, B \times D, A \times B, C \times E, \text{ and } B \times C$$

There are 8 main effects and 5 interactions, for a total of 13 factors. Thus, an array with at least 13 columns is required. The smallest available array is the L_{16}, which has 15 columns and 16 rows (for an array with only two-level factors, the number of columns is always 1 less than the number of rows).

We may note that factor A participates in two interactions, B in three, C

$$L_{16}\ (2^{15})$$

Run	1	2	3	4	5	6	7	8	9	10	11	12	13	14	15
1	1	1	1	1	1	1	1	1	1	1	1	1	1	1	1
2	1	1	1	1	1	1	1	2	2	2	2	2	2	2	2
3	1	1	1	2	2	2	2	1	1	1	1	2	2	2	2
4	1	1	1	2	2	2	2	2	2	2	2	1	1	1	1
5	1	2	2	1	1	2	2	1	1	2	2	1	1	2	2
6	1	2	2	1	1	2	2	2	2	1	1	2	2	1	1
7	1	2	2	2	2	1	1	1	1	2	2	2	2	1	1
8	1	2	2	2	2	1	1	2	2	1	1	1	1	2	2
9	2	1	2	1	2	1	2	1	2	1	2	1	2	1	2
10	2	1	2	1	2	1	2	2	1	2	1	2	1	2	1
11	2	1	2	2	1	2	1	1	2	1	2	2	1	2	1
12	2	1	2	2	1	2	1	2	1	2	1	1	2	1	2
13	2	2	1	1	2	2	1	1	2	2	1	1	2	2	1
14	2	2	1	1	2	2	1	2	1	1	2	2	1	1	2
15	2	2	1	2	1	1	2	1	2	2	1	2	1	1	2
16	2	2	1	2	1	1	2	2	1	1	2	1	2	2	1
	D	A	A	B	B	A	C	C	e	e	H	B	G	F	E
			×		×	×	×					×			
			D		D	B	E					C			

i

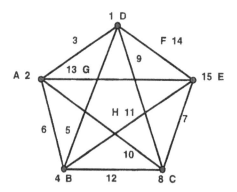

FIGURE 6-7a Assignment of a set of main effects and interactions to an L_{16} array.

in two, D in two, and E in one. This means that all these factors must be assigned to points, and that interactions are spread among many factors. Thus, the best linear graph choices in Figure 6-6 are *i* and *ii*. Before looking at the two possible solutions in Figure 6-7, the reader is encouraged to try to assign the factors to the array based on these graphs.

The array in Figure 6-7a is assigned according to graph *i*. It may be noted that main effects G and H are assigned to lines, which is acceptable as long as it is understood that they will be confounded with any interactions associated with these lines. It may also be noted that columns 9 and 10 are empty, and are designated with a small "*e*." Similar observations may be made about the array assigned according to graph ii in Figure 6-7b.

L_{16} (2^{15})

Run	1	2	3	4	5	6	7	8	9	10	11	12	13	14	15
1	1	1	1	1	1	1	1	1	1	1	1	1	1	1	1
2	1	1	1	1	1	1	1	2	2	2	2	2	2	2	2
3	1	1	1	2	2	2	2	1	1	1	1	2	2	2	2
4	1	1	1	2	2	2	2	2	2	2	1	1	1	1	1
5	1	2	2	1	1	2	2	1	1	2	2	1	1	2	2
6	1	2	2	1	1	2	2	2	2	1	1	2	2	1	1
7	1	2	2	2	2	1	1	1	1	2	2	2	2	1	1
8	1	2	2	2	2	1	1	2	2	1	1	1	1	2	2
9	2	1	2	1	2	1	2	1	2	1	2	1	2	1	2
10	2	1	2	1	2	1	2	2	1	2	1	2	1	2	1
11	2	1	2	2	1	2	1	1	2	1	2	2	1	2	1
12	2	1	2	2	1	2	1	2	1	2	1	1	2	1	2
13	2	2	1	1	2	2	1	1	2	2	1	1	2	2	1
14	2	2	1	1	2	2	1	2	1	1	2	2	1	1	2
15	2	2	1	2	1	1	2	1	2	2	1	2	1	1	2
16	2	2	1	2	1	1	2	2	1	1	2	1	2	2	1
	B	C	B	E	H	C	A	D	B	G	e	F	e	A	A
			X			x	X		x					x	
			C			E	D		D					B	

ii

FIGURE **6-7b**   Different assignment of the same set of main effects and interactions to an L_{16} array.

6.7   ARRAYS WITH MORE THAN TWO LEVELS

Sometimes, it is desirable to evaluate more than two levels of a factor. We have already seen an example of this with the L_{18} array in the previous chapter. Some others are shown in Appendix A. One of these may be chosen, but often it is necessary to evaluate some factors at two levels, some at three, some at four, etc. This is done by column *upgrading* or *downgrading*. This is not difficult, but it must be done according to some rules.

    Column upgrading and downgrading are based on the concept of *degrees of freedom*. Although the rules for determining degrees of freedom in a classical experiment are not difficult, they can be troublesome to remember. In Taguchi arrays, there is only one rule:

    The number of degrees of freedom for a factor in a Taguchi array is always equal to 1 less than the number of levels of that factor.

For example, a two-level factor always has $(2 - 1 =)$ 1 degree of freedom (dF); a four-level factor has $(4 - 1 =)$ 3 dF; an eight-level factor has 7 dF; etc.

    The number of degrees of freedom required to evaluate an interaction between two main effects is equal to the product of the degrees of freedom of the two factors. For example, if 2 two-level factors interact, the number of two-level columns required to evaluate their interaction is $(1 \times 1 =)$ 1. If a two-level factor interacts with a three-level factor, $(1 \times 2 =)$ 2 columns are required to evaluate the interaction.

6.8   HOW TO UPGRADE FROM TWO
TO FOUR LEVELS

In upgrading columns, it is critical to remember that the degrees of freedom must always be preserved. Thus, a single four-level column with 3 dF is equivalent to 3 two-level columns, each with 1 dF. In other words, to create a single four-level column, it is necessary to "consume" 3 two-level columns.

    The rules for upgrading from two- to four-level factors are explained here, using the L_8 array of Figure 6-8 as an example.

    1.   Choose a set of three interacting two-level columns, using the appropriate linear graph (see Figure 6-8a).
    2.   Choose any two of the three columns, ignoring the third (see Figure 6-8b).
    3.   Combine the levels in the chosen columns according to the scheme shown in Figure 6-8c.

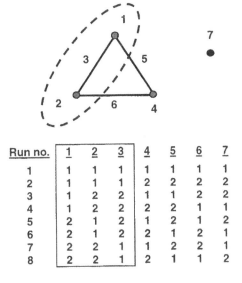

Run no.	1	2	3	4	5	6	7
1	1	1	1	1	1	1	1
2	1	1	1	2	2	2	2
3	1	2	2	1	1	2	2
4	1	2	2	2	2	1	1
5	2	1	2	1	2	1	2
6	2	1	2	2	1	2	1
7	2	2	1	1	2	2	1
8	2	2	1	2	1	1	2

FIGURE 6-8a Upgrading columns from two to four levels. Select any set of three interacting columns.

Run no.	1	2	3	4	5	6	7
1	1	1		1	1	1	1
2	1	1		2	2	2	2
3	1	2		1	1	2	2
4	1	2		2	2	1	1
5	2	1		1	2	1	2
6	2	1		2	1	2	1
7	2	2		1	2	2	1
8	2	2		2	1	1	2

FIGURE 6-8b Upgrading columns from two to four levels. Choose any two of the three columns selected in Figure 6-8b.

Old levels		New level
1	1	1
1	2	2
2	1	3
2	2	4

Run no.	1	2	3	4	5	6	7
1		1		1	1	1	1
2		1		2	2	2	2
3		2		1	1	2	2
4		2		2	2	1	1
5		3		1	2	1	2
6		3		2	1	2	1
7		4		1	2	2	1
8		4		2	1	1	2

FIGURE 6-8c Upgrading columns from two to four levels. Combine the columns according to the table on the left.

6.9　HOW TO UPGRADE FROM TWO TO EIGHT LEVELS

Upgrading from two to eight levels involves the same principles as upgrading from two to four levels, but is a little more complicated. For this example, an L_{16} array, shown in Figure 6-9, is used.

1. Choose a set of seven columns which contain three main effects, and all possible interactions among them, including the three-way interaction, as shown in Figure 6-9a.

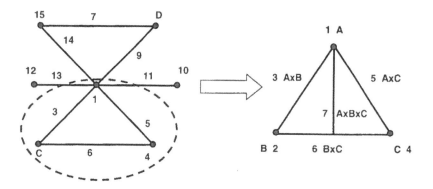

FIGURE 6-9a　Upgrading columns from two to eight levels. Choose a set of seven columns which contain three main effects and all possible interactions among them, including the three-way interaction.

Run	1	2	4		8	9	10	11	12	13	14	15
1	1	1	1		1	1	1	1	1	1	1	1
2	1	1	1		2	2	2	2	2	2	2	2
3	1	1	2		1	1	1	1	2	2	2	2
4	1	1	2		2	2	2	2	1	1	1	1
5	1	2	1		1	1	2	2	1	1	2	2
6	1	2	1		2	2	1	1	2	2	1	1
7	1	2	2		1	1	2	2	2	2	1	1
8	1	2	2		2	2	1	1	1	1	2	2
9	2	1	1		1	2	1	2	1	2	1	2
10	2	1	1		2	1	2	1	2	1	2	1
11	2	1	2		1	2	1	2	2	1	2	1
12	2	1	2		2	1	2	1	1	2	1	2
13	2	2	1		1	2	2	1	1	2	2	1
14	2	2	1		2	1	1	2	2	1	1	2
15	2	2	2		1	2	2	1	2	1	1	2
16	2	2	2		2	1	1	2	1	2	2	1

FIGURE 6-9b　Upgrading columns from two to eight levels. Choose the three main effects columns, ignoring the interaction columns.

Old levels			New level
1	1	1	1
1	1	2	2
1	2	1	3
1	2	2	4
2	1	1	5
2	1	2	6
2	2	1	7
2	2	2	8

FIGURE 6-9c Upgrading columns from two to eight levels. Use this scheme to combine the 3 two-level columns into a single new eight-level column.

Run	1	2	4	New	8	9	10	11	12	13	14	15
1				1	1	1	1	1	1	1	1	1
2				1	2	2	2	2	2	2	2	2
3				2	1	1	1	1	2	2	2	2
4				2	2	2	2	2	1	1	1	1
5				3	1	1	2	2	1	1	2	2
6				3	2	2	1	1	2	2	1	1
7				4	1	1	2	2	2	2	1	1
8				4	2	2	1	1	1	1	2	2
9				5	1	2	1	2	1	2	1	2
10				5	2	1	2	1	2	1	2	1
11				6	1	2	1	2	2	1	2	1
12				6	2	1	2	1	1	2	1	2
13				7	1	2	2	1	1	2	2	1
14				7	2	1	1	2	2	1	1	2
15				8	1	2	2	1	2	1	1	2
16				8	2	1	1	2	1	2	2	1

FIGURE 6-9d The new eight-level column is shown in the box.

2. Choose the three main effects columns, ignoring the remaining four interaction columns (see Figure 6-9b.)
3. Combine the levels in the chosen columns according to the scheme shown in Figure 6-9c.
4. The result is shown in Figure 6-9d.

6.10 HOW TO UPGRADE FROM TWO TO THREE LEVELS

Upgrading from two to three levels in a Taguchi array involves two steps:

1. Upgrade a set of 3 two-level columns to a single four-level column using the rules described above.
2. Downgrade the four-level column by replacing the "4's" with "3's."

$$L_8 \ (2^7)$$

Run no.	1	2	3	4	5	6	7
1	1	1	1	1	1	1	1
2	1	1	1	2	2	2	2
3	1	2	2	1	1	2	2
4	1	2	2	2	2	1	1
5	2	1	2	1	2	1	2
6	2	1	2	2	1	2	1
7	2	2	1	1	2	2	1
8	2	2	1	2	1	1	2

Run no.	1	2	3	4	5	6	7
1		1		1	1	1	1
2		1		2	2	2	2
3		2		1	1	2	2
4		2		2	2	1	1
5		3		1	2	1	2
6		3		2	1	2	1
7		3		1	2	2	1
8		3		2	1	1	2

FIGURE 6-10   Upgrading from two to three levels.

This process is illustrated in Figure 6-10. It may be noted that this is not a mathematically rigorous operation, since it involves only the substitution of 3's for 4's in the previously upgraded column. This is an inefficient use of a four-level column, and most of the time it is better to leave the four-level column alone and obtain additional information. Sometimes, however, it is not possible to do so, and the two- to three-level upgrade is desirable. There are efficient three-level columns, however, such as the L_9, the L_{18}, and the L_{27} (see Appendix A).

$$L_9 \ (3^4)$$

Run no.	1	2		3	4
1	1	1		1	1
2	1	2		2	2
3	1	3	=1'	3	3
4	2	1		2	3
5	2	2		3	1
6	2	3	=1'	1	2
7	3	1		3	2
8	3	2		1	3
9	3	3	=1'	2	1

FIGURE 6-11   Downgrading from three to two levels.

An example of just the downgrading step is shown in Figure 6-11, where a three-level column is downgraded to an inefficient two-level column.

6.11   USING COMBINATION DESIGNS

Another way to manipulate an array is to use a *combination design*. This is illustrated in the L_9 array of Figure 6-12. Suppose we want to evaluate 3 three-level factors A, B, and C and 2 two-level factors X and Y. This requires 8 degrees of freedom, which is the number of dF in the L_9.

We begin by assigning factors A, B, and C to columns 1, 2, and 3, respectively, as shown in Figure 6-12. We then identify the possible combinations of the two-level factors as shown:

Combination	Designation
X_1Y_1	$(XY)_1$
X_2Y_1	$(XY)_2$
X_1Y_2	$(XY)_3$
X_2Y_2	(Not used)

We then assign the XY combination to column 4 as shown in Figure 6–12. We now have an array in which each run has five factors: A in column 1, B in column 2, C in column 3, and X and Y in column 4. For example, run 5 has the following factor levels: $A_2B_2C_3(XY)_1$.

The results of the experiment are analyzed by treating (XY) as a single

$L_9\ (3^4)$

Run no.	1	2	3	4
1	1	1	1	1
2	1	2	2	2
3	1	3	3	3
4	2	1	2	3
5	2	2	3	1
6	2	3	1	2
7	3	1	3	2
8	3	2	1	3
9	3	3	2	1
	A	B	C	XY

FIGURE 6-12   Combination treatment design in which two factors, X and Y, are evaluated in a single column.

factor, calculating its effects at three levels. The effects of X and Y are then determined as follows:

- The effect of X with Y constant is $(XY)_1 - (XY)_2$, where $(XY)_1 = X_1Y_1$, and $(XY)_2 = X_2Y_1$.
- Likewise, the effect of Y with X constant is $(XY)_1 - (XY)_3$, where $(XY)_1 = X_1Y_1$, and $(XY)_3 = X_1Y_2$.

6.12 NESTED FACTORS

Another useful manipulation is the nested factor, in which the relationship between two factors is evaluated as a factor by itself. Both combination designs and nested factors are beyond the capability of most of the software currently available, so these techniques require manual analysis. This is not conceptually difficult, but it is time consuming. Therefore, these methods are not used widely.

In this chapter, some of the possible ways that arrays can be manipulated have been illustrated. Many other manipulations are available; in fact, it is possible to do almost anything that the degrees of freedom allow to customize an array to fit a particular experimental need. For reasons of space and clarity, those manipulations are considered beyond the scope of this book. If further capabilities are required, almost any of the excellent texts referenced at the end of Chapter 4 can be recommended.

In Chapter 7, a complete experimental design process is described; and in Chapter 8, a complete analysis of variance (ANOVA) is performed on the results.

Chapter 7

Designing an Experiment

7.1 INTRODUCTION

In this chapter, we discuss the design of a manufacturing experiment involving both main effects and interactions, which have been discussed in previous chapters. The problem selected for exposition is the reduction of solder joint defects in the wave solder process for component attachment to an electronic circuit card assembly. The experiment is set up in this chapter, and Chapter 8 presents analysis of variance of the results.

7.2 WHERE DO THE ORTHOGONAL ARRAYS COME FROM?

Before discussing the experiment, it is appropriate to discuss the generation and use of the arrays. This subject is touched upon in Chapter 6, and more detail is included here.

Table 7-1 is a full factorial 2^4 array to evaluate four factors A, B, C, and D at two levels each. It has 16 rows and 15 columns, and contains all possible interactions among the four main effects. A fractional factorial array can be extracted from this array by eliminating some combinations of levels which are of relatively low interest. For example, the four-way $A \times B \times C \times D$ interaction would rarely be considered significant. If we therefore eliminate all rows in which

TABLE 7-1 A Full Factorial 2^4 Array to Evaluate Four Factors

Run	1	2	3	4	5	6	7	8	9	10	11	12	13	14	15
1	1	1	1	1	1	1	1	1	1	1	1	1	1	1	1
2	1	1	1	1	1	1	1	2	2	2	2	2	2	2	2
3	1	1	1	2	2	2	2	1	1	1	1	2	2	2	2
4	1	1	1	2	2	2	2	2	2	2	2	1	1	1	1
5	1	2	2	1	1	2	2	1	1	2	2	1	1	2	2
6	1	2	2	1	1	2	2	2	2	1	1	2	2	1	1
7	1	2	2	2	2	1	1	1	1	2	2	2	2	1	1
8	1	2	2	2	2	1	1	2	2	1	1	1	1	2	2
9	2	1	2	1	2	1	2	1	2	1	2	1	2	1	2
10	2	1	2	1	2	1	2	2	1	2	1	2	1	2	1
11	2	1	2	2	1	2	1	1	2	1	2	2	1	2	1
12	2	1	2	2	1	2	1	2	1	2	1	1	2	1	2
13	2	2	1	1	2	2	1	1	2	2	1	1	2	2	1
14	2	2	1	1	2	2	1	2	1	1	2	2	1	1	2
15	2	2	1	2	1	1	2	1	2	2	1	2	1	1	2
16	2	2	1	2	1	1	2	2	1	1	2	1	2	2	1
	A	A	B	B	A	A	C	C	A	A	B	B	A	A	D
	×		×		×	×	×	×	×	×	×	×	×	×	
	B		C		C	B	D		B	C	D	C	D	B	
	×		×		×				×					×	
	C		D		D				D					C	
	×														
	D														

this interaction is at level 2, the interaction is eliminated from the array, and the array is reduced to eight rows, as shown in Table 7-2, which is called a *half-factorial* array. Table 7-2 has the following pairs of identical columns:

Columns	Factors
2, 3	A, B × C × D
4, 5	B, A × C × D
6, 7	A × B, C × D
8, 9	C, A × B × D
10, 11	A × C, B × D
12, 13	B × C, A × D
14, 15	D, A × B × C

When the experimental results are analyzed, we will not know whether we are seeing the effects of A or B × C × D, etc. These pairs of factors are said

TABLE 7-2 Full Factorial 2^4 Array, with Level 2 of the A $\times$ B $\times$ C $\times$ D Interaction Eliminated

Run	1	2	3	4	5	6	7	8	9	10	11	12	13	14	15
1	1	1	1	1	1	1	1	1	1	1	1	1	1	1	1
2	1	1	1	1	1	1	1	2	2	2	2	2	2	2	2
3	1	1	1	2	2	2	2	1	1	1	1	2	2	2	2
4	1	1	1	2	2	2	2	2	2	2	2	1	1	1	1
5	1	2	2	1	1	2	2	1	1	2	2	1	1	2	2
6	1	2	2	1	1	2	2	2	2	1	1	2	2	1	1
7	1	2	2	2	2	1	1	1	1	2	2	2	2	1	1
8	1	2	2	2	2	1	1	2	2	1	1	1	1	2	2
	A	A	B	B	A	A	C	C	A	A	B	B	A	A	D
	×	×	×		×	×	×		×	×	×	×	×	×	
	B	C	C		C	B	D		B	C	D	C	D	B	
	×	×			×				×					×	
	C	D			D				D					C	
	×														
	D														

to be *confounded*, or *aliased*. If this amount of confounding is acceptable, the array can be reduced to the L_8 array shown in Table 7-3. The determination of whether or not this is acceptable is, of course, a matter of engineering judgment, not statistics.

The L_8 array shown in Table 7-3 is capable of evaluating three main effects and all their interactions without confounding. If four main effects are investigated, one of them will be confounded with the three-way interaction or one of the two-way interactions. As the number of main effects is increased, the amount of confounding becomes more severe. Since the L_8 contains seven columns, seven main effects could be evaluated, but it would be heavily confounded. To evaluate seven factors in a full factorial would require 2^7, or 128 runs. Since the L_8 is only one-sixteenth of this, it is called a *sixteenth factorial*. This level of confounding should be done only if the experimenter is sure that there are no interactions.

The generation of arrays is of great interest to some experimenters but, paradoxically, it is not terribly relevant to the successful use of DoE. Success can be attained by applying the arrays correctly, even with little understanding of how they are generated.

From the above discussion, it is apparent that the cost of a designed experiment can be estimated accurately before work is begun, if the desired main effects and their interactions are known. The cost of adding or subtracting factors also can be estimated accurately. This knowledge is not always a blessing, however, since some unenlightened managers may resist committing to the known expense

TABLE 7-3 Full Factorial 2^4 Array, with the A × B × C × D Interaction and Redundant Columns Eliminated, Resulting in an L_8 Array

Run	2	4	6	8	10	12	14
1	1	1	1	1	1	1	1
2	1	1	1	2	2	2	2
3	1	2	2	1	1	2	2
4	1	2	2	2	2	1	1
5	2	1	2	1	2	1	2
6	2	1	2	2	1	2	1
7	2	2	1	1	2	2	1
8	2	2	1	2	1	1	2
	A	B	A	C	A	B	A
			×		×	×	×
			B		C	C	B
							×
							C

of a designed experiment before work is begun. Instead, they tacitly choose the option of "working on" the problem with no definite plan; and they get no definite results.

While the mechanics of a designed experiment are important, and rules governed by statistics are critical to success, they are not the main barriers to the successful application of Design of Experiments. The primary barriers to success are (1) insufficient planning for the experiment before the actual data collection process begins, and (2) insufficient preparation for the logistical challenges of preparing the samples and running them through the prescribed treatment combinations. Most texts on Design of Experiments are very good at explaining the statistical tools and mechanics of a designed experiment; in fact, they are more thorough and comprehensive than this text in that regard. In this chapter, we deal with some of the practical matters not contained in other texts.

7.3 GETTING STARTED

Like a good building project, a good DoE project must have a good foundation to be successful. In order to build a good DoE foundation, all the relevant information must be made available. Seldom is there sufficient knowledge, experience, and judgment present in the mind of a single individual to bring all this information to bear on a problem. Therefore, a team of individuals representing all relevant disciplines must be assembled to design and conduct the experiment.

In the wave solder example presented here, a team made up of a manufacturing engineer, an assembler, a production supervisor, and a quality engineer was assembled. Since this was the first DoE project for all of them, a DoE facilitator was included in the early discussions. Approximately 30 hours were spent in structured discussions to complete the experimental plan. While this may seem excessive, it was well rewarded in the end. The group should not be in a hurry to conclude the discussions in order to "get to the real work." Some time spent in unstructured discussions early in the process is useful to give all participants the confidence necessary to make honest suggestions.

It is good to remember that the purpose of the discussion is to produce suggestions that will lead to the *solution* of the problem, not necessarily those which will reveal the *cause* of the problem. (Remember the thick film resistor and the transistor gain examples.)

There are many good techniques for structuring brainstorming discussions to set up a DoE project. One such technique is known as nominal group theory (NGT). In this process, each participant is given a turn to suggest which factors should be considered, according to a set of rules, such as:

1. All participants are considered equal.
2. Each participant is allowed to make one suggestion per turn.
3. All suggestions should be short, succinct, and concise.
4. Passing is allowed, and a person who has passed on one turn is allowed to make a suggestion on following turns.
5. Absolutely *no* comments on others' suggestion are allowed.
6. The suggestion process continues until no more suggestions are made.
7. A prior commitment by all (including management) is made to implement the results of the process.

After all suggestions are listed, a facilitator groups the similar ideas to consolidate the options. Then the group selects the ones most likely to be successful by voting. Usually this is done by giving each person several votes.

An interesting feature of NGT is that there is very seldom less than unanimity of agreement on the results obtained. Since everyone contributes equally, all are equally committed to making the project a success.

7.4 FACTORS AND LEVELS FOR THE WAVE SOLDER EXPERIMENT

In a process as complicated as wave soldering, there are literally hundreds of factors which can impact the results. Even a cursory review of the literature on this subject reveals that many different factors have been reported as critical. In view of this, it may seem unreasonable to expect to select, evaluate, and optimize the 8, 16, or 32 main effects and interactions which will lead to significant improvement. This leads to two observations:

1. Even though most processes and designs are indeed affected by dozens, or even hundreds, of factors, experience has shown in almost every case that significant improvement can be achieved by evaluating only the relative few that can be fitted into a designed experiment.
2. The critical few factors do not have to be the same in every instance. For example, one manufacturer may optimize its process by conducting a 16-run experiment with a list of factors that is almost completely different from that of another manufacturer with the same process. They will both obtain satisfactory results.

The above observations may explain why two different manufacturers may produce products that meet the specifications, but are different from each other. It is also a strong caution against the practice of "copying" the results of one process into another one. Most veteran manufacturers have had the experience of acquiring experimental results from another manufacturer, and of then being completely unsuccessful in copying the conclusions into their own process. This is because all of those other variables which were not accounted for behaved differently. The lesson is: *Do your own work!* It's not only the right thing to do, but the only way to be successful. Further discussion of this topic is contained in Ref. 1.

The above is only one of several possible ways to structure the discussion of DoE setup. The important point is that some form of structured discussion must take place.

The factors selected for the experiment under discussion here were:

A. *Prebake conditions (two levels).* The circuit cards were baked in an oven prior to the process to dry them. The two levels of this factor were two different prebake temperatures.
B. *Flux density (two levels).* The density of the solder flux used in the process was varied by diluting it to two different levels with alcohol.
C. *Conveyor speed (two levels).* The two levels corresponded to two different speeds at which the boards passed through the solder wave.
D. *Preheat conditions (two levels).* Before the circuit cards enter the solder wave, they pass through an infrared preheater. Two different preheat settings are the two levels of this factor.
E. *Cooling time (two levels).* Since it was thought that the cooling rate of the cards might affect defect rates, two different cooling rates were investigated.
F. *Omega (two levels).* This is the agitation of the wave by ultrasonic

vibration to break up the dross on the surface. Two different settings on the ultrasonic apparatus were used as the two levels of this factor.

G. *Solder temperature (two levels).* This is the temperature of the solder in the wave, and two temperatures were used as the levels.

All the above factor levels had numerical values, but they are not used here in order to focus attention on the analytical process. Instead, "level 1" and "level 2" are used.

In addition to the above seven main effects, the following interactions were of interest:

Prebake × flux density	A × B
Prebake × conveyor speed	A × C
Flux density × conveyor speed	B × C
Prebake × preheat	A × D
Flux density × preheat	B × D
Conveyor speed × preheat	C × D

7.5 ASSIGNMENT OF FACTORS TO ARRAY COLUMNS

The seven main effects and six interactions have the following degrees of freedom:

Main effects: 7 factors × [(2 − 1 =) 1 dF/factor] = 7 dF

Interactions: 6 factors × [(1 × 1 =) 1 dF/factor] = 6 dF.

Thus, the total number of degrees of freedom in the experiment is 13, and at least 14 runs are required. Since there is no L_{14} array, we must go to the next larger one, the L_{16}, which has 15 dF. This array and some of its linear graphs are shown in Appendix A. The reader is encouraged to try to assign the factors to it before reading the solution below.

The first step is to use the linear graph. Since the interactions are spread almost evenly among the factors, an "equi-axed" linear graph is used. The experimenters chose the one labeled *i* in the L_{16} section of Appendix A, and the main effects are assigned as shown in Figure 7-1. (There are, of course, other correct solutions.) Factors E and F are assigned to lines instead of points, and thus are confounded with interactions. Since the subject interactions are not considered likely, however, this should not be a problem. Table 7-4 shows the L_{16} array with the main effects and interactions assigned according to Figure 7-1. It also may

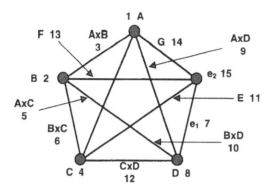

FIGURE 7-1   Linear graph for assignment of main effects and interactions for the wave solder experiment.

TABLE 7-4 L_{16} Array Showing Assignment of Main Effects, Interactions, and Empty Columns for the Wave Solder Experiment

Run	1	2	3	4	5	6	7	8	9	10	11	12	13	14	15
1	1	1	1	1	1	1	1	1	1	1	1	1	1	1	1
2	1	1	1	1	1	1	1	2	2	2	2	2	2	2	2
3	1	1	1	2	2	2	2	1	1	1	1	2	2	2	2
4	1	1	1	2	2	2	2	2	2	2	2	1	1	1	1
5	1	2	2	1	1	2	2	1	1	2	2	1	1	2	2
6	1	2	2	1	1	2	2	2	2	1	1	2	2	1	1
7	1	2	2	2	2	1	1	1	1	2	2	2	2	1	1
8	1	2	2	2	2	1	1	2	2	1	1	1	1	2	2
9	2	1	2	1	2	1	2	1	2	1	2	1	2	1	2
10	2	1	2	1	2	1	2	2	1	2	1	2	1	2	1
11	2	1	2	2	1	2	1	1	2	1	2	2	1	2	1
12	2	1	2	2	1	2	1	2	1	2	1	1	2	1	2
13	2	2	1	1	2	2	1	1	2	2	1	1	2	2	1
14	2	2	1	1	2	2	1	2	1	1	2	2	1	1	2
15	2	2	1	2	1	1	2	1	2	2	1	2	1	1	2
16	2	2	1	2	1	1	2	2	1	1	2	1	2	2	1
	A	B	A	C	A	B	e_1	D	A	B	E	C	F	G	e_2
			×		×	×			×	×		×			
			B		C	C			D	D		D			

be noted that there are two empty columns, labeled e_1 and e_2. Since only 13 factors were evaluated, this was inevitable. The empty columns are used to evaluate the overall noise in the experiment. It was decided to conduct three repetitions of the experiment, for a total of 48 circuit cards.

7.6 EFFECTS

There is only one effect, or response, to be evaluated in this experiment: the number of defects per board. They were determined by visual inspection. Both mean and signal-to-noise values are calculated for this effect.

7.7 LOGISTICS

After the experiment is designed, it would seem to be a simple matter to make the samples and conduct the experiment. In principle, this is correct. In practice, however, it can be very complicated to keep track of the various combinations of factor levels for each group of samples. Due to the nature of a designed experiment, the same set of samples are divided in several different ways at various

FIGURE 7-2 A benchtop laid out in grid pattern for samples in an L_{18} experiment. Note the experimental flow diagram above the samples. (Photo courtesy of ELDEC Corp., used by permission.)

points. This can become very confusing, and mistakes can be made if care is not exercised. Often, it is a good idea to dedicate a bench top or cabinet to store the samples between runs, and to lay out a grid of tape or boxes labeled with the array designations, as shown in Figure 7-2.

The next step after collecting of data is the analysis, which is the subject of Chapter 8.

REFERENCE

1. L. W. Condra, Value-Added Management with Design of Experiments, Chapman and Hall, London, 1995.

Chapter 8

Analyzing DoE Results

8.1 INTRODUCTION

We are now at a thrilling point in a well-designed experiment: the arrival of the data. We are beginning to reap tangible rewards from our planning. In this chapter, we conduct an analysis of variance on the data collected from the wave solder experiment, which was designed and conducted in Chapter 7. While data collection is the most visible part of an experiment, and may appear to an observer to be the major part of the effort, it is often the quickest and easiest part. The emphasis in DoE is to put considerable effort into designing and setting up the experiment, in order to be efficient in the costly enterprise of building and processing samples.

After the data from a properly designed experiment are collected, it is not unusual to discover information that was not previously expected. As they obtain experience, experimenters develop instincts for planning experiments which yield significant data well beyond that which was planned, and which is useful to other individuals and other departments. In the author's experience, it is possible to design and conduct experiments which will provide useful data for months and even years of future investigation. Methods for planning and executing this process are discussed in Chapter 22.

8.2   CALCULATING THE MEAN AND SIGNAL-TO-NOISE RATIO FOR THE WAVE SOLDER EXPERIMENT

The response in the wave solder experiment is the number of defects per card, as determined by visual inspection. The results for each of the three cards for each run of the L_{16} array are shown in Table 8-1. The equation used to calculate $\bar{Y}$ for each run is

$$\bar{Y} = \frac{Y_1 + Y_2 + \cdots + Y_n}{n} \tag{8-1}$$

Since the number of defects per card is a smaller-is-better criterion, it is calculated by

$$S/N = -10 \log \left(\frac{Y_1^2 \, Y_2^2 + \cdots + Y_n^2}{n} \right) \tag{8-2}$$

Sample calculations for $\bar{Y}$ and S/N for the first four runs of the array are shown in Table 8-2. The reader is encouraged to complete Table 8–1 before proceeding. Results of calculations for all runs are shown in Table 8-3. From this table, we

TABLE 8-1   Results of the Wave Solder Experiment

| Run no. | Defects per board | | | $\bar{Y}$ | S/N |
	Y_1	Y_2	Y_3		
1	13	30	26		
2	4	16	11		
3	20	15	20		
4	42	43	64		
5	14	15	17		
6	10	17	16		
7	36	29	53		
8	5	9	16		
9	29	0	14		
10	10	26	9		
11	28	173	19		
12	100	129	151		
13	11	16	11		
14	17	2	17		
15	53	70	89		
16	23	22	7		

TABLE 8-2 Sample Calculations for $\bar{Y}$ and S/N

$$\bar{Y} = \frac{Y_1 + Y_2 + \cdots + Y_n}{n}$$

$$\bar{Y} \text{ run } 1 = \frac{13 + 30 + 26}{3} = 23.0$$

$$\bar{Y} \text{ for run } 2 = \frac{4 + 16 + 11}{3} = 10.3$$

$$\bar{Y} \text{ for run } 3 = \frac{20 + 15 + 20}{3} = 18.3$$

$$\bar{Y} \text{ for run } 4 = \frac{42 + 43 + 64}{3} = 49.7$$

$$S/N = -10 \log \left(\frac{Y_1^2 + Y_2^2 + \cdots + Y_n^2}{n} \right)$$

S/N for run $1 = -10 \log (13^2 + 30^2 + 26^2) = -27.6$
S/N for run $1 = -10 \log (4^2 + 16^2 + 11^2) = -21.2$
S/N for run $1 = -10 \log (20^2 + 15^2 + 20^2) = -25.3$
S/N for run $1 = -10 \log (42^2 + 43^2 + 64^2) = -34.1$

TABLE 8-3 Completed $\bar{Y}$ and S/N Table for the Wave Solder Experiment

Run no.	Defects per board			$\bar{Y}$	S/N
	Y_1	Y_2	Y_3		
1	13	30	26	23.0	−27.6
2	4	16	11	10.3	−21.2
3	20	15	20	18.3	−25.3
4	42	43	64	49.7	−34.1
5	14	15	17	15.3	−23.7
6	10	17	16	14.3	−23.3
7	36	29	53	39.3	−32.2
8	5	9	16	10.0	−20.8
9	29	0	14	14.3	−25.4
10	10	26	9	15.0	−24.6
11	28	173	19	73.3	−40.2
12	100	129	151	126.7	−42.2
13	11	16	11	12.3	−21.9
14	17	2	17	12.0	−22.9
15	53	70	89	70.7	−37.2
16	23	22	7	17.3	−25.5

can see that we did indeed produce a wide variation in results among the runs and, from this point of view, the experiment was a success. Run 8 produced the best results, and run 12 was clearly the worst.

The next step is to calculate $\overline{\overline{Y}}$ and $\overline{S/N}$ for each main effect and interaction. A blank response table for this is shown in Table 8-4, and sample calculations are shown in Table 8-5.

TABLE 8-4   Blank Response Table for the Wave Solder Experiment

Factor	Level	$\overline{\overline{Y}}$	$\overline{S/N}$
A: Prebake	A_1		
	A_2		
B: Flux Density	B_1		
	B_2		
	$(A \times B)_1$		
	$(A \times B)_2$		
C: Conveyor Speed	C_1		
	C_2		
	$(A \times C)_1$		
	$(A \times C)_2$		
	$(B \times C)_1$		
	$(B \times C)_2$		
	$e1_1$		
	$e1_2$		
D: Upper preheat	D_1		
	D_2		
	$(A \times D)_1$		
	$(A \times D)_2$		
	$(B \times D)_1$		
	$(B \times D)_2$		
E: Cooling Time	E_1		
	E_2		
	$(C \times D)_1$		
	$(C \times D)_2$		
F: Omega	F_1		
	F_2		
G: Solder temperature	G_1		
	G_2		
	$e2_1$		
	$e2_2$		

TABLE 8-5 Sample Calculations for the Response Table for the Wave Solder Experiment

Effects for $\overline{\overline{Y}}$:

A_1: $\overline{\overline{Y}}$ = average of $\overline{Y}$'s for all runs in which A is at level 1

$$= \frac{23.0 + 10.3 + 18.3 + 49.7 + 15.3 + 14.3 + 39.3 + 10.0}{8} = 22.5$$

A_2: $\overline{\overline{Y}}$ = average of $\overline{Y}$'s for all runs in which A is at level 2

$$= \frac{14.3 + 15.0 + 73.3 + 126.7 + 12.3 + 12.0 + 70.7 + 17.3}{8} = 42.7$$

B_1: $\overline{\overline{Y}} = \frac{23.0 + 10.3 + 18.3 + 49.7 + 14.3 + 15.0 + 73.3 + 126.7}{8} = 41.3$

B_2: $\overline{\overline{Y}} = \frac{15.3 + 14.3 + 39.3 + 10.0 + 12.3 + 12.0 + 70.7 + 17.3}{8} = 23.9$

Effects for : $\overline{S/N}$

A_1: $\overline{S/N}$: = average of $\overline{S/N}$'s for all runs in which A is at level 1

$$= \frac{27.6 - 21.2 - 25.3 - 34.1 - 23.7 - 23.3 - 32.2 - 20.8}{8} = -26.0$$

A_2: $\overline{S/N}$: = average of $\overline{S/N}$'s for all runs in which A is at level 1

$$= \frac{-25.4 - 24.6 - 40.2 - 42.2 - 21.9 - 22.9 - 37.2 - 25.5}{8} = -30.0$$

B_1: $\overline{S/N}$: $= \frac{-27.6 - 21.2 - 25.3 - 34.1 - 25.4 - 24.6 - 40.2 - 42.2}{8} = -30.1$

B_2: $\overline{S/N}$ $= \frac{-23.7 - 23.3 - 32.2 - 20.8 - 21.9 - 22.9 - 37.2 - 25.5}{8} = -25.9$

The rule for interactions is: If both main effects participating in the interaction are at the same level, the interaction is at level 1; if the participating main effects are at different levels, the interaction is at level 2. In other words,

$1 \times 1 = 1$

$1 \times 2 = 2$

$2 \times 1 = 2$

$2 \times 2 = 1.$

The subscript at the end of the interaction is the level of the interaction, e.g., $(A \times B)_1$ indicates level one of the interaction, and not level 1 of either of the participating main effects.

The reader is urged to complete the table and compare results with those in Table 8-6.

At this point, we are in a position to draw some conclusions about the results of the experiment. The first place to look is in the empty columns, e1 and e2. Significant variation in these columns would indicate that we did not account

TABLE 8-6 Completed Response Table for the Wave Solder Experiment

Factor	Level	$\bar{\bar{Y}}$	$\overline{S/N}$
A: Prebake	A_1	22.5	−26.0
	A_2	42.7	−30.0
B: Flux density	B_1	41.3	−30.1
	B_2	23.9	−25.9
	$(A \times B)_1$	26.7	−27.0
	$(A \times B)_2$	38.5	−29.0
C: Conveyor speed	C_1	14.6	−23.8
	C_2	50.7	−32.2
	$(A \times C)_1$	43.9	−30.1
	$(A \times C)_2$	21.4	−25.9
	$(B \times C)_1$	25.0	−26.8
	$(B \times C)_2$	40.3	−29.2
	$e1_1$	38.4	−28.6
	$e1_2$	26.9	−27.4
D: Upper preheat	D_1	33.3	−29.2
	D_2	31.9	−26.8
	$(A \times D)_1$	33.4	−28.0
	$(A \times D)_2$	31.9	−28.0
	$(B \times D)_1$	22.8	−26.4
	$(B \times D)_2$	42.4	−29.6
E: Cooling time	E_1	36.3	−27.9
	E_2	29.0	−28.1
	$(C \times D)_1$	33.6	−27.7
	$(C \times D)_2$	31.7	−28.3
F: Omega	F_1	33.6	−28.9
	F_2	31.6	−27.1
G: Solder temperature	G_1	43.8	−30.6
	G_2	21.5	−25.4
	$e2_1$	30.5	−28.7
	$e2_2$	34.7	−27.3

for all the important factors when we set up the experiment, and that some of their effects are showing up here. Since the differences in levels for both e1 and e2 appear to be small, there is no significant factor operating in either of these columns.

If we use the rule of thumb that a difference of 3 dB in the signal-to-noise ratio is significant, we see that four main effects and two interactions are significant: A, B, C, G, A $\times$ C, and B $\times$ D. This leads to the conclusion that the following levels of main effects should be used:

A_1 B_2 C_1 G_2

Although the interactions A $\times$ C and B $\times$ D are significant, they cannot be set directly because their levels are determined by the levels of their participating main effects.

From Table 8-6, we see that the best level for A $\times$ C is level 2, indicating that both A and C should not be set at the same level. Since A_1 and C_1 are the preferred levels for those main effects, we must make a choice between optimizing them and optimizing their interaction. In cases like this, it is the usual practice of most experimenters to optimize the main effects, and we will do so here. That is by no means a rule, however, and each case must be decided on its individual merits. $(B \times D)_1$ is the preferred level for that interaction, and since the preferred levels of B and D are both 2, it is possible to optimize both B and D, as well as their interaction.

8.3   ANALYSIS OF VARIANCE

Analysis of variance (ANOVA) is somewhat complicated, and the calculations can appear formidable, but it is not conceptually difficult. The process is presented here in a step-by-step procedure which the reader can learn and use for future reference. The format is such that the process can be "copied" into other applications using data from other experiments.

For ease of presentation, only the $\bar{Y}$ results are considered here. ANOVA can be applied just as easily to S/N, but those calculations are arithmetically more difficult, and are omitted here. It is at this point that the subject of software for DoE usually comes up, since it is capable of taking the drudgery out of planning and analyzing DoE projects. Software is discussed at the end of this chapter.

Following the procedure used throughout this book, sample calculations are presented, and the reader is encouraged to perform further calculations to become familiar with the process.

Table 8-7 is a completed ANOVA table, and we shall step through it column by column. The first column, labeled "Factor level," is a listing of the factors and levels from the response table.

TABLE 8-7 Completed ANOVA Table for the Wave Solder Experiment

Factor level	$\bar{Y}$ total	dF	S_x	Pool?	dF_e	S_e	V_x	V_e	F	S'_x	ρ, (%)
A_1	180.3	1	1626.78	no			1626.78		13.22	1503.74	9.4
A_2	341.7										
B_1	330.7	1	1213.36	no			1213.36		9.86	1090.32	6.8
B_2	191.3										
$(A \times B)_1$	213.7	1	560.11	no			560.11		4.55	437.07	2.7
$(A \times B)_2$	308.3										
C_1	116.7	1	5208.03	no			5208.03		42.33	5084.99	31.9
C_2	405.3										
$(A \times C)_1$	351.0	1	2025.00	no			2025.00		16.46	1901.96	11.9
$(A \times C)_2$	171.0										
$(B \times C)_1$	200.0	1	930.25	no			930.25		7.56	807.21	5.1
$(B \times C)_2$	322.0										
$e1_1$	307.0	1	529.00	yes	1	529.00	529.00	529.00			
$e1_2$	215.0										

Source											
D₁	266.7	1	8.03	yes	1	8.03	8.03	8.03			
D₂	255.3	1	9.00	yes	1	9.00	9.00	9.00			
(A × D)₁	267.0	1		no					12.47	1410.99	8.9
(A × D)₂	255.0										
(B × D)₁	182.7	1	1534.03	yes		1534.03					
(B × D)₂	339.3										
E₁	290.3	1	215.11	yes	1	215.11	215.11	215.11			
E₂	231.7										
(C × D)₁	268.7	1	14.69	yes	1	14.69	14.69	14.69			
(C × D)₂	253.3										
F₁	269.0	1	16.00	yes	1	16.00	16.00	16.00			
F₂	253.0										
G₁	350.0	1	1980.25	no		1980.25			16.09	1857.21	11.7
G₂	172.0										
e2₁	244.3	1	69.44	yes	1	69.44	69.44	69.44		1845.60	11.6
e2₂	277.7								123.04		
e total					7		861.28	1062.61	123.04		
Total	15939.08	15						1062.61			

8.4   TOTAL VARIATION

The second column of Table 8-7, labeled "$\overline{Y}$ total," is the sum of all the $\overline{Y}$'s for each level of each factor. For example, the $\overline{Y}$ total for A_1 is the sum of all the $\overline{Y}$'s for which factor A is at level 1. From Table 8-3,

$$A_1 = 23.0 + 10.3 + 18.3 + 49.7 + 15.3 + 14.3 + 39.3 + 10.0 = 180.3$$
$$A_2 = 14.3 + 15.0 + 73.3 + 126.7 + 12.3 + 12.0 + 70.7 + 17.3 = 341.7$$

As we go down this column, we see that the sum of the $\overline{Y}$ totals is the same for each factor (discounting rounding effects). This is because we are using the same data for each factor; but we are grouping the runs differently for each factor. This is an important point, because it means we are using all the data from the entire experiment in the analysis of each factor.

8.5   DEGREES OF FREEDOM

The third column of Table 8-7, labeled dF, is the number of degrees of freedom. For a Taguchi experiment, the number of degrees of freedom of a factor is always equal to the number of levels of that factor, minus 1. All the two-level factors of this experiment have 1 degree of freedom.

8.6   CALCULATING THE SOURCE VARIATION

The source variation of a factor is the quantitative measure of the magnitude of its effect due to changes in its level. It is calculated by the equation

$$S_x \left(\frac{X_1^2 + X_2^2}{n/2} \right) - \left(\frac{T^2}{n} \right) \tag{8-3}$$

where

$\qquad X_1$ = sum of data for factor X at level 1
$\qquad X_2$ = sum of data for factor X at level 2
$\qquad n$ = total number of data points
$\qquad T$ = total of all the data

It is apparent from this equation that, the greater the difference in the $\overline{Y}$'s between the two levels of a factor, the greater is S_X. For factor A:

$\qquad A_1 = 180.3$
$\qquad A_2 = 341.7$
$\qquad n = 16$
$\qquad T = A_1 + A_2 = 180.3 + 341.7 = 522.0$

and

$$S_A = \frac{180.3^2 + 341.7^2}{16/2} - \frac{522.0^2}{16}$$

$$= 1626.78$$

Calculation of S_X for the remainder of the factors is left as an exercise for the reader.

8.7   POOLING

In the fourth column of Table 8-7, we list whether or not the variation of the factor is pooled. To understand this, we must back away from our data for a time, and consider the overall purpose of ANOVA. We are trying to determine if individual factors of an experiment are significant by comparing their variation with the overall variation of the data from the experiment. In order to do this, we need a good estimate of that overall variation, which is sometimes called *error variation*.

The error variation can be estimated in two ways. The first is to use empty columns. If a column has not been assigned a factor, and if no unknown cause is confounded with it, its variation should be random, and approximately equal to that of the entire data base. In this experiment, we have two empty columns.

The second way to estimate error variation is to use columns which have been assigned factors which were suspected to be significant, but which are shown by the results to be insignificant. In general, if the variation of a factor is less than that of an error column or if it is significantly lower than that of some other columns with factors in them, it can be considered to be random, and it can be *pooled* with other insignificant factors and error columns to provide a data base for estimating the random variation of the experiment. The decision of whether or not to pool data from a particular factor can be somewhat subjective, however, and sometimes it is decided not to pool certain data even though their variation is less than that of an error column.

In experiments containing no error columns, a judgment is made to pool some of the least significant factors.

In this experiment, all the S_X's were compared with those of the larger of the source variations of the two error columns. The source variations of the two error columns are 529.00 ($e1$) and 69.44 ($e2$). If a given S_X was less than the S_X for either of the error columns, it was pooled. In this experiment, all factors with S_X less than 529.00 were pooled. There were seven such factors:

$e1, D, A \times D, E, C \times D, F,$ and $e2$

These factors are designated by a "yes" in the column labeled "Pool?" in Table 8-7. Their values are then transferred to the columns labeled dF$_e$ (degrees of

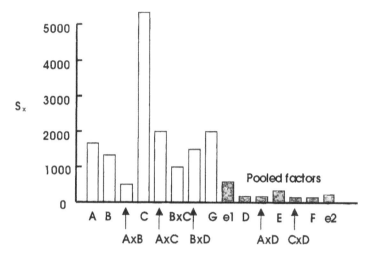

FIGURE 8-1   Graphical illustration of pooling factors for the wave solder experiment. The shaded columns represent pooled factors; the unshaded columns represent significant factors.

freedom of the error) and S_e (source variation of the error). The total degrees of freedom for the error terms is 7, and the total source variation of the error is 861.28.

Pooling is illustrated graphically in Figure 8-1, which is a sort of Pareto chart of the S_X values for all of the main effects and interactions. The lengths of the bars are proportional to their contribution to the variation of the process being optimized. These results can be used to quantify the importance of each factor, and to prioritize activities for improvement.

8.8   VARIANCE OF THE SOURCE

The *variance of the source* V_X is the variation of the source corrected for the number of degrees of freedom, according to the equation

$$V_X = \frac{S_X}{dF_X}$$

(8-4)

Since we are dealing in this experiment only with two-level factors, all source variances are equal to their source variations, e.g.,

$$V_A = \frac{1626.78}{1} = 1626.78$$

$$V_B = \frac{1213.36}{1} = 1213.36$$

The V_X values are shown for all factors in Table 8-7, but the V_e values are shown only for the pooled factors. It should also be noted that the total error variation at the bottom of the V_e column is equal to $S_e/7$, or 123.04.

8.9   THE F-TEST

The F-ratio of the source is calculated for significant, or unpooled, factors only. It is used in the F-test, which is a statistical test for significance. The equation is

$$F_X = \frac{V_X}{V_e} \qquad (8\text{-}5)$$

V_e is the total error variance, shown at the bottom of the V_e column in Table 8-7. For this experiment,

$$F_A = \frac{1626.78}{123.04} = 13.22$$

$$F_B = \frac{1213.36}{123.04} = 9.86$$

The F ratio is illustrated in Figure 8-2, which shows a dotted line drawn at the e total value of 123.04. The F ratio for each of the unpooled factors is the

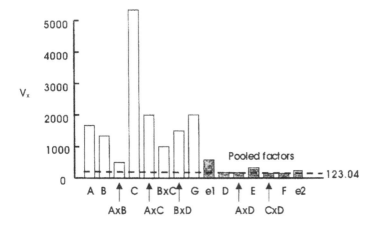

FIGURE 8-2   Graphical representation of the variance of the source calculation of the F value for the wave solder experiment.

ratio of the length of its bar above the dotted line to the length below the dotted line. From Figure 8-2, it is apparent that, the greater the V_X, the greater is the F ratio.

The F ratios are then compared with the values from the F tables shown in Appendix B. In this experiment, the number of degrees of freedom in the numerator is 1, and that for the denominator is 7. The F value for 95% confidence is 5.59. From the column labeled F in Table 8-7, we can see that the following factors are significant at this level:

A, B, C, A × C, B × C, B × D, and G

The F value for 99% confidence is 12.25, and at this level A, C, A × C, B × D, and G are significant.

8.10   PURE VARIATION OF THE SOURCE

The *pure variation of the source* is calculated only for unpooled factors. It is the variation of a factor with the portion due to error variance removed, and is calculated by

$$S'_X = S_X - (dF \cdot V_e) \qquad (8\text{-}6)$$

For this experiment,

$$S'_A = 1626.78 - (1)(123.04) = 1503.74$$
$$S'_B = 1213.36 - (1)(123.04) = 1090.32.$$

8.11   PERCENT CONTRIBUTION OF THE SOURCE TO TOTAL VARIATION

The percent contribution of the source to the total variation also is also calculated only for unpooled factors. It is designated by the Greek letter ρ, and is calculated by

$$\rho_X = \left(\frac{S'_X}{S'_T}\right) \times (100\%) \qquad (8\text{-}7)$$

For this experiment,

$$\rho_A = 1503.74/15939.08 = 9.4\%$$
$$\rho_B = 1090.32/15939.08 = 6.8\%$$

It can be seen from the ρ column of Table 8-7 that 31.9% of the total variation in the experiment was due to factor C, and that the interaction A × C and factor G were the next two largest contributors. Overall, 88.4% of the variation was accounted for by the factors investigated. By this measure, the experiment could be considered a success.

8.12 CONCLUSIONS OF THE WAVE SOLDER EXPERIMENT

Upon completion of the analysis of variance, we now have four different quantitative measures of the effects of the various factors:

1. The mean, or $\overline{\overline{Y}}$ of the number of defects
2. The signal-to-noise ratio of the number of defects
3. The F ratio of the number of defects
4. The percent contribution to the total variation

(We also could have calculated F ratios and percent contributions using the signal-to-noise values.)

Fortunately in this case, the factors and levels we chose for evaluation are so significant that there is no ambiguity in choosing the optimum values for our process. They are

A_1, B_2, C_1, D_2, E_2, F_2, and G_2

The preferred levels of interactions are

$(A \times B)_1$, $(A \times C)_2$, $(B \times C)_1$, and $(B \times D)_1$.

If the preferred levels of the main effects are used, the only interaction which will be optimized is $B \times D$. Since in this case we attach more significance to main effects, the other interactions will not be optimized.

The optimized factors and their percent contributions to the overall variation are

A: Prebake conditions	9.4%
B: Flux density	6.8%
C: Conveyor speed	31.9%
D: Preheat conditions	Pooled
E: Cooling time	Pooled
F: Omega	Pooled
G: Solder temperature	11.7%
H: B × D	8.9%

By optimizing these factors, we should be able to account for 68.7 percent of the variation in the process.

The two largest contributors to defect rates were solder temperature and conveyor speed. These two conditions can be optimized by setting them at levels which increase the heat input to the board, i.e., slower conveyor speed and higher solder temperature. Other less significant factors also showed the same tendency. The result was that, when all of them were put at their optimum levels, the cards were discolored by excessive heat. Thus the confirmation run was made at conditions slightly less heat intensive than those suggested by the experiment.

In the confirmation run, the conditions suggested by the experiment, with the slight modifications described above, were compared with previous production settings. The results showed a 43% reduction in defects. Control chart readings over the next several months indicated an annual cost savings rate equal to approximately three times the cost to conduct the experiment. Since this was the first DoE project for all of the participants, it was considered highly successful.

8.13   AUTOMATED DESIGN OF EXPERIMENTS

It is clear that designing, setting up, and analyzing an experiment, while not conceptually difficult, is a time-consuming process, subject to arithmetic errors and repetitive in nature. Since humans are incredibly intelligent, incredibly slow, and incredibly inaccurate, and since computers are incredibly stupid, incredibly fast, and incredibly accurate, these tasks are ideal for a computer.

One of the quickest and least expensive ways to use a computer for DoE is to develop a spreadsheet for the analysis. The author has done so on several occasions, and with a little experience, spreadsheets for analysis of a typical L_{16} experiment can be completed in between 20 and 40 hours. While this is a way to avoid the cost of purchased software, it is an inefficient use of engineering time and talent. There are many excellent companies who have DoE software on the market. They are too numerous to mention here, but can be found easily by searching the internet.

Most of the software packages available for DoE will take the experimenter's inputs for desired factors, levels, and responses, and will produce a customized array to fit specific needs. The packages also will produce data sheets to be filled out during data collection. After the data are collected and fed back to the computer, the software packages conduct the calculations. For anyone who is serious about using DoE, this type of software will quickly become a necessity. The reader is cautioned, however, against moving to the software solution too quickly. It is suggested that some hands-on experience of the nature described in the last few chapters be gained before completely trusting the software. Also, some software packages are limited in size, and if a very large experiment (above about 40 runs) is contemplated, the software may not be able to handle it.

With this analysis, we have covered most of the basic skills of DoE. The examples, however, have been limited to parameter optimization of process design and improvement. In Chapter 9, we look at some product design examples.

Chapter 9

DoE in Product Design

9.1 INTRODUCTION

Up to this point, the DoE examples all have been directed toward developing and optimizing manufacturing processes. These examples are the easiest to understand, and therefore are most commonly used in teaching. This is in a sense unfortunate, because it gives some people the impression that DoE is a tool only for process development and improvement, and they do not try to apply it elsewhere. In fact, DoE can be used in a variety of applications, such as concurrent supplier and customer product development, marketing, etc. It was pointed out in Chapter 1 that the major costs of a product are committed even before the processes to manufacture it are defined, and the best opportunities for cost, time, quality, and reliability improvement are in product design. Product design, therefore, is the focus of this chapter.

DoE can be used in product design in exactly the same way as it is used in process design and improvement. The important design factors are determined and evaluated, and those which produce the lowest-cost, highest quality, and most reliable product are selected.

Product design is viewed by many as a discipline where intuition and inspiration are the most valuable commodities, and efforts to systematize and organize these skills are viewed as attempts to stifle creativity. Experience has shown, however, that the most creative and productive designers have been made even more effective with the use of tools such as DoE. It is appropriate here to restate a point made earlier: DoE does not make the user more intelligent; it merely helps organize the user's intelligence and make the user more efficient.

In this chapter, we review three examples of the use of DoE in product design: a mechanical one and two electrical ones.

9.2   THE FUEL RAIL DESIGN EXAMPLE

An automotive parts manufacturer wanted to optimize the design of a thermoplastic fuel rail, in order to maximize its strength, and thus its ability to withstand the mechanical stresses imposed upon it in service* [1]. The first prototype brackets were so weak that they could be snapped off by hand.

It was decided to improve the design by using finite element analysis (FEA) to evaluate several different design features that might impact the strength of the fuel rail. The first design to be evaluated was produced by the best judgment of the design engineers. The design was modeled, and a "worst-case" static load was applied. The results were disappointing. Ten algebraic maxima were obtained, and all of them were considered too high. (Algebraic maxima in FEA are simply points of relative maximum stress, and are located in different spots with different designs.) Improvements were required.

The usual approach to this type of problem is to continue to try various options and use FEA to analyze the results until a satisfactory result is obtained. Since many design features could be changed (in this case seven were involved), and since each analysis is expensive and time consuming (even with FEA), the prospects for success using the traditional approach were not good. The designers decided to use DoE to improve the efficiency of their design process.

9.2.1   Factors in the Fuel Rail Design Array

Seven design features were chosen as factors in the fuel rail design using a Taguchi L_8 array. Used in this manner, the L_8 is a screening array, since only main effects and no interactions were evaluated. Figure 9-1 shows the seven factors, and they are listed along with their levels in Table 9-1. Table 9-2 shows their assignment to the L_8 array. Two-dimensional sketches of the fuel rail for the eight runs are shown in Figure 9-2.

Most of the factors represent *design decisions*, that is, the levels are the presence or absence of a given feature, such as a radius, or a supporting rib, or a bolt pad, etc. These are examples of *system design* decisions. Only factor *E*,

---

* This example is reported in Ref. 1, and is used by permission of the publisher, the American Supplier Institute. The project was conducted by Flex Technologies, Inc.

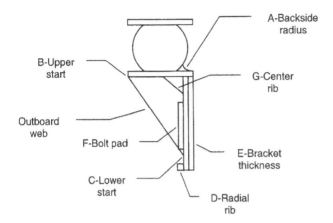

FIGURE 9-1 Schematic drawing of the fuel rail, with the features chosen as factors in the fuel rail design experiment. (© 1987 American Supplier Institute, used by permission.)

the bracket wall thickness, is a purely parametric factor, which can be considered a *parameter design* decision. In general, system and parameter design features fit well into inner arrays for product design, while tolerance design features are best dealt with in outer arrays. (Inner and outer arrays are described in Chapter 10.)

Since this was a main effects experiment, the factors were assigned arbitrarily to the columns. The array is heavily confounded (the orthogonal columns in the L_8 array are 1, 2, and 4, and the other columns contain potential interactions

TABLE 9-1 Factors and Levels for the Fuel Rail
Design Experiment

Factor	Level 1	Level 2
A: Backside radius	None	5.0 mm
B: Upper start of outboard web	Rail edge	C/L
C: Lower start of outboard web	Bottom	C/L
D: Radial rib on bracket face	None	2.0 mm
E: Bracket wall thickness	2.0 mm	4.0 mm
F: Bolt pad on bracket	2.0 mm	None
G: Center rib on bracket	Yes	No

Source: From Ref. 1. © 1987 American Supplier Institute, used by permission.

TABLE 9-2 Assignment of Factors to the L_8 Array for the Fuel Rail Design Experiment

Run no.	A	B	C	D	E	F	G
1	1	1	1	1	1	1	1
2	1	1	1	2	2	2	2
3	1	2	2	1	1	2	2
4	1	2	2	2	2	1	1
5	2	1	2	1	2	1	2
6	2	1	2	2	1	2	1
7	2	2	1	1	2	2	1
8	2	2	1	2	1	1	2

Source: From Ref. 1. © 1987 American Supplier Institute, used by permission.

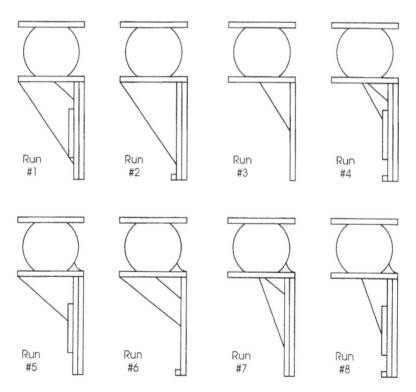

FIGURE 9-2 Schematic drawings of the seven runs of the fuel rail design experiment. (© 1987 American Supplier Institute, used by permission.)

among them), but in this case, the designers decided that the likelihood of interactions was small.

9.2.2 Results and Analysis of the Fuel Rail Design Experiment

The conditions for the eight experimental FEA runs were defined by the array. These eight runs were conducted using "worst case" loading, and in each run the experimenters recorded the 10 largest algebraic maxima, or locations in the design where mechanical stresses were highest. These locations were not the same for all runs, since there were large design variations among runs. The results are shown in Table 9-3.

Table 9-4 shows the responses from the experiment, using the signal-to-noise ratio as the evaluation criterion. Since lower stresses were the most desirable, the S-type, or smaller-is-better, formula was used to calculate signal-to-noise ratio:

$$S/N = -10 \log \left(\frac{Y_1^2 + Y_2^2 + \cdots + Y_n^2}{n} \right) \qquad (9\text{-}1)$$

In this case, the ten Y's observed for each run were the 10 algebraic maxima.

From Table 9-4, we see that run 2 produced the highest (least negative) S/N, and run number 3 produced the lowest. A response table is shown as Table 9-5, where $\overline{S/N}$ is shown for each level of each factor.

From Table 9-5, it may be seen that factors B, C, and E were the largest contributors to the variation in the results, because the differences between the two levels were greatest for these factors. This was confirmed by an analysis of variance, the results of which are shown in Table 9-6. The ANOVA was con-

TABLE 9-3 Ten Highest Algebraic Maxima (in psi) for Each Run of the Fuel Rail Design Experiment

Run	1	2	3	4	5	6	7	8	9	10
1	89	89	87	84	83	81	77	77	76	76
2	72	71	70	68	67	62	61	61	60	60
3	334	312	310	268	259	256	244	238	231	224
4	129	128	125	120	120	118	118	116	113	113
5	117	117	115	112	111	106	104	104	103	100
6	153	149	142	139	139	135	134	124	122	123
7	122	113	112	112	111	110	104	101	100	99
8	170	152	143	133	131	123	119	118	115	109

Source: From Ref. 1. © 1987 American Supplier Institute, used by permission.

TABLE 9-4 Signal-to-Noise Ratio for Each of the Runs of the Fuel Rail Design Experiment

Run no.	A	B	C	D	E	F	G	S/N
1	1	1	1	1	1	1	1	−38.28
2	1	1	1	2	2	2	2	−36.32
3	1	2	2	1	1	2	2	−48.63
4	1	2	2	2	2	1	1	−41.49
5	2	1	2	1	2	1	2	−40.76
6	2	1	2	2	1	2	1	−42.69
7	2	2	1	1	2	2	1	−40.71
8	2	2	1	2	1	1	2	−42.44

Source: From Ref. 1. © 1987 American Supplier Institute, used by permission.

TABLE 9-5 Signal-to-Noise Response Table for the Fuel Rail Experiment

Factor	A	B	C	D	E	F	G
Level 1	−164.82	−158.05	−157.75	−168.38	−172.04	−163.07	−163.27
Level 2	−166.60	−173.37	−173.67	−163.04	−159.38	−168.35	−168.15
Total	−331.42	−331.42	−331.42	−331.42	−331.42	−331.42	−331.42

Source: From Ref. 1. © 1987 American Supplier Institute, used by permission.

TABLE 9-6 Analysis of Variance Table for the Fuel Rail Design Experiment

Factor	dF	S_X	V_X	S'_X	ρ (%)
A	1	0.40	0.40	pooled	
B	1	29.34	29.34	26.73	29.2
C	1	31.38	31.38	29.08	31.8
D	1	3.56	3.56	pooled	
E	1	20.03	20.03	17.43	19.1
F	1	3.48	3.48	pooled	
G	1	2.98	2.98	pooled	
e'	4	10.42	2.61	18.24	19.9
Total	7	91.48		91.48	100.0

Source: From Ref. 1. © 1987 American Supplier Institute, used by permission.

ducted in exactly the same manner as that discussed in Chapter 8, with the exception that there were no error columns in this experiment. The error term e' was obtained by pooling results from the four factors judged least significant. The three most significant factors, B, C, and E, accounted for over 80% of the total variation. On this basis, the experiment can be considered successful.

9.2.3 Selection of Factor Levels for the Fuel Rail Design Experiment

The actual selection of levels for the factors evaluated was based on results obtained from the experiment, along with other engineering information available to the designers. These selections, and reasons for them, are shown in Table 9-7. Note that only levels for factors B and C were selected purely on the basis of experimental results. Some other factor levels were selected on the basis of cost. Factor E, bracket wall thickness, was one of the more significant factors. Even though level 2, (4.0 mm) yielded the best results, level 1 (2.0 mm) was chosen because 4.0 mm was subsequently found to be too thick for consistently good molding.

In their selection of factor levels for the design of the fuel rail, the design engineers exhibited good judgment in using all the information available to them, and not just the results from the experiment. They had sufficient knowledge and

TABLE 9-7   Selection of Factor Levels for the Fuel Rail Design Experiment

Factor	Level	Description	Reason
A	2	5.0 mm backside radius	Increased strength in some types of loads
B	1	Upper start of webs at outboard position	Best condition as identified by the experiment
C	1	Lower start of webs at bottom position	Best condition as identified by the experiment
D	1	No radial rib on bracket face	Rib did not affect stress and is expensive to tool
E	1	2.0 mm thick brackets	4.0 mm is too thick for the material, sufficient stress reduction obtained by picking best levels of other factors
F	2	No bolt pad on bracket	Pad did not affect stress and is expensive to tool
G	2	No center rib	Rib did not affect stress and would add complexity to design

Source: From Ref. 1. © 1987 American Supplier Institute, used by permission.

control of the statistics of the analysis to be able to decide when to let the statistics dictate their actions, and when to let other engineering knowledge override the statistics.

Using the factor levels selected above, a confirmation run was conducted, with the following algebraic maxima: 87.53, 87.21, 86.96, 86.01, 84.52, 83.56, 81.45, 77.46, 76.43, and 75.05. The signal-to-noise ratio for the confirmation run was -38.35. The use of the new design resulted in significant financial benefit to the manufacturer.

9.3   THE ANALOG CIRCUIT DESIGN EXAMPLE

An electronics manufacturer applied DoE to the design of an analog application specific integrated circuit (ASIC), to determine both the optimum design topology and the design parameters of a specific circuit [2]. Since this is a somewhat specialized field, the terms used in this experiment may not be familiar to a wide range of readers, but the project illustrates the application of DoE to a discipline which is usually dominated by intuition, as opposed to the systematic approach. In other words, if DoE can be successful in analog circuit design, it can be successful almost anywhere.

9.3.1   Factors and Levels for the Analog Circuit Design Experiment

Six main effects, at two levels each, and nine potential interactions were chosen for evaluation, and are shown in Table 9-8. The factors indicated by asterisks were those which changed the topology of the circuit, as opposed to the parametric factors. In a gross manner, the asterisked factors may be considered system design factors, and the rest may be considered parametric factors. (The analog design process is a bit more complicated than is indicated here, and other qualifi-

TABLE **9-8**   Main Effects and Levels for the Analog Design Experiment

Factor	Level 1	Level 2
A: Input bias current	1 μA	100 μA
B: Size of input transistors	1 × unit transistor size	4 × unit transistor size
C*: Cascode the input transistors	Yes	No
D: Ratio of input transistors to load transistors	1 : 1	1 : 4
E*: Type of input transistor	p channel	n channel
F*: Cascode the load transistors	Yes	No

cations should be made for our readers who are analog designers. To the overwhelming relief of our readers who are not analog designers, those qualifications are considered beyond the scope of this book.)

An L_{16} array was chosen for this experiment. The six main effects require 6 degrees of freedom, and the nine interactions used up the remaining available dF in the array. Although interactions were originally thought to be of great interest, results showed that none were significant, and they are not discussed further.

9.3.2 Effects of the Analog Design Experiment

Four effects were measured in this experiment:

Input offset voltage (nominal-is-best)
Unity gain bandwidth (bigger-is-better)
Low frequency gain, open loop (bigger-is-better)
Power consumption (smaller-is-better)

The best result for this experiment is a circuit in which the lowest power consumption and an input offset voltage near zero are achieved for the highest unity gain bandwidth and low frequency open loop gain. Results were obtained by simulation.

9.3.3 Results and Analysis of the Analog Design Experiment

As mentioned earlier, the interactions among the factors were insignificant. The same was found to be true for the variations in input offset voltage. Therefore the response table (Table 9-9) shows only main effects responses for bandwidth, gain, and power consumption. Since simulation was used, only one sample per run was necessary. Therefore, the responses were the values observed in a single sample.

For two factors, cascoding the load and the type of input transistor, the same level optimized all the effects. The best choices were to cascode the load transistors, and to use p-channel input transistors.

Cascoding the input transistors provided a small benefit in gain, but both power consumption and bandwidth were improved by not doing so. Therefore, it was decided not to cascode the input transistors. This decision also simplified the design.

Bias current accounted for 91% of the variation in power consumption and, in order to minimize it, the bias current was set at the lower level. This large effect on power consumption allowed the designers to choose levels of the size of the input transistors and ratio of the input transistors to the load transistors that optimized the bandwidth.

TABLE **9-9**   Response Table for the Analog Circuit Design Experiment

Factor	Level	Bandwidth		Gain		Power consumption	
		Mean	%	Mean	%	Mean	%
A: Input bias current	1	402	70.6	134	39.4	0.01	91.0
	2	5255		110		3.06	
B: Size of input	1	3695	9.0	111	31.4	1.90	5.2
transistor	2	1962		132		1.17	
C: Cascode input	1	2154	5.5	125	3.2	1.58	0.1
transistor	2	3503		118		1.50	
D: Ratio of input to	1	3818	11.7	124	0.9	1.84	3.7
load transistor	2	1839		120		1.23	
E: Type of input	1	3322	2.9	128	9.6	1.51	0.0
transistor	2	2335		116		1.5	
F: Cascode load	1	2963	0.2	129	15.5	1.52	0.0
transistor	2	2694		115		1.55	

The factor levels chosen as a result of this experiment were:

Bias current	10 μA
Input transistor size	1×
Cascode the input	No
Channel type	P
Ratio of input to load	1 : 1
Cascode the load	Yes

A confirmation run was made with these levels, and the following performance characteristics were obtained:

Input offset voltage	191.110 μV
Bandwidth	4.272 MHz
Gain	117.7 dB
Power consumption	0.451 mW
Phase margin	45.5°

Although individual runs of the experiment had some individual responses which were better than those of the confirmation run, none of the experimental runs produced a combination of responses that were judged to be as good as the confirmation run.

9.4   PRODUCT DESIGN FACTORS

As mentioned above, both system design and parameter design factors were used in the examples reported here. There is certainly nothing incorrect in combining both types in a product design experiment. Some designers, however, prefer to conduct a preliminary screening experiment, in which only design decisions, or system design factors, are evaluated first. After these decisions are made, a second experiment is made to evaluate different parametric levels of the factors chosen in the screening experiment.

The third type of design factor, tolerance design, is best evaluated in an outer array. Therefore, discussion of tolerance design is deferred until after outer arrays are discussed in Chapter 10.

9.5   THE DIGITAL INPUT BOARD EXPERIMENT

A third example of using DoE in product design is the digital input board [3], in which the experimenters wanted to set the low and high threshold voltages at levels that would assure no false readings. Figure 9-3 illustrates the high and low threshold voltages. Ideally, there would be no difference between them, but each of them has a range of values because of variations in environmental and operating conditions. The environmental factor that affects threshold voltage is the temperature, and the operating factors are the supply voltage, the frequency, and the wave shape. Because of variations in these factors, there must be a "gap" between the high and the low threshold voltages, to avoid a false signal. The purpose of this experiment was to

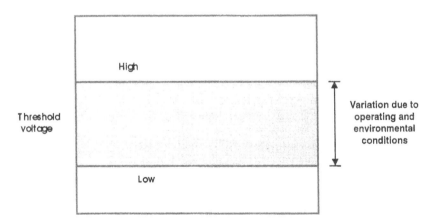

FIGURE 9-3   Schematic illustration of the high and low threshold voltage separation in the digital input board example. (© 1990, John Wiley & Sons, used by permission.)

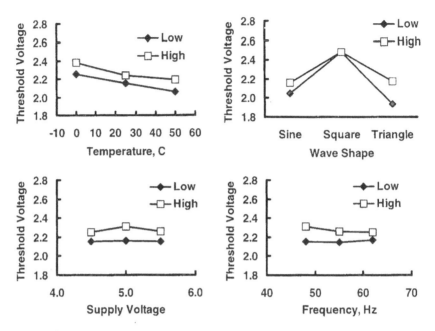

FIGURE 9-4   Response plots for the digital input board experiment. (© 1990, John Wiley & Sons, used by permission.)

set the high threshold voltage high enough, and the low threshold voltage low enough, so that a false signal would never be observed.

Table 9-10 shows the Taguchi L_9 experimental array used to evaluate each of the three factors at three levels. The low and high threshold voltages are shown for each of the nine runs. Table 9-11 is a response table for each of the four factors.

To provide a visual aid in evaluating the effects of the factors, the data in the response table were plotted, and the results are shown in Figure 9-4. From the response table and the plots, the following levels appear to produce the most extreme results:

Factor	Low threshold voltage	High threshold voltage
Temperature	50°C	0°C
Supply voltage	5.5 V	5.0 V
Frequency	55 Hz	48 Hz
Wave shape	Triangular	Square

TABLE 9-10   Experimental Array and Results for the Digital Input Board Experiment

Run	Temp., °C	Supply voltage	Frequency, Hz	Wave shape	Threshold Voltage	
					Low	High
1	0	4.5	48	sine	2.146	2.281
2	0	5.0	62	square	2.605	2.605
3	0	5.5	55	triangle	2.022	2.261
4	25	4.5	62	triangle	1.945	2.093
5	25	5.0	55	sine	2.040	2.148
6	25	5.5	48	square	2.476	2.476
7	50	4.5	55	square	2.370	2.370
8	50	5.0	48	triangle	1.846	2.171
9	50	5.5	62	sine	1.954	2.048

Source: From Ref. 3. © 1990, John Wiley & Sons, used by permission.

TABLE 9-11   Response Table for the Digital Input Board Experiment

Factor	Level	Value	Low threshold voltage	High threshold voltage
Temperature	1	0°C	2.258	2.382
	2	25°C	2.154	2.239
	3	50°C	2.057	2.196
Supply voltage	1	4.5 V	2.154	2.248
	2	5.0 V	2.164	2.308
	3	5.5 V	2.151	2.262
Frequency	1	48 Hz	2.156	2.309
	2	55 Hz	2.144	2.260
	3	62 Hz	2.168	2.249
Wave shape	1	Sine	2.047	2.159
	2	Square	2.484	2.484
	3	Triangle	1.938	2.175

Source: From Ref. 3. © 1990, John Wiley & Sons, used by permission.

The experimenters used the above results to calculate the values at which to set the low and high threshold voltages. To do so, they calculated the overall mean of the low threshold voltage for all the runs (2.156 V) and used the following equations:

Low threshold voltage = overall mean value

+ (50°C value − overall mean)

+ (5.5 V value − overall mean)

+ (55 Hz value − overall mean)

+ (triangular value − overall mean)

= 2.156 + (2.057 − 2.156) + (2.151 − 2.156)

+ (2.144 − 2.156) + (1.938 − 2.156)

= 1.822 V

and

High threshold voltage = overall mean value

+ (0°C value − overall mean)

+ (5.0 V value − overall mean)

+ (48 Hz value − overall mean)

+ (square value − overall mean)

= 2.273 + (2.382 − 2.273) + (2.308 − 2.273)

+ (2.309 − 2.273) + (2.484 − 2.273)

= 2.665 V

Based on the experimental results, and on the calculations made from them, the low threshold voltage was set at 1.822 V, and the high threshold voltage was set at 2.665 V.

9.6 FURTHER USE OF DoE IN PRODUCT DESIGN

In this chapter, the use of DoE in product design is introduced by considering three examples. Two of the three examples involved the use of computer simulation to produce the experimental results. In recent years, the combination of computer simulation and DoE has proven to be quite powerful. In particular the use of finite element analysis to produce the "experimental" results has be used extensively.

In the third example, the results and analysis of a designed experiment were used, not just to select the best value of a factor level, but to calculate the expected results when certain decisions were made. A number of other methods to perform similar calculations are available, such as response surface methodology and residual analysis. Those methods are not discussed in this elementary text, but are included in a number of standard references, e.g., Ref. 4. Reference 5 contains an excellent presentation of the combination of DoE and finite element analysis, and analysis of the DoE results using residual analysis and regression analysis. It illustrates these methods by the use of a design example.

9.7   CONCURRENT ENGINEERING

Before leaving this subject, a brief discussion of concurrent engineering is appropriate. This currently popular topic is often presented as just putting the design and manufacturing engineers in the same room. If concurrent engineering is so limited, its failure is assured. Manufacturing and engineering personnel must have the proper tools to be successful at concurrent engineering. Because it applies to both product and process design (and as we shall see later, to quality and reliability), DoE is an excellent concurrent engineering tool. With DoE, both design and process factors can be evaluated in the same experimental array. Producing the product design first, and then developing the manufacturing processes, is inefficient, even if it is done iteratively. A designed experiment with both product and process design factors is a true concurrent engineering tool.

In addition to its obvious technical benefits, the use of DoE for concurrent design has the following additional benefits:

1. It provides a vehicle (excuse?) for design and manufacturing engineers to work together on a project.
2. It improves communication by providing a common language for designers and manufacturing engineers.
3. It provides a means for both types of engineers to learn the other's jobs.
4. It helps those in different disciplines to appreciate the challenges (and the people) in other disciplines.
5. It is a management tool to aid in the administration of product and process development.

REFERENCES

1. J. Quinlan, Application of Taguchi Methods to Finite Element Analysis, Proceedings of the Fifth Symposium on Taguchi Methods, American Supplier Institute, Detroit, MI, 1987.
2. K. Rodenhiser, private communication, 1992.
3. G. R. Bandurek, H. L. Hughes, and D. Crouch, The Use of Taguchi Methods in Perfor-

mance Demonstrations, Quality and Reliability Engineering International, vol. 6, pp. 121–131, 1990.

4. D. Montgomery, Design and Analysis of Experiments, 3rd ed., John Wiley & Sons, New York, 1991.

5. A. Dasgupta, M. Pecht, and B. Matthieu, Design-of-Experiment Methods for Computational Parametric Studies in Electronic Packaging, Finite Elements in Analysis and Design, vol. 30, pp. 125–146, 1998.

Chapter 10

Managing the Uncontrollable Factors

10.1 INTRODUCTION

Most engineers and other manufacturing personnel are used to thinking in terms of the factors that are under their control in a product, process, or other work-related activity. As a matter of fact, it has always been considered unfair to expect people to be responsible for factors beyond their control. The fact is, however, that most of the factors we have to deal with in any enterprise are not under our control. It is pointless to protest this fact: our task is to be successful, even if we cannot control or even understand all of the factors in our business.

The point could be made that the difference between success and failure is the way we deal with factors we cannot control. Swings in the economy are beyond the ability of economists to control, predict, or in many cases, to explain even after they have occurred, but successful businesses find ways to be successful any way. Farmers complain constantly about the weather, but the successful ones find ways to produce profits even in widely varying conditions. All businesses are subject to unpredictable swings in their markets, but the good ones learn what they can about them, and find ways to adapt. Natural hazards and disasters can strike any time or place, but those who are prepared to deal with them increase their chances to survive.

Up to this point, our examples have dealt with factors that could be controlled easily by the experimenter. One of the major strengths of designed experiments, and of Taguchi arrays in particular, is their ability to deal with uncontrollable factors. That is the subject of this chapter. DoE is especially valuable in dealing with uncontrollable factors in reliability improvement; that topic is touched upon here and in Chapter 11, but the main discussion will have to be delayed until we have covered reliability in Section III, and accelerated testing in Section IV.

Some specific techniques for dealing with uncontrollable factors are discussed here, but this principle should always be kept in mind: our task is to deal with the controllable factors in such a way that the uncontrollable factors do not keep us from accomplishing our goals.

10.2   TYPES OF UNCONTROLLABLE FACTORS

There are several types of uncontrollable factors. We must deal with them all, but it is helpful to know that they are not all alike. Some of them are discussed in this section.

10.2.1   Factors That the Experimenter Does Not Want to Control

Some factors are well known to the experimenter, and their effects even may be quantified, but for various reasons they are not controlled on a continuous basis. One of the most common such factors is the variation in environmental conditions in a factory. In many operations, it is known, for instance, that humidity can have a strong affect on product quality, especially at different times of the year. It is financially impossible, however, to upgrade the facility to provide the necessary humidity control, so the manufacturer must accommodate this variation by other means.

Another known, but uncontrollable factor, is the variation in attention of the workers. Many manufacturers have known for years that their worst-quality products are made on Mondays and Fridays, when their workers are not as attentive as on other days. An obvious response would be to not work on Mondays and Fridays, but this would cause other, worse problems.

10.2.2   Factors That Are Known, But Cannot Be Controlled

Some factors are well known, but there is no physical means of controlling them. The weather is an example of this type of factor to a farmer. For a manufacturer, material properties can be this type of factor; e.g., it is hard to find a material that is a good conductor of heat, and also an electrical insulator, or one which is both strong and flexible.

TABLE **10-1** Factors and Levels
for the Baseball Experiment

Seven factors that may affect batting average	Level 1	Level 2
1. Stance	Open	Closed
2. Time of game	Day	Night
3. Weight of bat	Heavy	Light
4. Weight of player	<200 lb	>200 lb
5. Temperature	Hot	Cold
6. Location of game	Home	Away
7. Glasses	Yes	No

10.2.3   Factors That Are Unknown

By far the most frustrating factors are those which are not known. For obvious reasons, no examples are cited here.

10.3   THE BASEBALL EXAMPLE*

To illustrate the Design of Experiments approach to managing the uncontrollable factors, we shall use a nonmanufacturing example: that of a baseball player who wishes to improve his batting average.

The player approaches the problem in the typical DoE fashion by listing all the factors that might affect his batting average. His list of 7 two-level factors is shown in Table 10-1.

In theory, all these factors are under the control of the player. In practice, however, he considers it unwise to tell his manager that he will play only day games, or home games, or on days when it is warm. For this experiment, then, the controllable and uncontrollable factors are as shown in Table 10-2.

The task of the experimenter is to evaluate the effects of the controllable factors in such a way as to optimize performance, given the fact that the uncontrollable factors also will have their effects. He does this by evaluating the controllable factors in an *inner array*, and the uncontrollable factors in an *outer array*. This is shown in Table 10-3.

In Table 10-3, the inner array is an L$_8$, and it is arranged in the usual way.

---

* Most of the examples in this book are real ones, based on work performed by the author or other experimenters. This one is entirely fictional. Obviously, I prefer not to use "made-up" examples, but many of my students have told me that this one has been valuable to them in understanding uncontrollable factors. Therefore, it is discussed extensively here.

TABLE 10-2 Controllable and Uncontrollable Factors

Controllable Factors
1. Stance
2. Weight of bat
3. Weight of player
4. Glasses

Uncontrollable Factors
1. Time of game
2. Temperature
3. Location of game

TABLE 10-3 Inner and Outer Arrays for the Baseball Experiment

								Inner Array		Outer Array				
			A		A	B		2	2	1	1	E
			×		×	×		2	1	2	1	F
Run	A	B	B	C	C	C	D	1	2	2	1	G
1	1	1	1	1	1	1	1					
2	1	1	1	2	2	2	2					
3	1	2	2	1	1	2	2					
4	1	2	2	2	2	1	1					
5	2	1	2	1	2	1	2					
6	2	1	2	2	1	2	1					
7	2	2	1	1	2	2	1					
8	2	2	1	2	1	1	2					
Avg.												

Factor	Level 1	Level 2	Factor	Level 1	Level 2
A: Stance	Open	Closed	E: Time	Day	Night
B: Weight of bat	Heavy	Light	F: Temperature	Hot	Cold
C: Weight of player	<200 lb	>200 lb	G: Location	Home	Away
D: Glasses	Yes	No			

The controllable factors (both main effects and their interactions) have been assigned to columns according to the linear graph for this array.

The outer array is where the uncontrollable factors are evaluated. From Table 10-3, it may be seen that this is an L_4 array that has been turned on its side. The factors and levels are assigned exactly as they would be to an L_4 inner array. Factor E, time, for instance, has been assigned to column 1, which is now row 1 of the outer array in Table 10-3. (From now on, we will refer to columns and rows of the outer array as they appear in Table 10-3, i.e., juxtaposed from the usual way of thinking of them.) Using this approach, *each run of the outer array is now a separate effect of the inner array*.

We are now able to evaluate the effects of the inner and outer array factors separately and, most importantly, to choose the inner array factors in such a way as to minimize the effects of the outer array factors. To say it in a different way: our goal is not to control the outer array factors; instead, we want to control the inner array factors in a way that we optimize the results, regardless of the variation in the outer array factors.

The reader is advised to take a few minutes to contemplate the possible ways to evaluate data from this experiment. It may be seen, for example, that if only the results of column 1 of the outer array are used, we would choose inner array factors that would optimize the player's batting average for cold night games at home.

10.4   RESULTS OF THE BASEBALL EXPERIMENT

Being resourceful, and a bit lazy (both admirable traits of a good DoE practitioner), our baseball player has figured out that, by analyzing statistics from prior years, he can complete his experiment using existing data.[‡] The results are shown in Table 10-4.

The responses are the player's batting averages for the various combinations of inner and outer array factors, and are shown in Table 10-5.

The player's overall average for the time period analyzed was .300, but it varied widely over the range of conditions considered. From the inner array response table, it may be seen that three main effects: A (stance), B (bat), and C (weight), had strong effects; while factor D (glasses), and all the interactions, had minimal effects. To optimize his batting average across the range of outer

---

[‡] This practice is not limited to baseball. Most manufacturing operations have a wealth of data available to them, and many good experiments have been conducted using existing data. Once an organization gets used to using designed experiments on a routine basis, and to thinking in DoE terms, it can plan to collect data routinely, which can be used subsequently to analyze operations in a DoE format.

TABLE 10-4. Inner and Outer Arrays for the Baseball Experiment, with Results

			Inner Array							Outer Array		
			A		A	B	2	2	1	1	E
			×		×	×	2	1	2	1	F
Run	A	B	B	C	C	C	D	1	2	2	1	G
1	1	1	1	1	1	1	1	.346	.336	.345	.342	.342
2	1	1	1	2	2	2	2	.302	.301	.321	.317	.310
3	1	2	2	1	1	2	2	.316	.328	.338	.320	.326
4	1	2	2	2	2	1	1	.298	.310	.315	.306	.307
5	2	1	2	1	2	1	2	.289	.295	.310	.299	.298
6	2	1	2	2	1	2	1	.271	.274	.284	.280	.277
7	2	2	1	1	2	2	1	.266	.286	.291	.274	.279
8	2	2	1	2	1	1	2	.254	.265	.267	.255	.260
Avg.								.293	.299	.309	.300	.300

Factor	Level 1	Level 2	Factor	Level 1	Level 2
A: Stance	Open	Closed	E: Time	Day	Night
B: Weight of bat	Heavy	Light	F: Temperature	Hot	Cold
C: Weight of player	<200 lb	>200 lb	G: Location	Home	Away
D: Glasses	Yes	No			

TABLE 10-5 Response Table
for the Baseball Experiment

Inner Array		Outer Array	
A_1	.3213	E_1	.3040
A_2	.2788	E_2	.2961
B_1	.3070	F_1	.2993
B_2	.2931	F_2	.3008
$(A \times B)_1$	.2980	G_1	.2959
$(A \times B)_2$	.3021	G_2	.3041
C_1	.3113		
C_2	.2888		
$(A \times C)_1$	.3013		
$(A \times C)_2$	.2988		
$(B \times C)_1$	.3020		
$(B \times C)_2$	.2986		
D_1	.3015		
D_2	.2986		

array conditions, the player should use an open stance and a heavy bat. He should keep his weight down, and should wear glasses. The glasses factor is an interesting one in that the effect is small, and there may be a risk of betraying a human frailty if he wears glasses. It might be a reasonable choice to not wear glasses, but for this discussion, we shall assume the player implements all the levels of the factors as suggested by the experimental results.

Two of the outer array factors had effects, but here we are not looking to choose optimum outer array conditions. Instead, we want to use the outer array results to provide insight into the inner array factors.

10.5   CALCULATING THE OPTIMUM EFFECTS

It is possible to calculate the player's expected batting average if he chooses the optimum levels of the inner array factors. This can be done with the equation

$$\hat{Y} = \bar{\bar{Y}} + (A_1 - \bar{\bar{Y}}) + (B_1 - \bar{\bar{Y}}) + (C_1 - \bar{\bar{Y}}) + (D_1 - \bar{\bar{Y}}) \qquad (10\text{-}1)$$

where $\hat{Y}$ = predicted optimum batting average, $\bar{\bar{Y}}$ = overall average of all the experimental data, and A_1, B_1, C_1, and D_1 are the responses at the optimum levels of all the main effects. In this case,

$$\bar{\bar{Y}} = .300, \quad A_1 = .321, \quad B_1 = .307, \quad C_1 = .311, \quad D_1 = .302$$
$$\hat{Y} = .300 + .021 + .007 + .011 + .002 = .341.$$

Thus the player could expect to bat .341 by using the optimum levels of the inner array factors.

The player could further improve his average if he were to use the optimum outer array conditions. If he were to do so, his average would be

$$\hat{Y} = \bar{\bar{Y}} + (E_1 - \bar{\bar{Y}}) + (F_2 - \bar{\bar{Y}}) + (G_2 - \bar{\bar{Y}}) \qquad (10\text{-}2)$$

$$\hat{Y} = .341 + .004 + .001 + .004 = .350$$

Conceivably, the player could improve his average to .350 by operating under optimum conditions of both the inner and outer array. This is of academic interest only, however, since the player has already decided that the outer array conditions will not be controlled. Even so, improvement from .300 to .341 is significant.

10.6   USING OUTER ARRAYS FOR CONCURRENT ENGINEERING

Outer arrays are a powerful tool for concurrent engineering. In Chapter 9, we saw how to combine design and process factors in an inner array. The point was made that both system and parameter design factors could be evaluated concurrently in an inner array. To use outer arrays most effectively for concurrent design, they should contain factors over which the experimenter cannot or does not wish to exercise control. Tolerance design factors fit this definition.

Tolerances are routinely placed on most design specifications. Usually, however, they are not subject to real control by the designer without some cost. Standard machine shop tolerances, for example, are well established. To obtain tighter tolerances, the machine shop must take extraordinary measures, and the customer must be prepared to make extraordinary expenditures. In electronics design, tighter component tolerances are usually obtained by inspection, and selection of those components that have a tighter range of outputs, which also involves extra expense. From the cost standpoint, then, tolerances are often an uncontrollable factor. Primarily for this reason, design evaluations are often conducted at only one point in the tolerance range for each design dimension or other value.

By placing tolerances of key design parameters in the outer array, it is possible to evaluate their effects on the overall design. This is usually done by making the "plus" and "minus" values of the tolerance the two levels of the outer array factor, and allowing them to become an operator on the nominal value. This approach allows the investigator to evaluate the effects of varying several different tolerances simultaneously. On the bases of these evaluations, it is possible to decide which, if any, tolerances should be tightened, or if other measures should be taken.

Another type of uncontrollable factor is how the customer will use the product. The operating environment, which includes temperature, humidity, mechanical stresses, etc., is often beyond the control or even the knowledge of the product manufacturer. Likewise, the customer may handle or store the product under a wide range of conditions. It is apparent that this type of factor is getting us closer to the use of Design of Experiments for reliability improvement, which is introduced in chapter 11.

Chapter 11

Application of DoE
to Reliability Assurance

11.1 INTRODUCTION

In this last chapter in the section on Design of Experiments, we introduce the general concepts of applying DoE to reliability assurance. The application of DoE to reliability assurance requires an understanding of reliability principles and accelerated testing methods, so a complete treatment of this subject will have to wait until after we have covered those subjects in Chapters 12 through 20.

11.2 QUALITY AND RELIABILITY

It is appropriate here to review briefly some of the points made about quality and reliability in Chapter 2. We defined a reliable product as follows:

> A reliable product is one that does what the user wants it to do, when the user wants it to do so.

Appreciation of this definition should have a profound effect on the way manufacturers and designers do their jobs. Traditionally, a product or process is conceived, prototypes are built, and if some time-zero functional and quality requirements are met, production is allowed to begin. The reliability dimension, however, says that the functional and quality requirements must be met, not only at time zero in the production facility but for some time in the future, and in a variety of environmental and operating conditions, not all of which can be predicted exactly during design and production.

If we view quality and reliability, not as separate disciplines, but as the same discipline with different time and use conditions, we are making a good beginning. In the same way that DoE can be used for time-zero functional and quality optimization, it also can be used for time-future functional and quality optimization.

11.3 RELIABILITY AS A RESPONSE

To evaluate reliability in a designed experiment, some type of reliability metric must be used as a response. To do this, the experimenter must understand the use conditions of the product which are likely to impact its lifetime or function in the future, and select levels of design, manufacturing, and material factors which make the product robust with respect to these conditions. For example, if a paint manufacturer knows that ultraviolet light degrades paint in service, an experiment might be conducted to select the paint composition that is least likely to be affected by UV light. An inner array of different compositions and manufacturing conditions could be developed. The outer array could include various lengths of time of exposure to UV light; or a standard time of exposure to several different intensities of UV light; or both. The response would be the degradation of the color or other condition of the paint. The best combination of inner array factors would be that which showed the smallest slope on a plot of paint condition vs. time of exposure, or of paint condition vs. intensity of exposure.

An example of the use of DoE in a manner almost identical to that described in the previous paragraph is the automotive interior plastic experiment [1], conducted jointly by General Motors and the Dow Chemical Company. In that experiment, the controllable factors were:

1. Type of vehicle (sedan, station wagon, hatchback)
2. Color of plastic (blue, brown, maroon, red, green, gray, black)
3. Type of plastic (ABS, PP, AES, ASA, PE, PBT, acetal, nylon, PC, PC/ABS, PVC)
4. Location in the vehicle (window, wheel well, middle of door, bottom of door, roof, back shelf, arm rest, window pillar, dash board)

The uncontrollable factors were:

1. Type of exposure (field use, static soak facing either north or south)
2. Location (Texas, Arizona)

The response was the degree to which the plastic degraded during exposure.

11.4 EXAMPLES OF THE APPLICATION OF DoE TO RELIABILITY

Design of Experiments has been applied to reliability work for at least the last 40–50 years, and the theory of this application is well documented in the literature [2–6]. There are several ways to use DoE in reliability improvement; two of them are described in this chapter, and a third is described in Chapter 21, after the chapters on reliability and accelerated testing. The simplest approach is to fix the design and manufacturing variables of a product, and then to use DoE to generate an array of accelerated environmental and/or operating stresses. Samples of the product are subjected to combinations of stresses defined according to this array, and the effects of each stress on the reliability of the product are quantified. This approach is illustrated in Section 11.4.1 by the analog circuit board experiment [7].

A second, more proactive approach to the use of DoE in reliability work is to use it to optimize design and manufacturing variables for reliability of the product. Here, an array of design, manufacturing, and/or assembly factors are used to produce samples which are then subjected to some type of accelerated test, with reliability as the response. The levels of the factors that produce the most reliable samples are then used in the design and manufacture of the product. This approach is illustrated in Section 11.4.2 by the ceramic capacitor experiment [8].

The third approach is the most comprehensive; and it is a combination of the first two. Here, the manufacturing and/or design factors are considered controllable factors, and are evaluated in an inner array. The environmental and/or operating stresses are considered uncontrollable factors, and are evaluated in an outer array. This approach is illustrated in Chapter 21 by the surface mount capacitor experiment.

11.4.1 The Analog Circuit Board Experiment: Environmental and Operating Stresses as Factors, and Reliability as a Response

In the analog circuit board experiment, it was desired to test the sensitivity of a four-channel digital-to-analog converter to variations in four external conditions: (1) ambient temperature, (2) supply voltage, (3) resistance of loads (both voltage and current outputs), and (4) interference from other channels. The four external conditions were used as the factors, and they were evaluated at two levels each. Output variation was the effect; and two Taguchi L_8 experiments were conducted: one for voltage outputs and one for current outputs. The two arrays were identical with the exception of the load, and are shown along with the responses for each run in Tables 11-1 and 11-2. The variation in output was calculated by using this form of the Taguchi signal-to-noise ratio:

$$\text{Inaccuracy} = 10 \log (\text{variance of errors}) \tag{11-1}$$

TABLE 11-1 Experimental Array and Responses for the Voltage Outputs of the Analog Circuit Experiment

Run	Temperature, °C	Supply voltage	Other channel	Load, ohms	Voltage, inaccuracy of output
1	0	−6	Off	Open	4.565
2	0	−6	On	500	5.650
3	0	+6	Off	500	4.786
4	0	+6	On	Open	4.795
5	70	−6	Off	500	4.616
6	70	−6	On	Open	4.751
7	70	+6	Off	Open	4.717
8	70	+6	On	500	4.861

Source: From Ref. 7. © 1990 John Wiley & Sons, used by permission.

The effects of each factor were calculated by comparing the results from all the runs in which the factor was set at level 1 with those from all the runs in which the factor was set at level 2. For example, the effect of temperature was calculated by comparing the average response of runs 1–4 with the average response of runs 5–8; the effect of supply voltage was obtained by comparing the results of runs 1, 2, 5, and 6 with those of runs 3, 4, 7, and 8. Response tables for voltage and current outputs are shown in Tables 11-3 and 11-4, respectively. The investigators also plotted these results, and the plots may be found in Ref. 7.

TABLE 11-2 Experimental Array and Responses for the Current Outputs of the Analog Circuit Experiment

Run	Temperature, °C	Supply voltage	Other channel	Load, ohms	Voltage, inaccuracy of output
1	0	−6	Off	Short	9.12
2	0	−6	On	600	46.53
3	0	+6	Off	600	8.68
4	0	+6	On	Short	21.06
5	70	−6	Off	600	9.07
6	70	−6	On	Short	10.39
7	70	+6	Off	Short	9.80
8	70	+6	On	600	10.16

Source: From Ref. 7. © 1990 John Wiley & Sons, used by permission.

TABLE 11-3 Response Table for Voltage Outputs of
the Analog Circuit Board Experiment

Factor	Level	Mean value of responses
Temperature	0	4.6990
	70	4.7363
Supply voltage	−6	4.6455
	+6	4.7898
Other channel	Off	4.6710
	On	4.7643
Load	Open	4.7070
	500	4.7283

All factors had large effects, which was not expected by the investigators. In considering possible causes of this phenomenon, it was noted that the L_8 array contains seven degrees of freedom. This means that there is confounding among different sets of two-factor interactions, as follows:

(Other channel—load) is confounded with (temperature—supply voltage)
(Supply voltage—load) is confounded with (temperature—other channel)
(Supply voltage—other channel) is confounded with (temperature—load)

None of the two-factor interactions are confounded with any of the main effects, so that was not believed to be a source of the anomaly. There was, however,

TABLE 11-4 Response Table for Current Outputs of the
Analog Circuit Board Experiment

Factor	Level	Mean value of responses
Temperature	0	21.348
	70	9.855
Supply voltage	−6	18.778
	+6	12.425
Other channel	Off	6.168
	On	22.035
Load	Short	12.593
	600	18.610

potential confounding of a three-factor interaction with one of the main effects. Further experiments confirmed that this was indeed the case; i.e., the observed effect of load variation was actually the effect of the three-way interaction among the other three factors. Although such a three-factor interaction is possible, the likelihood of its occurrence in service is remote, since that would require all three factors to be near an extreme limit of their variation simultaneously. The product was thus considered reliable with respect to expected variations in the external conditions investigated.

11.4.2 The Ceramic Capacitor Experiment

An electronics assembly company wanted to guarantee the reliability of ceramic capacitors in its product [8]. The capacitors were purchased from various suppliers and assembled into the product by the manufacturer. Analysis of field failures showed that the most common failure mechanism was the propagation of micro cracks introduced into the product during production of the capacitors by the suppliers. It was hypothesized that the rate and amount of crack propagation could be affected by various assembly conditions, such as number of solder reflows or conformal coating material. It could also be affected by environmental conditions during use by the customer, such as variations in the ambient operating temperature and pressure (due to altitude in the aerospace product), or possibly by humidity.

An experiment was designed in which the following controllable factors were evaluated in the inner array:

A: Capacitor manufacturer
B: Capacitor assembly temperature
C: Type of conformal coating of the assembly
D: Number of solder reflow cycles in assembly

The uncontrollable factors were the product operating conditions, four of which were used as outer array factors:

1. Elevated temperature and humidity in combination
2. Temperature cycling
3. Vibration followed by temperature cycling
4. Vibration, followed by altitude (pressure) cycling, followed by temperature cycling

The factors and levels are shown in the experimental array of Table 11-5. This is an L_{16} array, which has 15 degrees of freedom. Three four-level factors were evaluated, along with a two-level factor and an empty column. The two-level column was actually a four-level column in which only two levels were used.

TABLE 11-5 Assignment of Factors and Levels to the Inner and Outer Arrays for the Capacitor Reliability Experiment

Run	Inner Array					Outer Array			
	A	B	C	D	E	1	2	3	4
1	1	1	1	1	1				
2	1	2	2	1	2				
3	1	3	3	2	3				
4	1	4	4	2	4				
5	2	1	2	2	4				
6	2	2	1	2	3				
7	2	3	4	1	2				
8	2	4	3	1	1				
9	3	1	3	2	2				
10	3	2	4	2	1				
11	3	3	1	1	4				
12	3	4	2	1	3				
13	4	1	4	1	3				
14	4	2	3	1	4				
15	4	3	1	2	1				
16	4	4	2	2	2				

Inner Array Factors	Outer Array Factors
A: Assembly method (4 levels)	1. Temperature—humidity
B: Capacitor manufacturer (4 levels)	2. Temperature cycling
C: Conformal coat material (4 levels)	3. Vibration—temperature cycling
D: Number of solder reflow cycles (2 levels)	4. Vibration—altitude—temperature cycling
E: Empty	

Source: From Ref. 8. © 1992 IEEE, used by permission.

The measured response was the number of failures due to increases in capacitor leakage. For each run of the experiment, leakages of four different groups of capacitors were measured after exposure for a specified time to the four different sets of environmental conditions. If the leakage was above a certain value, the capacitor was called a failure.

The results are shown in Table 11-6. Since all the initial sample sizes were equal, the greater the number of failures, the worse the result. From Table 11-6, then, it is possible to determine which environmental conditions had the greatest impact on the reliability of the product, and also to determine which factors,

TABLE 11-6 Results of the Capacitor Reliability Experiment[a]

Run	Inner Array					Outer Array			
	A	B	C	D	E	1	2	3	4
1	1	1	1	1	1	0	2	3	2
2	1	2	2	1	2	0	0	0	0
3	1	3	3	2	3	1	0	0	0
4	1	4	4	2	4	0	0	1	0
5	2	1	2	2	4	3	1	5	4
6	2	2	1	2	3	0	0	0	0
7	2	3	4	1	2	1	0	0	0
8	2	4	3	1	1	0	0	0	0
9	3	1	3	2	2	1	4	2	3
10	3	2	4	2	1	0	0	0	0
11	3	3	1	1	4	0	1	1	1
12	3	4	2	1	3	0	0	0	0
13	4	1	4	1	3	1	2	3	3
14	4	2	3	1	4	0	0	0	1
15	4	3	1	2	1	1	1	0	3
16	4	4	2	2	2	0	0	0	3

Inner Array Factors	Outer Array Factors
A: Assembly method (4 levels) B: Capacitor manufacturer (4 levels) C: Conformal coat material (4 levels) D: Number of solder reflow cycles (2 levels) E: Empty	1. Temperature—humidity 2. Temperature cycling 3. Vibration—temperature cycling 4. Vibration—altitude—temperature cycling

[a] The results, shown in the outer array columns, are the number of defects.
Source: From Ref. 8. © 1992 IEEE, used by permission.

controllable by the assembler, produced a product that was least sensitive to these environments.

The response table (Table 11-7) shows that the outer array condition that produced the most failures was environmental condition 4: the combination of vibration, altitude, and temperature cycling. It also shows that for all the outer array conditions, the largest inner array contributor to failures was the manufacturer of the capacitor. The contribution of this factor was so strong that, in one case, it was the only factor that mattered. With this knowledge, the assembler was able to conduct further analysis of the incoming product from all manufactur-

TABLE 11-7 Response Table for the Capacitor Reliability Experiment

Factor and level	1. Temperature—humidity		2. Temperature cycle		3. Vibration—temperature cycling		4. Vibration—altitude—temperatue cycling	
	No. failures	ρ, %	No. failures	ρ, %	No. failures	ρ, %	No. failures	ρ, %
Assembly method		9.5		8.9		Pooled		21.4
A_1	1		2		4		2	
A_2	4		1		5		4	
A_3	1		5		3		4	
A_4	2		3		3		10	
Capacitor manufacturer		38.1		75.1		78.5		45.7
B_1	5		9		13		12	
B_2	0		0		0		1	
B_3	3		2		1		4	
B_4	0		0		1		3	
Conformal Coat		14.3		Pooled		Pooled		2.9
C_1	0		3		4		6	
C_2	4		2		5		7	
C_3	2		4		2		4	
C_4	2		2		4		3	
No. reflows		17.5		Pooled		Pooled		11.4
D_1	2		5		7		7	
D_2	6		6		8		13	
Empty		Pooled		Pooled		Pooled		Pooled
E_1	1		3		3		5	
E_2	2		4		2		6	
E_3	2		2		3		3	
E_4	3		2		7		6	
e		20.6		16.0		21.5		18.6

Source: From Ref. 8. © 1992 IEEE, used by permission.

TABLE **11-8**   Factors and Levels for the Automatic Welding System Experiment

Factor	Current level (1)	Alternate level (2)
A: Rail material	A-6061T aluminum alloy	SS41 carbon steel
B: Bearing roller material	Steel	Plastic
C: Weight of welding unit	30 kg	35 kg
D: Pressure between rollers and rails	50 psi	100 psi
E: Debris removal interval	1 hr	2 hrs
F: Dust concentration	50,000 ppm	10,000 ppm
G: Speed of movement	10 m/sec	15 m/sec

Source: From Ref. 9. © 1999 John Wiley & Sons, used by permission.

ers, to isolate some key features that produced the desired results, and to implement tests which guaranteed the reliability of incoming capacitors.

11.4.3   The Automatic Welding System Experiment

The automatic welding system experiment [9]§, is an example of a *degradation experiment*. In the automatic welding system, the welding unit moves back and forth on two aluminum rails; and wear of the rails limits the lifetime of the system. Seven design and operating factors were selected for evaluation, and their levels are shown in Table 11-8. The degradation metric was the amount of wear observed on the rails, observed at intervals of 50 weld cycles. Table 11-9 shows the assignment of the factors to a Taguchi L_8 array. Also shown in Table 11-9 are the results, in the form of signal-to-noise ratios, for observations at 500 and 1000 cycles.

Table 11-10 is a response table for the automatic welding experiment. The preferred levels for each of the factors are indicated by asterisks. It may be noted that, for all factors except *C*, the weight of the welding unit, the preferred levels are the same for 500 and 1000 cycles. The weight is not very significant, however, since there is very little difference between the levels at either 500 or 1000 cycles.

11.4.4   The Light-Emitting Diode Experiment

Another example of a degradation experiment is the light-emitting diode (LED) experiment [10]. An LED is composed of a light-emitting semiconductor chip

---

§ In addition to the analyses described here, this experiment included additional analyses that are beyond the scope of this chapter. The reader is encouraged to review this reference after reading the later chapters of this book.

TABLE 11-9 Assignment of Factors to the Experimental Array for the Automatic Welding Experiment[a]

Run	C	D	B	E	F	G	A	S/N 500 cycles	S/N 1000 cycles
1	1	1	1	1	1	1	1	18.81	11.46
2	1	1	1	2	2	2	2	14.73	6.81
3	1	2	2	1	1	2	2	18.80	11.80
4	1	2	2	2	2	1	1	18.20	11.07
5	2	1	2	1	2	1	2	20.69	13.62
6	2	1	2	2	1	2	1	18.17	10.98
7	2	2	1	1	2	2	1	18.78	12.59
8	2	2	1	2	1	1	2	11.81	4.95

[a] Also shown are results, in the form of signal-to-noise ratios for wear after 500 and 1000 cycles.
Source: From Ref. 9. © 1999 John Wiley & Sons, used by permission.

TABLE 11-10 Response Table for S/N After 500 and 1000 Cycles for the Automatic Welding Experiment[a]

Factor and level	S/N at 500 cycles	S/N at 1000 cycles
C_1	17.63*	10.28
C_2	17.36	10.53*
D_1	18.10*	10.71*
D_2	16.89	10.10
B_1	16.03	8.95
B_2	18.96*	11.86*
E_1	19.27*	12.36*
E_2	15.72	8.45
F_1	16.89	9.79
F_2	18.09*	11.02*
G_1	17.37	10.27
G_2	17.62*	10.54*
A_1	18.48*	11.52*
A_2	16.50	9.29

[a] Preferred levels are indicated by asterisks.
Source: Adapted from Ref. 9.

TABLE 11-11   Factors, Levels, and Noise Array for the Light-Emitting Diode Experiment

Run	A: Chip vendor	B: Epoxy type	C: Lead frame	Noise level		Estimated loss
1	−	−	−	−	+	5.286
2	+	−	−	−	+	4.683
3	−	+	−	−	+	7.863
4	+	+	−	−	+	7.958
5	−	−	+	−	+	6.896
6	+	−	+	−	+	6.009
7	−	+	+	−	+	6.875
8	+	+	+	−	+	7.072

Source: From Ref. 10. © 1996 John Wiley & Sons, used by permission.

attached by silver epoxy to a lead frame that carries the electrical current. The controllable factors for this experiment were (1) the semiconductor chip vendor, (2) the type of epoxy material, and (3) the lead frame design. Since one of the goals of the experiment was to produce a design that was robust with respect to the placement of the chip on the lead frame, chip location was chosen as a noise factor. The effect, or reliability metric, was the degradation of luminous flux observed during operation of the LED. The goal of the experiment, then, was to select the levels of chip vendor, epoxy, and lead frame design to minimize degradation in luminosity, regardless of where the chip was placed on the lead frame.

A full-factorial array was chosen for this experiment; and it is shown in Table 11-11. (Note that classical DoE notation is used.) Thirty samples were made for each of the eight treatment combinations; and since there were two noise levels, a total of 480 samples were made. Observations of the degradation in luminosity were made at 10 time intervals throughout the experiment.

The results of the experiment also are shown in Table 11-11 as the estimated loss. The values shown are the average of those for the "−" and "+" levels of the noise factor. Since this was a full factorial experiment, the results are the response table, and the optimum levels of the factors are shown in run 1, with run 2 also yielding good results.

11.5   DoE AND RELIABILITY

The above examples are somewhat elementary, and we cannot go further in the application of DoE to reliability because we have not yet laid the groundwork in reliability and accelerated testing to allow us to discuss it in more detail. The examples do, however, illustrate in principle that DoE can be applied to reliability evaluations.

TABLE 11-12 Example Arrays for Concurrent Design, Process, and Reliability Evaluation

	Inner Array								Outer Array			
		A			A	B		2	2	1	1	E
		×			×	×		2	1	2	1	F
Run	A	B	B	C	C	C	D	1	2	2	1	E × F
1	1	1	1	1	1	1	1					
2	1	1	1	2	2	2	2					
3	1	2	2	1	1	2	2					
4	1	2	2	2	2	1	1					
5	2	1	2	1	2	1	2					
6	2	1	2	2	1	2	1					
7	2	2	1	1	2	2	1					
8	2	2	1	2	1	1	2					

Factor	Level 1	Level 2	Factor	Level 1	Level 2
A: Time (manufacturing process)	Short	Long			
B: Temperature (manufacturing process)	High	Low	E: Temperature (operating)		
C: Size (design feature)	Large	Small	F: Humidity (operating)		
D: Pressure (manufacturing process)	High	Low			

In the capacitor example, responses were measured in four different sets of environmental conditions. This is a way of evaluating reliability as a response. It may be noted that the set of environmental conditions evaluated here is not a true outer array. Instead, it is just four different combinations of environmental factors evaluated independently. A true outer array of environmental factors is shown as an example in Table 11-12. This type of experiment offers many opportunities for the effective evaluation of reliability in a timely and cost-efficient manner, and after our discussion of reliability in Section III, and of accelerated testing in Section IV, we shall explore some of those opportunities.

11.6 SUMMARY

Although the use of reliability as a response in a designed experiment is not new, it has not been used widely by reliability engineers in the last few decades. The combination of inner arrays of design and manufacturing conditions with outer arrays of environmental and operating conditions is a potentially powerful tool, and its exploitation for reliability improvement is only beginning. The use of DoE for reliability improvement is not intended to replace good reliability theory, such as failure distribution and acceleration models, but it can help organize and analyze this type of knowledge. Because of its efficiency, DoE also makes it realistic to conduct reliability experiments that would not be considered without such a tool.

REFERENCES

1. T. Sutton and C. Wurst, Prediction of Automotive Interior Plastic Weathering Performance with Accelerated Weathering Techniques, SAE Technical Paper 930629, Society of Automotive Engineers, Warrendale, PA, 1993.
2. W. Weibull, Statistical Design of Fatigue Experiments, Journal of Applied Mechanics, 19(1):109–113, March 1952.
3. M. Zelen, Factorial Experiments in Life Testing, Technometrics 1(3): 269–289, August 1959.
4. G. C. Derringer, Considerations in Single and Multiple Stress Accelerated Life Testing, Journal of Quality Technology, 14(3):130–134, July, 1982.
5. D. J. Hannaman, N. Zamani, J. Dhiman, and M.G. Buehler, Error Analysis for Optimal Design of Accelerated Tests, Proceedings of the International Reliability Physics Symposium, IEEE, 1990, pp. 55–60.
6. M.J. Luvalle, A Note on Experiment Design for Accelerated Life Tests, Microelectronics and Reliability 30(3):591–603, 1990.
7. G.R. Bandurek, H.L. Hughes, and D. Crouch, The Use of Taguchi Methods in Performance Demonstrations, Quality and Reliability Engineering International 6(2): 121–131, 1990.
8. L.W. Condra, G.M. Johnson, M.G. Pecht, and A. Christou, Evaluation of Manufacturing Variables in the Reliability of Surface Mount Capacitors, IEEE Transactions on Components, Hybrids, and Manufacturing Technology 15(4):542–552, August 1992.
9. Y. Jeng, Improving the Lifetime of an Automatic Welding System, Quality and Reliability Engineering International 15(4): 47–55, 1999.
10. C. Chaio and M. Hamada, Robust Reliability for Light Emitting Diodes Using Degradation Measurements, Quality and Reliability Engineering International 12:89–94, 1996.

Chapter 12

Reliability Measurement

12.1 PROBABILISTIC AND DETERMINISTIC RELIABILITY

The story is told of a census bureau worker who visited a small backwoods region and was impressed by the large number of elderly people he saw. He became so interested that he began asking people their ages, and found that many of them were indeed quite old. He sought out the local physician and asked him what the death rate was in the region. The physician replied, "The same as it is anywhere else, sir, one to a person."

This story illustrates two different approaches to the field of reliability. On the one hand is the demographer, whose main interest is in measuring the rate of failure on a broad statistical basis. He takes a *probabilistic* approach to reliability. On the other is the physician, whose primary concern is finding and reducing the causes of individual failures, and thereby increasing the lifetimes of individuals. His view of reliability is *deterministic*. The difference in viewpoint often leads to some spirited discussions among reliability professionals, who often are unaware that they are approaching the subject with different premises.

Both of the above points of view have their place in reliability, and in actual practice they often use the same methods. The difference is worth noting, however, since it can lead to serious misunderstanding, and sometimes unnecessary conflict, if it is not recognized. Table 12-1 summarizes some of the major

TABLE **12-1** Probabilistic and Deterministic Reliability Methods

Probabilistic reliability	Deterministic reliability
1. Concerned with overall failure rates, regardless of individual mechanisms or causes.	Concerned with individual failure mechanisms and their distributions in time.
2. Primary activities are system level failure measurement and prediction for maintenance scheduling, spares provision, etc.	Primary activities are investigations of causes of failure at subsystem levels.
3. Interested in system level reliability improvement through failure accommodation, redundancy, maintenance, reduction of failure effects, etc.	Interested in improving reliability at the component level by understanding failures at the structural level, and removing or minimizing the causes.
4. Work with failure distributions that may include combinations of several causes and mechanisms.	Work with distributions of single failure causes and mechanisms, or of probabilities of random events which overstress the components.

differences between the two approaches, and it is easy to see the importance of both.

In Section 2.4, we saw that there are three ways to categorize reliability methods by intent:

Methods to measure and predict failures
Methods to accommodate failures
Methods to prevent failures

Both deterministic and probabilistic reliability professionals are interested in methods to measure and predict failures, although they may use them in different ways. Probabilistic reliability engineers are more interested in failure accommodation methods, while deterministic reliability engineers are more interested in methods to prevent failures.

This book is about the use of Design of Experiments to improve reliability and, since DoE is more applicable to deterministic reliability methods, we will concentrate on methods of failure prevention. Generally, this implies a concentration on the reliability of the components of a system, instead of the reliability of the overall system.

First, however, we must explore the common ground of reliability measurement and prediction. That is the subject of this chapter, and of Chapter 13.

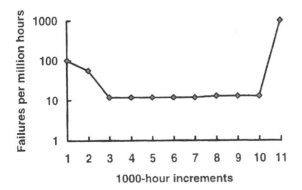

FIGURE 12-1 Failure rate vs. time for a single lot of lightbulbs without replacement.

12.2 THE FAILURE RATE CURVE

Consider the case of a lightbulb manufacturer* who produces a lot of 1000 lightbulbs. The bulbs are designed for a life of between 10,000 and 11,000 hours but, due to a defect in some of the seals, 100 of them will wear out in the first 1000 hours. Because of defective filaments, 50 more will expire in the second 1000 hours; and due to various other defects, 10 more will burn out in each 1000 hour increment until the design life of 10,000 hours is reached.

If the user of this 1000 piece lot places the bulbs in service, and does not replace them as they burn out, the resulting failure rate vs. time will be as shown in Figure 12-1. This curve has been calculated by dividing the number of failures in each 1000 hour increment by the total number of operating hours (bulbs × hours) for each increment. (These calculations are explained later in greater detail.)

Consider a second lightbulb example, in which a manufacturer produces 1000-piece lots, each with a single defect which causes 100 of the bulbs to burn out in the first 1,000 hours of operation, and none thereafter until wearout. The design life of these bulbs is also 10,000 hours. If such a lot is placed in service every thousand hours, and if bulbs are replaced as they fail, the failure rate vs.

---

* Lightbulbs have always been a favorite example for reliability engineers. In the first edition of this work, I suggested that, in view of the popularity of lightbulb jokes, i.e., "How many _____ does it take to change a lightbulb?" it is curious that reliability engineers have not been featured. Since then, I have conducted an extensive, unscientific survey, and the best answer is, "We don't know; reliability engineers never have enough data to make a conclusive decision."

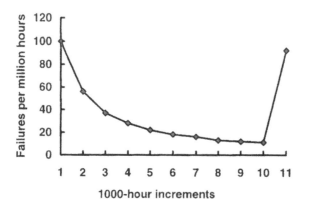

FIGURE 12-2   Failure rate vs. time curve for 1000-piece lightbulb lots introduced to service at 1000 hour intervals, each lot with a single defect causing 100 of each lot to fail in the first 1000 hours, with replacement.

time curve of Figure 12-2 results. (It may be noted that, in Figure 12-1, failure rate is plotted on a log scale, while in Figure 12-2, a linear scale is used. There is no established convention for this, and both are commonly used.)

The examples of Figures 12-1 and 12-2 are hypothetical, but they are typical of real situations. Figures 12-3 and 12-4, replotted from Refs. 1 and 2, show real examples of failure rate vs. time data. The general shape of all of these curves is quite familiar to reliability engineers. It has an "early-life" phase, in which the

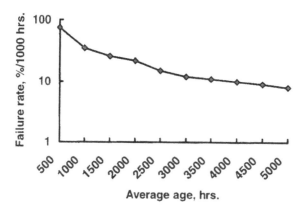

FIGURE 12-3   Failure rate for commercial airline computers under warranty. (From Ref. 1. © 1979 IEEE, used by permission.)

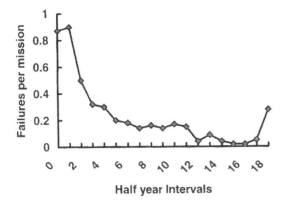

FIGURE 12-4 Failure rates for spacecraft operation in orbit. (From Ref. 2. © 1987 IEEE, used by permission.)

failure rate decreases with time; a "useful-life" phase, in which the failure rate is somewhat constant; and sometimes a "wearout" phase, in which the failure rate increases. It is commonly called the "bathtub curve" because its shape resembles the profile of a bathtub. An idealized form of the bathtub curve is shown in Figure 12-5, as the sum of three smaller curves.

It is interesting to note that, although the two curves in Figures 12-1 and 12-2 have common characteristics, such as a high initial rate, which decreases asymptotically to a relatively constant rate, and then an abrupt rise; they represent physical phenomena which are entirely different. For example:

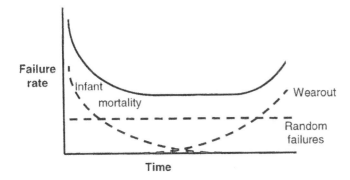

FIGURE 12-5 Bathtub curve for reliability, shown as the sum of the infant mortality, random failure, and wearout curves.

1. The lightbulbs of Figure 12-1 contain 10 different types of defects, and therefore different failure mechanisms in each 1000 hour increment. The lightbulbs of Figure 12-2 have a single defect, which causes failures due to a single mechanism only in the first 1000 hours.
2. A single lot of lightbulbs is considered in Figure 12-1, while 10 different lots are considered in Figure 12-2. Furthermore, the lots in Figure 12-2 are placed in service at different times.
3. The failures are not replaced in Figure 12-1, while they are replaced in Figure 12-2.

The theoretical basis for the bathtub curve was established in a classic paper by Proschan [3]. There is discussion, and some disagreement, among professionals about the nature and use of the bathtub curve, and some even doubt its existence [4–6]. This argument can be very subjective. The curve of Figure 12-5 is stylistic. In real situations, the infant mortality region is fairly well established, although its slope can vary widely. The useful life region in real life situations usually trends downward, although gradually; it often has little humps in it, and sometimes it is difficult to observe at all. The wearout region is fairly well established for mechanical products, but there is little evidence for it in high-technology electronic products.

The story used to introduce this chapter is not entirely whimsical, because the bathtub curve was originally developed to describe the death rate of humans. It was observed that the death rate is high in the early stages of life, due to birth defects. In the useful-life region, the death rate should be relatively constant, since death is primarily due to random events, such as accidents or violence. For older people, wearout dominates, and the death rate increases due to common fatal illnesses. Some data for the failure rate of carbon-based units (humans) are shown in Table 12-2, and plotted in Figure 12-6.

The bathtub curve for reliability has three distinct regions, distinguished by the three different slopes of the solid curve of Figure 12-5. The first section is usually called the region of infant mortality. This region is characterized by a sharply decreasing failure rate. The failures in this region usually are due to serious defects introduced during the manufacturing process. There may be many such defects, each of which afflicts a small subpopulation, and each of which has its own distribution in time, and the smaller curve shown in Figure 12-5 actually may be the sum of several even smaller curves. Failures in this region are often called intrinsic failures, since they result from causes within the item.

The second region is one in which the failure rate is relatively constant. This is called the useful-life region, and it is popular with probabilistic reliability professionals because, being constant, it can be used to predict failure rates. A common term for expressing a constant failure rate is mean-time-between-failures

TABLE **12-2** Survival Rates for Carbon-Based Units (Humans)

Age	Survivors	Failures	Age	Survivors	Failures
0	10,000,000	0	52	8,610,244	152,062
1	9,929,200	70,800	54	8,431,654	178,590
2	9,911,725	17,475	56	8,222,990	208,664
3	9,896,659	15,066	58	7,980,171	242,819
4	9,882,210	14,449	60	7,698,678	281,493
5	9,868,375	13,835	62	7,374,352	324,326
6	9,855,053	13,322	64	7,003,907	370,445
7	9,842,241	12,812	66	6,584,596	419,311
8	9,829,840	12,401	68	6,114,070	470,526
9	9,817,749	12,091	70	5,591,994	522,076
10	9,805,870	11,879	72	5,025,837	566,157
12	9,781,958	23,912	74	4,431,782	594,055
14	9,756,737	25,221	76	3,826,877	604,905
16	9,728,950	27,787	78	3,221,866	605,011
18	9,698,230	30,720	80	2,626,354	595,512
20	9,664,994	33,236	82	2,058,523	567,831
22	9,630,039	34,955	84	1,542,763	515,760
24	9,593,960	36,079	86	1,100,119	442,644
26	9,557,155	36,805	88	741,456	358,663
28	9,519,442	37,713	90	468,156	273,300
30	9,480,358	39,084	91	361,347	106,809
32	9,439,447	40,911	92	272,534	88,813
34	9,396,358	43,089	93	200,054	72,480
36	9,350,279	46,079	94	142,173	57,881
38	9,299,482	50,797	95	97,147	45,026
40	9,241,359	58,123	96	63,019	34,128
42	9,173,375	67,984	97	37,769	25,250
44	9,093,740	79,635	98	19,313	18,456
46	9,000,587	93,153	99+	857	857
48	8,891,204	109,383			
50	8,762,306	128,898			

(MTBF). The failure distribution in this region is also the sum of a series of smaller failure distributions. These distributions represent failures due to

1. Defects of subpopulations which, while not as severe as those in the infant mortality region, cause failures before wearout or
2. Random environmental or operating events which can overstress an item

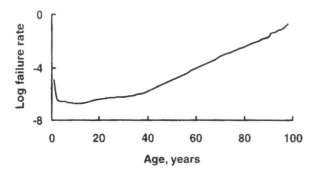

FIGURE 12-6   Failure rate vs. time curve for carbon-based units (also called the human death rate).

In most cases, random events are assumed to dominate this region, and failures here are often called extrinsic because they result from events external to the item. (As noted earlier, occurrences of lengthy flat regions of the bathtub curve are rare; usually, this region may be represented by a gradual downward slope which ends in an abrupt rise, as wearout takes over.)

The third region is characterized by an increasing failure rate, and is called wearout. This is the region in which a single major failure mechanism progresses to the point where it causes failure of all the remaining items. It is possible, but unlikely, that more than one smaller curve will be included in the wearout region. These failures are commonly considered to be intrinsic.

Probabilistic reliability engineers use the constant failure rate portion of the bathtub curve for failure rate prediction in the scheduling of system maintenance, and for spare parts provisioning. The infant mortality section is used to justify the use of "burn-in" or "run-in" processes to reduce early failures in service.

The bathtub curve does not describe how long an individual item will last, nor does it distinguish among various failure mechanisms. To do this, we must consider the individual failure distributions which contribute to the bathtub curve. These will be called the deterministic reliability curves.[‡]

[‡] It is acknowledged that these smaller curves can also be probabilistic, i.e., each curve could represent several different failure mechanisms. If we continue to refine the data, however, we ultimately get to the point where a single curve represents a single failure mechanism. For instructional purposes, we assume this point has already been reached. Each of the smaller curves also could represent the probability of a random event causing a failure. To the extent that the reliability engineer can predict, control, or minimize the effects of random events, their probabilities fit our definition of deterministic curves.

So far, we have used the term "failure rate" rather loosely. In reliability terms, the failure rate is either a *hazard rate*, which is the likelihood of failure of an unrepairable item, or the *rate of occurrence of failures* (ROCOF), which is applied to repairable systems. It is true that individual unrepairable items are either failed or unfailed, and therefore do not have failure rates; however, it is also true that groups of such items do have failure rates. Since the concentration of this text is on deterministic failures of unrepairable items, we use the term "failure rate" to mean "hazard rate" unless indicated otherwise.

12.3   DETERMINISTIC RELIABILITY CURVES

Each individual failure mechanism of an item or set of items has its own distribution in time, which may or may not begin at time zero. For example, if a group of products has a subset with cracks of a certain size, the cracks will grow during the service life of the product, and cause failures of the subset according to their own distribution before wearout of the entire lot. If the same lot has a different subset with somewhat smaller cracks, they will fail later than the first according to a different distribution, but still before wearout. Another subset with a different defect will fail according to another distribution, etc.

These individual distributions may also represent the probability of random events which can overstress an item and cause immediate failure. Examples of such events are electrical surges, dropping or damaging an item by a maintenance worker, gusts of wind, etc. Usually, the probability of such random events is considered constant throughout the useful life region.

These individual distributions which are the concern of the deterministic reliability engineer. They usually can be described statistically as some sort of probability of failure vs. time distribution. Deterministic methods are sometimes called science-based, or physics-of-failure, methods.

12.4   LIFE DISTRIBUTIONS

The basic tool of the reliability engineer is the life distribution, which may be also called the failure distribution. Life distributions are mathematical descriptions of failures in time. They can be either probabilistic (a combination of smaller distributions of different failure mechanisms) or deterministic (a single distribution representing a single failure mechanism). The ordinate of a life distribution is usually some function of time, and the abscissa is usually some measure of failure probability or failure rate. Life distributions can have any shape, but some standard forms have been developed over the years, and reliability engineers generally try to fit their data to one of them so that conclusions can be drawn. Four such distributions are discussed in some detail in this chapter, due to their widespread use in reliability: normal, lognormal, Weibull, and exponential.

Before discussing these distributions, we must introduce some mathematical functions which describe various forms of life distributions. We also introduce some parameters which are useful in describing life distributions.

12.5 FORMS OF LIFE DISTRIBUTIONS

Life distributions can take on various forms, depending on how the user wants to analyze the data and draw conclusions. The most commonly used forms of life distribution are the probability density function (pdf), the cumulative density function (cdf), the reliability function, and the hazard rate.

12.5.1 The Probability Density Function (pdf)

The pdf for the for the normal distribution is shown in Figure 12-7a. This function is designated $f(t)$, and describes the probability that an item will fail at a given time t. Its general mathematical expression is

$$\int_{-\infty}^{\infty} f(t)\, dt = 1 \qquad 0 \le f(t) \le 1 \tag{12-1}$$

For the normal pdf, this probability begins at zero, increases to a maximum, and decreases again to zero, in a curve which is bell shaped.

12.5.2 The Cumulative Distribution Function (cdf)

This function describes the probability that an item will have failed by a given time t, and is designated as F(t). It is shown for the normal distribution in Figure 12-7b. Its general mathematical expression is

$$F(x) = \int_{-\infty}^{x} f(t)\, dt \tag{12-2}$$

where $F(x)$ is the probability that failure will occur before time x. The cdf begins at zero and starts to increase at the same point in time as the pdf. Since it is cumulative, it never shows a decrease, and reaches a maximum of 1, or 100%.

12.5.3 The Reliability Function

This function is designated $R(t)$. It represents the probability that an item will *not* have failed by a given time t, that is, that the item will be reliable. In its general form, it is equal to 1 minus the cdf, or

$$R(t) = 1 - F(t) \tag{12-3}$$

$R(t)$ for the normal distribution is shown in Figure 12-7c. It starts at 1 and ends at zero.

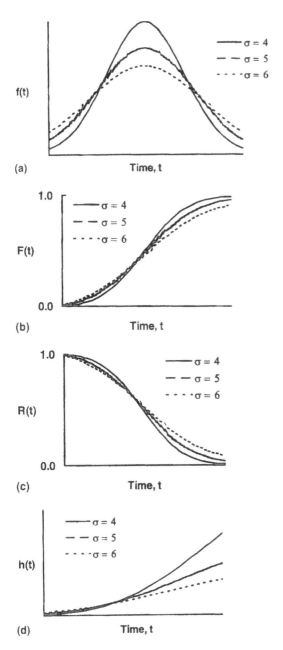

FIGURE 12-7 Probability density function of (a) the normal distribution; (b) cumulative density function; (c) reliability function; (d) failure rate curve.

12.5.4   The Hazard Rate

The hazard rate is the probability that an item will fail during a given time
interval, Δt, given that it is functioning at the beginning of that interval. It is
given by

$$\overline{\lambda(t)} = \frac{N_f}{N_u\,\Delta t} \tag{12-4}$$

where $\overline{\lambda(t)}$ is the average failure rate during the interval Δt, N_f is the number of
items that fail during Δt, and N_u is the number of good items entering Δt. As Δt
approaches zero, Eq. (12-1) becomes the *instantaneous failure rate*, which is also
called the *hazard rate*. The general expression for the hazard rate is

$$\lambda(t) = \frac{f(t)}{R(t)} \tag{12-5}$$

(This and other derivations of reliability equations are provided in detail in Ref.
6.) The hazard rate is also commonly designated $h(t)$. Figure 12-7d shows the
hazard rate function for the normal distribution. In general, when we refer to
failure rate in this book, we mean the hazard rate.

The above expressions show only the general forms of the pdf, cdf, $R(t)$,
and $\lambda(t)$. Each type of life distribution has its own specific form.

12.6   SIGNIFICANT PARAMETERS OF LIFE DISTRIBUTIONS

It is often cumbersome or impractical in reliability or accelerated testing work
to use an entire distribution as a variable, or a factor. Instead, we use parameters
of distributions which are important to our purpose. These parameters could be
almost any feature of a life distribution, but some parameters have become common, such as the mean, median, mode, and standard deviation. We are most
familiar with these terms as they relate to the normal distribution. They are calculated in different ways, but have the same meaning, for other distributions.

A common terminology in reliability work is the time at which a certain
percentage of items fail. t_{50}, t_{16}, and t_{75}, for instance, are the times by which 50,
16, and 75% of the items in a sample have failed. The median of a normal distribution is t_{50}.

The *location parameter* of a distribution locates it in time. For a normal
distribution, the location parameter is the mean. Location parameters are also
called *measures of central tendencies*, *measures of central values*, and *measures
of location*.

The *shape parameter* provides a quantitative measure of the shape, or spread, of a distribution. For a normal distribution, the shape parameter is the standard deviation. Shape parameters are also called *measures of variation*.

12.7 FOUR IMPORTANT LIFE DISTRIBUTIONS

As mentioned earlier, there are many statistical probability distribution functions, among which are the normal, lognormal, Weibull, exponential, gamma, binomial, Poisson, chi-square, etc. Of these, the first four are widely used in reliability and are discussed in some detail here.

12.7.1 The Normal Distribution

The pdf, cdf, reliability, and hazard functions for the normal distribution are shown graphically in Figure 12-7. Its important parameters are:

Mean (arithmetic average):	$t = \mu$
Median (t_{50}, or time at 50% failure):	$t = \mu$
Mode [highest value of $f(t)$]:	$t = \mu$
Location parameter:	μ
Shape parameter:	σ
s (estimate of σ):	$t_{50} - t_{16}$

The probability density function for the normal distribution is

$$f(t) = \frac{1}{\sigma \sqrt{2\pi}} \exp\left[\left(-\frac{1}{2}\right)\left(\frac{t - \mu}{\sigma}\right)^2\right] \tag{12-6}$$

The cumulative density function is

$$F(t) = \frac{1}{\sigma \sqrt{2\pi}} \int_0^t \exp\left[\left(-\frac{1}{2}\right)\left(\frac{x - \mu}{\sigma}\right)^2\right] dx \tag{12-7}$$

The reliability is

$$R(t) = 1 - F(t) \tag{12-8}$$

The hazard rate is

$$\lambda(t) = \frac{\exp\left[\left(-\frac{1}{2}\right)\left(\frac{t - \mu}{\sigma}\right)^2\right]}{\int_t^\infty \exp\left[\left(-\frac{1}{2}\right)\left(\frac{x - \mu}{\sigma}\right)^2\right] dx} \tag{12-9}$$

The normal distribution is not as common in reliability work as the other three distributions in this section. It is included here because of its familiarity.

12.7.2   The Lognormal Distribution

The important parameters of the lognormal distribution are:

Mean: $t = \exp\left[\left(\mu + \dfrac{\sigma^2}{2}\right)\right]$

Median (t_{50}): $t = e^{\mu}$

Mode: $t = \exp\left(\mu - \sigma^2\right)$

Location parameter: e^{μ}

Shape parameter: σ

s (estimate of σ): $\ln\dfrac{t_{50}}{t_{16}}$

The pdf, cdf, reliability, and hazard functions for the lognormal distribution are similar to those of the normal distribution, except that they are functions of the log of time, instead of time. They may be seen graphically in Figure 12-7, if the time axes are considered to be logarithmic instead of linear.

The pdf for the lognormal distribution is

$$f(t) = \frac{1}{\sigma t \sqrt{2\pi}} \exp\left[\left(-\frac{1}{2}\right)\left(\frac{\ln t - \mu}{\sigma}\right)^2\right] \tag{12-10}$$

The cdf is

$$F(t) = \frac{1}{\sigma\sqrt{2\pi}} \int_0^t \frac{1}{x} \exp\left[\left(-\frac{1}{2}\right)\left(\frac{\ln x - \mu}{\sigma}\right)^2\right] dx \tag{12-11}$$

The reliability is

$$R(t) = 1 - F(t) \tag{12-12}$$

The hazard rate is

$$\lambda(t) = \frac{f(t)}{1 - F(t)} \tag{12-13}$$

The lognormal distribution is very common in reliability work. It is used to describe failure distributions due to accumulated damage, i.e., in which a defect gets progressively worse in service until it finally causes failure. Crack propagation, oxide growth, and wear are examples of this.

12.7.3 The Weibull Distribution

The Weibull distribution was first reported in 1951 [8], and it is one of the most flexible and widely used of all the reliability distribtutions. Reference 8 reports its application to (1) yield strength of a Bofors steel; (2) size distribution of fly ash; (3) fiber strength of Indian cotton; (4) length of Cyrtoideae; (5) fatigue life of a St-37 steel; (6) height of adult males born in the British Isles; and (7) breadth of beans of Phaseolus Vulgaria. If anything, the Weibull distribution is used too widely, and often inappropriately. Waloddi Weibull, for whom the distribution is named, says this about it:

> The author has never been of the opinion that this function is always valid. On the contrary, he very much doubts the sense of speaking of the "correct" distribution . . . the only practicable way of progressing is to choose a simple function, test it empirically, and stick to it as long as none better has been found. [8]

The pdf, cdf, reliability, and hazard functions for the Weibull distribution are shown graphically in Figure 12-8. Its three most important parameters are:

Location, or time-delay, parameter: γ

Shape parameter, or slope: β

Scale parameter: η

The Weibull pdf is

$$f(t) = \frac{\beta}{\eta}\left(\frac{t-\gamma}{\eta}\right)^{\beta-1}\exp\left[-\left(\frac{t-\gamma}{\eta}\right)^{\beta}\right] \tag{12-14}$$

The cdf is

$$F(t) = 1 - \exp\left[-\left(\frac{t-\gamma}{\eta}\right)^{\beta}\right] \tag{12-15}$$

The reliability is

$$R(t) = \exp\left[-\left(\frac{t-\gamma}{\eta}\right)^{\beta}\right] \tag{12-16}$$

The hazard rate for the Weibull distribution is

$$\lambda(t) = \frac{\beta}{\eta}\left(\frac{t-\gamma}{\eta}\right)^{\beta-1} \tag{12-17}$$

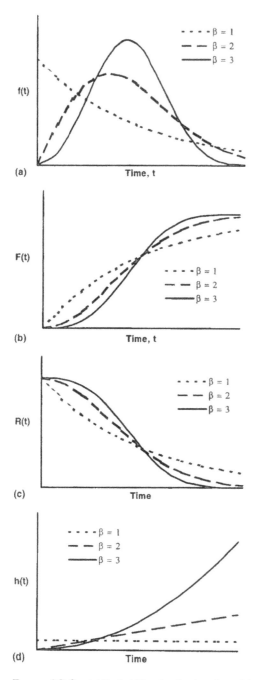

FIGURE 12-8   (a) Probability density function of the Weibull distribution; (b) cumulative density function; (c) reliability function; (d) reliability curve.

The Weibull distribution is one of the most common and most powerful in reliability, because of its flexibility in taking on many different shapes for various values of β. For example, it can be used to describe the three regions of the bathtub curve as follows:

Infant mortality (decreasing failure rate): $\beta < 1$

Useful life (constant failure rate): $\beta = 1$

Wearout (increasing failure rate): $\beta > 1$

The location parameter γ is used to move the distribution forward or backward in time. If the distribution begins at $t = 0$, then $\gamma = 0$, and the distribution becomes the *two-parameter Weibull*, which is very common in reliability work.

12.7.4 The Exponential Distribution

The exponential distribution is a special case of the Weibull distribution, occurring when the Weibull slope β is equal to 1. This condition describes a constant failure rate condition. The pdf, cdf, reliability, and hazard rate are shown graphically in Figure 12-8 as the curves for which $\beta = 1$. The pdf for the exponential distribution is

$$f(t) = \lambda \exp[-\lambda t] \qquad (12\text{-}18)$$

Its cdf is

$$F(t) = 1 - \exp[-\lambda t] \qquad (12\text{-}19)$$

Its reliability is

$$R(t) = \exp[-\lambda t] \qquad (12\text{-}20)$$

Its hazard rate is

$$\lambda(t) = \lambda \qquad (12\text{-}21)$$

Since the failure rate is constant for the exponential distribution, its inverse is defined as the mean-time-between failures, or MTBF.

$$\text{MTBF} = \frac{1}{\lambda} = \theta \qquad (12\text{-}22)$$

From a statistical point of view, the exponential distribution usually is applicable only to failures which are totally random. If the failure rate is constant, its probability of failure does not change with time, or it is independent of past history. An item which fails according to the exponential failure rate is as good the moment before failing as it was at the time it was put in service. This is true of an

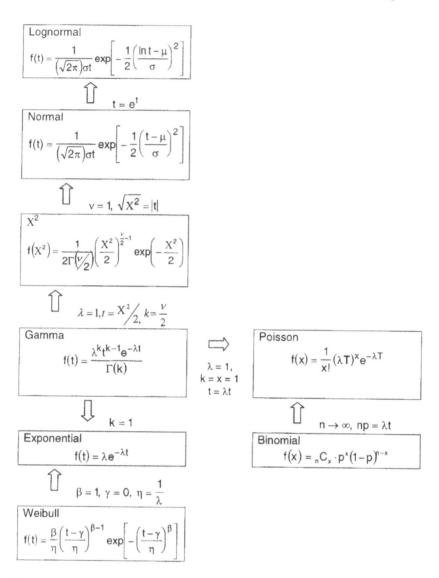

FIGURE 12-9   Relationships among reliability distribution equations.

electrical fuse, for instance, or a shear pin on a hay baler. The cause of failure is external, such as an electrical surge or a sudden mechanical load. A common cause of such failures in electronic products is handling damage.

From a reliability point of view, the exponential distribution usually does not represent a single mechanism. Instead, it is the sum of several other distributions, which may be both random and phenomenological. Since most products are subject to these mixed-mode failures, the exponential distribution often is used empirically to describe and predict their reliability. This is an example of the probabilistic approach.

12.8 OTHER DISTRIBUTIONS

In addition to the four distributions presented here, other distributions may be used in reliability work. Although they are not discussed in detail in this book, Figure 12-9 shows the mathematical relationships among some of them.

The normal, lognormal, Weibull, and exponential are examples of *continuous* distributions, because the parameters may take on any value on a continuous scale. By contrast, the parameters of a *discrete* distribution may take on only discrete values from a certain set of numbers. Perhaps the best-known discrete distribution is the well-known binomial distribution. Atlhough it is not discussed in detail here, Ref. 9 is an excellent exposition on the use of the binomial distribution.

12.9 SUMMARY

In later chapters, we probe further into the individual failure causes and distributions which make up the constant failure rate distribution, in our quest for deterministic failure prevention. We are not yet quite ready to do this, however, and in Chapters 13 and 14 we consider some methods to measure, calculate, and predict reliability.

REFERENCES

1. A. G. Bezat and L. L. Montague, The Effect of Endless Burn-in on Reliability Growth Projections, Reliability and Maintainability Symposium Proceedings, IEEE, 1979.
2. H. Hecht and E. Fiorentino, Reliability Assessment of Spacecraft Electronics, Reliability and Maintainability Symposium Proceedings, IEEE, 1987.
3. F. Proschan, Theoretical Explanation of Observed Decreasing Failure Rate, Technometrics, 5:375–383, 1963.
4. K. L. Wong and D. L. Lindstrom, Off the Bathtub and onto the Roller-Coaster Curve, Reliability and Maintainability Symposium Proceedings, IEEE, 1988.

5. K. L. Wong, What is Wrong with the Existing Reliability Prediction Methods? Quality and Reliability Engineering International, 6:251–257, 1990.

6. D. J. Sherwin, Concerning Bathtubs, Maintained Systems, and Human Frailty, IEEE Transactions on Reliability, 46(2), June 1997.

7. D. Kececioglu, Reliability Engineering Handbook, vols. 1 and 2, Prentice Hall, Englewood Cliffs, NJ, 1991.

8. W. Weibull, A Statistical Distribution Function of Wide Applicability, Journal of Applied Mechanics, September 1951.

9. J. E. King, Binomial Statistics and Binomial Plotting Paper: the Ugly Ducklings of Statistics, Quality Engineering, vol. 7(5): 493–521, 1995.

Chapter 13

Graphical Reliability Methods

13.1 INTRODUCTION

Chapter 12 contains some relatively complicated equations. To the reliability engineer, as with any professional in his field, these equations are basic; to those in other disciplines, they can be intimidating. It is a formidable challenge to know when and how to apply the various reliability methods. Our task is to make the reliability equations user friendly, so that they can be learned and applied by those who are not reliability engineers.

Reliability data can be handled in two ways: (1) by using mathematical equations derived from those in Chapter 12, and (2) by plotting them on graphs. Both methods are considered in this book. Graphical methods are easier to learn, and more instructive, and they are considered first.

13.2 PROBABILITY PLOTTING

A cynical definition of technical research is the manipulation of available data to the point where it can be plotted as a straight line. This is disconcertingly close to the truth in reliability and accelerated testing. The use of probability plots not only enables the user to estimate important parameters with reasonable accuracy, but also allows one to maintain perspective about what is being accomplished. As shown in Figures 12-7 and 12-8, all of the major forms of reliability

163

distributions, pdf, cdf, reliability, and hazard, can be plotted. The most practical of these, however, is the cumulative distribution function, cdf.

Figure 13-1 shows a plot of the cdf for the normal distribution. The normal probability distribution, or histogram, is turned on its side and projected onto a plot of cumulative percent (ordinate) vs. a linear scale (abscissa). In reliability work, the abscissa is some measure of time, such as days, thousands of hours, weeks, temperature cycles, etc. The location parameter, or mean μ, of the distribution is projected onto the cumulative distribution plot, and since μ is the median for the normal distribution, its value is plotted at the 50% point on the ordinate. Following are other percentile values for the normal distribution:

$$\mu - 1\sigma = 15.87\% \qquad \mu + 1\sigma = 84.13\%$$
$$\mu - 2\sigma = 2.28\% \qquad \mu + 2\sigma = 97.72\%$$
$$\mu - 3\sigma = 0.135\% \qquad \mu + 3\sigma = 99.865\%$$
$$\mu - 4\sigma = 0.003\% \qquad \mu + 4\sigma = 99.997\%$$

It may be seen that the $\mu - 1\sigma$ projection from the normal distribution intersects the straight-line probability plot at approximately 16%, and that the $\mu + 1\sigma$ projection intersects the plot at approximately 84%. The standard deviation of a

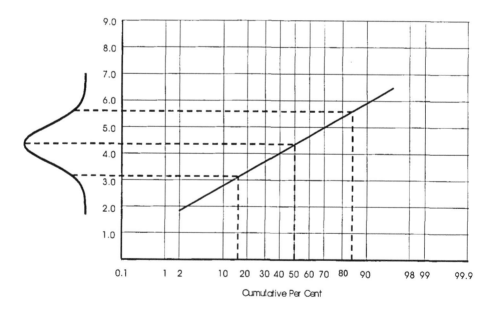

FIGURE 13-1   A normal probability distribution histogram and it relationship to the normal cumulative distribution.

normal distribution can be estimated as $(t_{50} - t_{16})$. A set of data plots as a straight line on normal probability only if it is normally distributed. If this is the case, it is possible to use graphical methods to extrapolate or interpolate to any value on the distribution. If it is desired to estimate very small failure percentages, which is common in reliability work, this can be an enormous advantage. (At this point, the author is obliged to caution the reader about the dangers of extrapolation. Because it is used to draw conclusions about what might or might not happen outside the range of the available data, it should be done only when the experimenter is reasonably certain of the behavior of the object of the experiment.)

    This type of plot also can be made for other distributions, and the lognormal, Weibull, and exponential are illustrated in this chapter. If data are distributed according to one of these functions, they will plot as a straight line on the appropriate plotting paper.

13.2.1   Normal Probability Plotting

A normal probability plot is a plot of the cumulative density function of the normal distribution [Eq. (12-7)]. In order to plot probabilities, it is necessary to have the proper paper. Sources of probability paper for various distributions are listed as Refs. 1–4 at the end of this chapter. A number of excellent software programs also are available for probability plotting, two of which are listed in Refs. 5 and 6. Before using the software, the reader is advised to use manual methods in order to get a feel for the processes used by the software.

    Normal probability plotting can be illustrated with the following times to failure (in thousands of hours) for a set of 20 samples of a manufactured product: 4.9, 5.2, 3.2, 5.6, 4.4, 4.7, 6.2, 4.9, 5.2, 4.0, 4.9, 5.8, 3.7, 6.5, 4.8, 3.4, 5.7, 5.3, 4.1, 4.4.

    The steps in probability plotting are:

1. Rank-order the data from the smallest to the largest value. If i is the rank order of a given item, and j is the total number of items, then in our example i ranges from 1 to 20, and j is equal to 20. The rank-ordered data are shown in the first column of Table 13-1.

2. Obtain an estimate E_i for the cumulative percent for each data point. Several equations have been used for this. One of them is the *midpoint plotting position*:

$$E_i = \frac{100(i - 0.5)}{j} \qquad (13\text{-}1)$$

A second equation is the *expected plotting position*:

$$E_i' = \frac{100i}{j + 1} \qquad (13\text{-}2)$$

TABLE 13-1 Cumulative Percents for Data in a 20-Piece Sample for Normal
Probability Plotting, as Estimated by Three Different Methods

Time to failure, (1000 hr)	Rank order	Cumulative percent			
		Midpoint plotting position	Expected plotting position	Median plotting position	Median rank
3.2	1	2.5	4.8	3.4	3.4
3.4	2	7.5	9.5	8.3	6.3
3.7	3	12.5	14.3	13.2	13.1
4.0	4	17.5	19.0	18.1	18.0
4.1	5	22.5	23.8	23.0	23.0
4.4	6	27.5	28.6	27.9	27.9
4.4	7	32.5	33.3	32.8	32.8
4.7	8	37.5	38.1	37.7	37.7
4.8	9	42.5	42.8	42.6	42.6
4.9	10	47.5	47.6	47.5	47.5
4.9	11	52.5	52.4	52.5	52.5
4.9	12	57.5	57.1	57.4	57.4
5.2	13	62.5	61.9	62.3	62.3
5.2	14	67.5	66.7	67.2	67.2
5.3	15	72.5	71.4	72.1	72.1
5.6	16	77.5	76.4	77.0	77.0
5.7	17	82.5	80.1	81.9	81.9
5.8	18	87.5	85.7	86.8	86.8
6.2	19	92.5	90.5	91.7	91.7
6.5	20	97.5	95.2	96.6	96.6

A third equation is the *median plotting position*:

$$E_i'' = \frac{100(i - 0.3)}{j + 0.4} \tag{13-3}$$

A fourth method is to use a ranks table. Ranks tables have been calculated such that the values given in them have a certain probability of
being less than the actual value. The most commonly used rank is the
median rank [7]. Appendix C shows median, 5%, and 95% ranks for
sample sizes up to 30. For a given i on a median rank table, the estimated cumulative percent has a probability of 50% of being less than
the true time to failure for the distribution. Confidence limits can be
obtained by plotting the 5% and 95% ranks along with the median
ranks.

All the above methods yield slightly different results. The differences are usually slight, however, compared with the inaccuracies in data collection and plotting of the distribution, so there is rarely any practical difference among them. The choice is left to the reader. Results for the four different methods for our data are shown in Table 13-1. It may be noted that the cumulative percents for the median plotting position and the median ranks method are nearly identical. The reader is encouraged to calculate a few of the values for practice.

3. Plot the data on probability paper. Each time to failure is plotted vs. its cumulative percent. Results for this example, using median ranks, are shown in Figure 13-2.

4. Draw a best-fit straight line through the plotted data points. This is almost always done by eyeball. If several distributions are plotted on the same paper, and their standard deviations are known to be the same, the best fit is a set of parallel lines.

From this plot, it may be seen that the mean t_{50} of the distribution is 4.8, and that mean minus 1 sigma (t_{16}) is 3.9. The estimated standard deviation is thus 0.7. It is possible to estimate the time to failure for any cumulative percent.

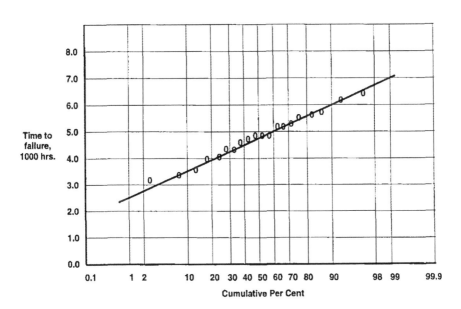

FIGURE 13-2   Probability plot of times to failure for a 20-piece sample of a manufactured item.

Probability plots are the best way to compare two or more distributions. They all can be plotted on the same paper, and comparisons can be made for mean, standard deviation, etc. A common comparison is that of time to a certain percent failure for different samples.

In this example, the time of each failure is known, and the data points have been plotted individually. In some cases, especially when the sample sizes are large, it is more convenient to group the data into time intervals and plot a single point for each interval. This is especially true when a reliability test is being conducted, and pass-fail tests are made only at certain time intervals.

13.2.2 Lognormal Probability Plotting

A lognormal probability plot is a plot of the lognormal cdf equation (12-11). Lognormal plotting is done in exactly the same manner as normal probability plotting, except that the logarithms of the times to failure are plotted. Usually, this is $\log_{10}$, but the natural logarithm also can be used. Plotting is made easier with the use of logarithmic probability paper.

To illustrate lognormal probability plotting, consider the following data from a mechanical fatigue test (the numbers are for cycles to failure): 100, 62, 52, 150, 120, 66, 82, 82, 140, 56, 100, 84, 44, 66, 130, 200, 72, 100, 140, 94.

The rank-ordered data and their median rank values are shown in Table 13-2, and the probability plot is shown in Figure 13-3. The logs of the mean and standard deviations may be estimated from this plot, and extrapolations may be made from it.

Many failure mechanisms have been shown to fit the lognormal probability distribution. Times to failure for data from a reliability test, or from the product

TABLE 13-2   Fatigue Test Data for a Lognormal Probability Plot

Cycles to failure	Rank order	Cumulative percent	Cycles to failure	Rank order	Cumulative percent
44	1	3.4	94	11	52.5
52	2	8.3	100	12	57.4
56	3	13.1	100	13	62.3
62	4	18.0	100	14	67.2
66	5	23.0	120	15	72.1
66	6	27.9	130	16	77.0
72	7	32.8	140	17	81.9
82	8	37.7	140	18	86.8
82	9	42.6	150	19	91.7
84	10	47.5	200	20	96.6

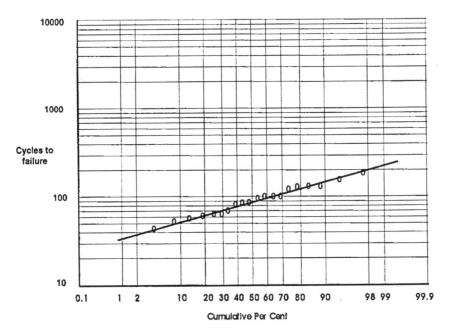

FIGURE 13-3 Lognormal probability plot of fatigue test data.

in service, often are so long that only a small percent of the total sample will fail in the time available for testing. If we know the shape of the life distribution, we can make reasonable estimates of the reliability of our sample.

13.2.3 Weibull Probability Plotting

A Weibull probability plot is a plot of the Weibull cdf equation (12-15). The Weibull distribution is quite complicated mathematically, but it can be plotted with no more difficulty than the normal or the lognormal distributions. Weibull plotting paper is available for this purpose. The forms of Weibull paper vary in the limits of their per cent scales, and in the number of cycles in the log scale.

As an exercise, we shall make a Weibull plot of the following 10 failure times (in hours) from a reliability test: 20, 26, 40, 45, 50, 60, 75, 90, 95, 110. The data are already rank-ordered and the median ranks method is used to produce the Weibull plot shown in Figure 13-4.

The Weibull shape parameter, or slope, can be obtained using the nomograph in the upper left corner of the plot. A line is drawn parallel to the best-fit line for the data through the target, and it intercepts the small horizontal scale

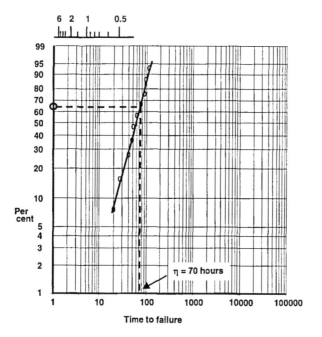

FIGURE **13-4**   Weibull probability plot.

at a point equal to the Weibull slope β. Recall that different values of β describe different regions on the bathtub curve:

$\beta < 1$: Decreasing failure rate (infant mortality)
$\beta = 1$: Constant failure rate (useful life)
$\beta > 1$: Increasing failure rate (wearout)

For the data in Figure 13-4, β is approximately equal to 2.0.

Estimation of the scale parameter, η, is also shown in Figure 13-4. It is obtained by drawing a horizontal line from the 63.2% failure point on the per cent failure axis until it intersects the distribution line. A vertical line drawn from this point intersects the time-to-failure axis at η. For our example, η is estimated at slightly over 70 hours. The derivation of 63.2% for η is as follows:

$$R(t) = \exp\left[-\left(\frac{t-\gamma}{\eta}\right)^{\beta}\right] \quad \text{with } \gamma = 0 \qquad (12\text{-}16)$$

$$\ln R(t) = -\left(\frac{t}{\eta}\right)^{\beta} \qquad (13\text{-}4)$$

$$\ln \frac{1}{R(t)} = \left(\frac{t}{\eta}\right)^{\beta} \tag{13-5}$$

$$\ln \ln \frac{1}{R(t)} = \beta \ln t - \beta \ln \eta \tag{13-6}$$

At $t = \eta$:

$$\ln \ln \frac{1}{R(t)} = 0 \tag{13-7}$$

$$\ln \frac{1}{R(t)} = 1 \tag{13-8}$$

$$\frac{1}{R(t)} = e \tag{13-9}$$

$$R(t) = \frac{1}{e} = 0.368 \tag{13-10}$$

$$F(t) = 1 - R(t) = 0.632 \tag{13-11}$$

If the plotted data appear to fit a smooth, but not straight, line on a Weibull plot, it may be possible to make it straight by adding or subtracting a constant time value to each data point. The value required to straighten the data is equal to γ, the location parameter of a Weibull distribution.

If the true distribution begins at $t = 0$, then $\gamma = 0$; if the true distribution does not begin at zero, then the displacement from zero is equal to γ. For instance, if it were found that a curved Weibull plot of data with a first failure of 100 hours could be made straight by subtracting 70 hours from each failure time, then it would be estimated that the true first failure would be near 30 hours.

There is much that can be done with Weibull plots and their interpretation, but that is beyond the scope of this book. An excellent presentation of these methods can be found in Ref. 8.

13.2.4   Exponential Probability Plotting

As pointed out in Chapter 12, the exponential distribution is a special case of the Weibull distribution, occurring when the Weibull slope β is equal to 1. The equation for the lognormal cdf is

$$F(t) = 1 - \exp[-\lambda t] \tag{12-19}$$

Since exponential distributions are plotted on semilog paper with the slope equal to λ, it is not the cumulative distribution function $F(t)$ that plots as a straight line, but the reliability function:

$$R(t) = 1 - F(t) = \exp(-\lambda t) \tag{12-20}$$

To illustrate exponential reliability plotting, consider the following times to failure: 16, 20, 30, 50, 55, and 90. These data are plotted in Figure 13-5.

This is an appropriate time to introduce the mean time between failures (MTBF), one of the most basic concepts in reliability engineering. In any given time interval, two things may happen to an item: it may fail, or it may not fail. Only the failures are countable, however, since it is meaningless to talk about the number of times an item does not fail. The probability that an item will fail is typically represented by the Poisson distribution; and the exponential distribution is used to describe the intervals between the failures. For repairable items, λ is called the failure rate, and $1/\lambda$ is called the MTBF.

The failure rate λ of an exponential distribution is constant; and the mean time between failures (MTBF) is also constant and equal to $1/\lambda$. (Note that, since the exponential distribution is a special case of the Weibull distribution, the derivation of η in the previous section also applies here.) The MTBF of an exponen-

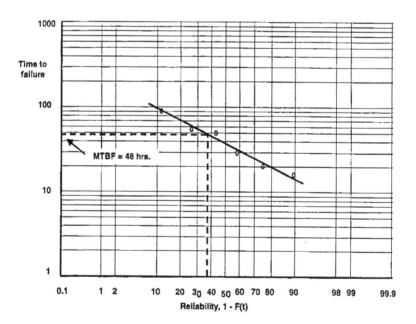

FIGURE 13-5   Exponential probability plot.

tial distribution corresponds to the point where 63.2% of the items have failed, or $R(t) = 0.368$. In Figure 13-5, this value is approximately 48 hours. (The MTBF of other distributions with nonconstant failure rates may be calculated, but it is meaningless if the time for which it is calculated is not specified. This is often the source of erroneous conclusions in reliability work.)

13.3 GOODNESS OF FIT FOR PROBABILITY PLOTS

The most common method of fitting lines to probability plots is by eyeball (unless, of course, plotting software is used). This is sufficiently accurate for most work, and mathematical curve fitting or tests for goodness of fit are rarely used. The confidence limits on a plot made with median ranks can be estimated by plotting other rank values, such as the 5% and 95% ranks, on the same plot.

There are at least three goodness-of-fit tests which can be used easily on probability plots. They are (1) the chi-square test, (2) the Kolmogorov-Smirnov test, and (3) the least-squares test. Although these methods are not difficult, they are not developed here. An excellent presentation of all three can be found in Ref. 10.

13.4 PRACTICAL RELIABILITY PLOTTING

The examples shown in the previous sections for plotting four important life distributions have been idealized to illustrate the methods. Reality, however, rarely fits our theories as well as textbook examples. Consider, for example the solder fatigue data shown in Table 13-3, from Ref. 9. These data have been plotted on normal, lognormal, and Weibull probability paper in Figures 13-6, to 13-8. All three plots

TABLE 13-3   Solder Fatigue Data

Failure no.	Cumulative percent	Cycles to failure, 60°C	Cycles to failure, 100°
1	6.7	185	7
2	16.2	242	46
3	25.8	254	52
4	35.5	280	82
5	45.1	305	90
6	54.8	353	100
7	64.5	381	101
8	74.1	504	105
9	83.8	556	112
10	93.3	687	151

Source: From Ref. 9. © 1988 IEEE, used by permission.

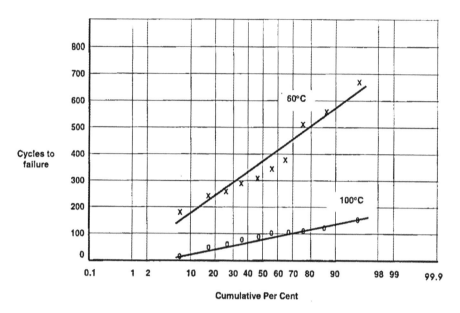

FIGURE 13-6   Data from Table 13-3, plotted on normal probability paper.

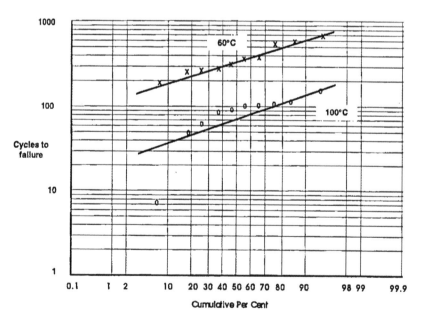

FIGURE 13-7   Data from Table 13-3, plotted on lognormal probability paper.

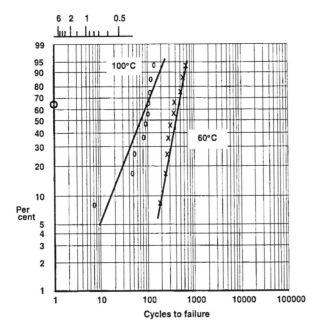

FIGURE 13-8   Data from Table 13-3, plotted on Weibull probability paper.

seem to fit the data equally well. (The author of Ref. 9 chose to use the Weibull plot.)

Another choice facing the practical reliability engineer is what to do when the data have obvious points of inflection, such as those shown in Figure 13-9. These plots "tail off" at the low end or the high end. The distributions represented by these plots have points that are sometimes called "outliers" at the ends. The upper curve has a few items with lifetimes so long that they do not fit the distribution; and the lower curve has a few items with lifetimes so short that they do not fit the distribution. The engineer has two choices at this point: (1) to replot the data on another type of distribution paper in an effort to obtain a better fit, or (2) to consider the data as representative of two or more distributions (such as two different failure mechanisms). This is an engineering, not a statistical, decision. This is the point where careful analysis of the failed samples is critical.

13.5   BIMODAL DISTRIBUTIONS

Consider the following hypothetical times-to-failure data, in hours, from a reliability test: 15, 20, 25, 30, 200, 300, 300, 400, 400, 400, 400, 500, 600, 600,

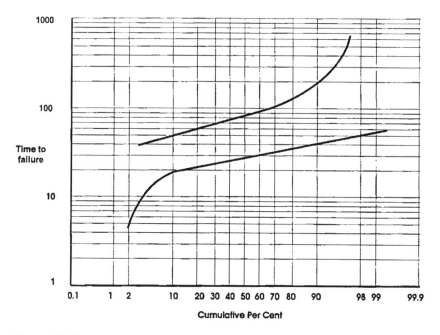

FIGURE **13-9** Lognormal distributions showing high-end and low-end tails.

600, 1500, 2000, 2000, 2000, 2500. A lognormal plot of these data is shown in Figure 13-10, with an obvious point of inflection. These data are from a bimodal distribution, or from two different distributions. Both segments have an apparent standard deviation of approximately 1.0, but have very different means. Since the point of inflection occurs at about 20%, we can assume that about 20% of the data have failed due to one mechanism, and the remaining 80% have failed due to a different mechanism (this must be confirmed by failure analysis). We can thus divide the data into two different sample sets, one with the lowest four data points, and the other with the highest sixteen data points. These two distributions have been plotted in Figure 13-11. The upper distribution has a t_{50} of approximately 700 hours, and the lower distribution has a t_{50} of about 20 hours. Notice that their standard deviations are both less than 1.

   This type of data can be very revealing. The lower 4 points might represent a defective subgroup or a group of items which has a more severe form of the defect which caused all of the items to eventually fail. This information could be used to improve the product by eliminating the cause of failure. It could also be used to set up an environmental stress screening process to expose the defective items before shipping.

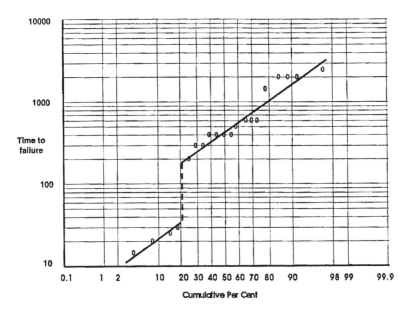

FIGURE 13-10 Lognormal plot of data showing a bimodal distribution.

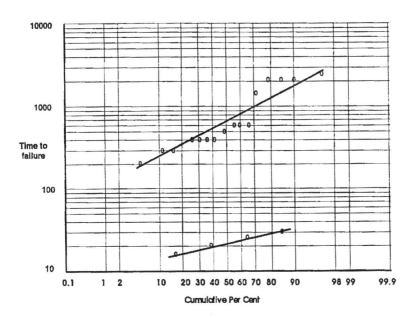

FIGURE 13-11 Lognormal plots of the two different distributions represented in the plot of Figure 13-10.

13.6 SUMMARY

In this chapter, we have considered some of the basic ways to handle data by using graphical techniques. Graphical methods are quite satisfying, since they provide "pictures" of the data in a form that allows the user to interpret the data intuitively. Graphical methods are not always the most useful, however, and in Chapter 14, some calculation methods are considered.

REFERENCES

1. CODEX Book Company, 22 High Street, P.O. Box 905, Brattleboro, VT 05302.
2. Technology Associates, 51 Hillbrook Drive, Portola Valley, CA 94028.
3. K+E (Keuffel & Esser Co.), 20 Whippany Road, Morristown, NJ 07960.
4. TEAM (Technical and Engineering Aids for Management), P.O. Box 25, Tamworth, NH 03886.
5. Weibullsmith, 536 Oyster Road, North Palm Beach, FL 33408–4328.
6. Reliasoft, 3037 North Dodge, Tucson, AZ 85716.
7. L. G. Johnson, The median ranks of sample values in their population with an application to certain fatigue studies, Industrial Mathematics, 2:1–9, 1951.
8. D. Kececioglu, Reliability Engineering Handbook, Prentice Hall, Englewood Cliffs, NJ, 1991.
9. J. H. Lau, G. Harkins, D. Rice, J. Kral, and B. Wells, Experimental and Statistical Analysis of Surface-Mount Technology PLCC Solder-Joint Reliability, IEEE Transactions on Reliability, 37(5):524–530, 1988.
10. P. D. T. O'Connor, Practical Reliability Engineering, 3rd ed., John Wiley & Sons, Baffins Lane, Chichester, England, pp. 56–60, 1991.

Chapter 14

Reliability Calculations: Constant Failure Rate

14.1 INTRODUCTION

Reliability calculations can be somewhat baffling to those who are being introduced to them. There are a variety of reasons for this, the most common of which are that the purpose of the calculations is not stated clearly, and that "shortcuts" are taken that often confuse the first-time user. In this chapter, some basic reliability calculations are presented, along with step-by-step illustrations of their application to various situations. If these methods are used systematically, the calculations are relatively easy.

14.2 MTBF CALCULATIONS

MTBF calculations may be made from test data or from field observations. Usually, they are based on the assumption that the failure rate is constant over the range for which the calculations or predictions are made. Many distributions have failure rates that are constant for certain conditions, but the exponential distribution is the only one for which it is always constant. The exponential distribution is therefore very popular because it allows easy MTBF calculations. Sometimes, it is used inappropriately for this reason, and applied to distributions to which it does not fit. Before performing any reliability calculations, the reader is advised to ascertain that the equations being used are appropriate to the data.

179

The calculations illustrated in this section are based on the assumption that the failure rate is constant, and that the failures are caused by random events, at least over the time period for which the calculations are made. If this is not the case, other methods must be used.

The most common reliability tests are conducted by placing a defined number of items on test, or by considering a defined number of items in service, for a defined period of time, and counting the number of failures during that time period. This type of test can be conducted in four different ways:

1. Failure-terminated with replacement
2. Time-terminated with replacement
3. Failure-terminated without replacement
4. Time-terminated without replacement

Combinations of the above approaches also are possible. Equations can be written to estimate the MTBF for each of these cases, but the calculation method is the same for all of them:

To estimate the MTBF for reliability tests in which the failure rate is considered constant, divide the total number of accumulated device hours by the total number of relevant failures.

Accumulated device hours include those achieved prior to failure by all items in the test. This includes both failed and unfailed items, as well as original and replacement items.

The general equation for the estimate of MTBF is

$$\text{MTBF}(e) = \frac{t_a}{r} \tag{14-1}$$

where t_a is the total number of device hours and r is the number of relevant failures.

The number of accumulated device hours, t_a, is obtained by summing the hours that all the items accumulate until they are removed for any cause. Causes of removal may be relevant failure, irrelevant failure, removal for non-test reasons, or termination of the test. A relevant failure is one that will be considered as a failure in the MTBF calculations, usually because it would be expected to occur in service. An irrelevant failure is one due to some spurious cause, such as test equipment failure, or inadvertent handling damage.

14.3 FAILURE-TERMINATED TESTS WITH REPLACEMENT OF FAILED ITEMS

The MTBF is estimated for tests in which failed items are replaced by

$$\text{MTBF}(e) = \frac{t_a}{r} = \frac{nt_c}{r} \tag{14-2}$$

where n is the number of items entering the test and t_c is the clock time of the test.

As an example, consider a failure-terminated test, with replacement, in which eight items are tested until the third failure occurs. If failures occur at 150, 400, and 650 hours:

$$\text{MTBF}(e) = \frac{(8)(650)}{3} = 1733 \text{ hours} \qquad (14\text{-}3)$$

In this example, even though two items failed prior to the end of the test, they were replaced immediately, and their t_a's plus those of their replacements each equaled t_c.

14.4 TIME-TERMINATED TESTS WITH REPLACEMENT

The equation used here is the same as (14-2), except that t_c is determined by the experimenter before the test starts, instead of by a failure. As an example, consider a variation of the above test, also in which eight items are placed on test, and three failures occur at 150, 400, and 650 hours. This test, however, is time-terminated at 1000 hours, and

$$\text{MTBF}(e) = \frac{(8)(1000)}{3} = 2667 \text{ hours} \qquad (14\text{-}4)$$

Since each failed item was replaced immediately, the sum of its hours before failure, and its replacement's hours to the end of the test is equal to t_c.

14.5 FAILURE-TERMINATED TESTS IN WHICH FAILED ITEMS ARE NOT REPLACED

Equation (14-1) applies to tests without replacement, but t_a is calculated differently:

$$t_a = \sum_{i=1}^{r} t_i + (n - r)t_c \qquad (14\text{-}5)$$

In words, this says that the accumulated device hours are equal to those accumulated by the failed items r, up to the point of failure, plus those accumulated by the unfailed items $(n - r)$ from the beginning to the end of the test.

Consider the example of a test in which 12 items are placed on test, and which is terminated at the occurrence of the fourth failure. Failures occur at 200, 500, 625, and 800 hours. In this case the t_a's for the four failed items are 200,

500, 625, and 800 hours, for a total of 2125 hours. The eight unfailed items accumulate a total of (8 × 800=) 6400 hours, and

$$\mathrm{MTBF}(e) = \frac{2125 + 6400}{4} = 2131 \text{ hours} \qquad (14\text{-}6)$$

14.6 TIME-TERMINATED TESTS WITHOUT REPLACEMENT

In this case, Eq. (14-5) is used for t_a, except that t_c is determined by the experimenter, instead of being dictated by the occurrence of a failure. As an example, consider a variation of the above test, in which 12 items are placed on test, also with four failures at 200, 500, 625, and 800 hours. In this case, however, the test is time-terminated at 1000 hours. The accumulated device hours for the four failed items are still 2125, but the accumulated device hours for the unfailed items are now (8 × 1000=) 8000 hours.

$$\mathrm{MTBF}(e) = \frac{2125 + 8000}{4} = 2554 \text{ hours} \qquad (14\text{-}7)$$

14.7 MTBF FOR MIXED REPLACEMENT AND NONREPLACEMENT

Suppose we want to test six items in a time-terminated test to a clock time of 1000 hours, but will replace any failed item only once. In this case, the first failure occurs at 300 hours, it is replaced, and the replacement fails after an additional 400 hours (clock time of 700 hours). The second failure occurs at 350 hours, it is replaced, and the replacement fails after an additional 500 hours (clock time of 850 hours). The third failure occurs at 600 hours, it is replaced, and the replacement does not fail before a clock time of 1000 hours. In this case, $n = 6$, $r = 5$, and

$$\mathrm{MTBF}(e) = \frac{700 + 850 + 1{,}000 + (6-3)1000}{5} = \frac{5550}{5} = 1100 \text{ hours} \qquad (14\text{-}8)$$

14.8 INTERVAL TESTING

In this example, we test 1000 lightbulbs from 0 to 1000 hours, and record failures at intervals of 100 hours (similar to the lightbulb example of Section 12.2). Failed bulbs are not replaced. Accumulated times for all failures within an interval are calculated on the assumption that the failures occurred the instant before the end of the interval.

TABLE 14-1 Failure Rate Calculations for Interval Testing

Time interval, hr	Interval no.	No. of items entering interval n	Accumulated hours t_a	No. of failures r	Failures per million hours r/t_a	No. of survivors, $n - r$
0–100	1	1000	100,000	60	600	940
101–200	2	940	94,000	52	553	888
201–300	3	888	88,800	39	439	849
301–400	4	849	84,900	36	424	813
401–500	5	813	81,300	28	348	785
501–600	6	785	78,500	26	331	759
601–700	7	759	75,900	25	329	734
701–800	8	734	73,400	25	341	709
801–900	9	709	70,900	23	324	686
901–1000	10	686	68,600	22	321	664

This example is treated as a series of tests over 10 intervals within which the failure rate is assumed to be constant; that is, the exponential failure rate is assumed for each 100 hour interval. The data for the test are shown in Table 14-1.

For each 100 hour interval in Table 14-1, the failure rate is calculated by dividing the accumulated time for that interval by the number of failures that occurred within it. Since failures occurred within each interval, the number of accumulated hours decreased in each interval. The failure rates, in failures per million hours, are plotted in Figure 14-1.

It may be noted that, although the failure rate is constant within each interval, the overall failure rate initially decreases, and then becomes constant, like the bathtub curve. As an exercise, the reader is encouraged to plot the overall failure rate on a Weibull plot, and to determine the appropriate Weibull parameters.

14.9 CONFIDENCE LIMITS ON MTBF ESTIMATES

Confidence limits can be placed on estimates of MTBF from test data by use of the chi-squared test. We shall designate the *confidence level* (CL) as $1 - \alpha$, and the *risk level* as α. If a large number of tests are conducted to estimate the true MTBF for a population, the confidence level is defined as the proportion of samples in which the true MTBF is contained within the confidence limits.

The use of the chi-square distribution for these calculations is based on the work of Epstein and Sobel [1].

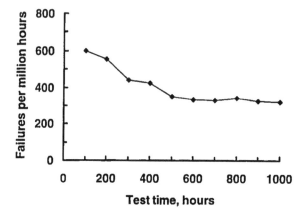

FIGURE 14-1 Plot of failure rate vs. time for the interval test data of Table 14-1.

The general equation for the upper and lower confidence limits of an estimate of MTBF is

$$\text{MTBF}(u, l) = \frac{2t_a}{\chi^2_{\gamma;\, dF}} \qquad (14\text{-}9)$$

where $\text{MTBF}(u, l)$ is the upper or lower confidence limit on the true MTBF, and χ^2_{γ} is the chi-square value for the appropriate confidence level and degrees of freedom. For confidence level $= 1 - \alpha$, and number of failures $= r$, γ and dF

TABLE 14-2 Values of γ and dF to Be Used in Eq. (14-9) or Estimating Lower and Upper Limits of MTBF

	MTBF (lower)		MTBF (upper)	
Test type	γ	dF	γ	dF
Two-sided, failure terminated	$\alpha/2$	$2r$	$1 - \alpha/2$	$2r$
One-sided, failure terminated	α	$2r$	$1 - \alpha$	$2r$
Two-sided, time terminated	$\alpha/2$	$2r + 2$	$1 - \alpha/2$	$2r$
One-sided, time terminated	α	$2r + 2$	$1 - \alpha$	$2r$
No failures observed	α	2		$2r$

The confidence interval is equal to $1 - \alpha$ and the number of failures is equal to r.

for the appropriate test are obtained from Table 14-2. Chi-squared values are listed in Appendix D. Some examples are considered in Sections 14.10 to 14.13.

14.10 TWO-SIDED CONFIDENCE LIMITS ON MTBF FOR A FAILURE-TERMINATED TEST

In this example, 10,500 device hours are accumulated, with six failures. We want to know the upper and lower limits on MTBF with 95% confidence. The following values apply:

$t_a = 10,500$

$CL = 1 - \alpha = 0.95$

$\alpha = 0.05$

$\alpha/2 = 0.025$

$1 - \alpha/2 = 0.975$

$r = 6$

We can calculate the 90% confidence limits on MTBF by using the above information, Table 14-2, and Appendix D, with the equations below.

$$MTBF(l) = \frac{2(10,500)}{\chi^2_{0.025; \, 12}} = \frac{21,000}{23.337} \approx 900 \text{ hours} \qquad (14\text{-}10)$$

and

$$MTBF(u) = \frac{2(10,500)}{\chi^2_{0.975; \, 12}} = \frac{10,500}{4.404} \approx 2,384 \text{ hours} \qquad (14\text{-}11)$$

14.11 ONE-SIDED CONFIDENCE LIMITS FOR A FAILURE-TERMINATED TEST

In this example, 16,000 device hours are accumulated, with four failures, and we want to know the lower and upper one-sided 90% confidence limits on MTBF. The following values apply:

$t_a = 16,000$

$CL = 1 - \alpha = 0.90$

$\alpha = 0.10$

$\alpha/2 = 0.05$

$1 - \alpha/2 = 0.95$

$r = 4$

The 90% confidence limits on MTBF are then

$$MTBF(l) = \frac{2(16,000)}{\chi^2_{0.10;\ 8}} = \frac{32,000}{13.362} = 2395 \text{ hours} \tag{14-12}$$

and

$$MTBF(u) = \frac{2(16,000)}{\chi^2_{0.90;\ 8}} = \frac{32,000}{3.490} \approx 9195 \text{ hours} \tag{14-13}$$

14.12 TWO-SIDED CONFIDENCE LIMITS FOR A TIME-TERMINATED TEST

In this example, we want to know the 95% confidence limits on MTBF for a test in which 12,400 device hours are accumulated, and nine failures occur. The following values apply:

$t_a = 12,500$

$CL = 1 - \alpha = 0.95$

$\alpha = 0.05$

$\alpha/2 = 0.025$

$1 - \alpha/2 = 0.975$

$r = 9$

The estimates are

$$MTBF(l) = \frac{2(12,500)}{\chi^2_{0.025;\ 20}} = \frac{25,000}{34.170} \approx 732 \text{ hours} \tag{14-14}$$

and

$$MTBF(u) = \frac{2(12,500)}{\chi^2_{0.975;\ 18}} = \frac{25,000}{8.231} \approx 3037 \text{ hours} \tag{14-15}$$

14.13 ONE-SIDED CONFIDENCE LIMITS FOR A TIME-TERMINATED TEST

In this test 21,000 device hours are accumulated with seven failures, and we want to know the 99% confidence limits on our estimate of MTBF. The following values apply:

$t_a = 21,000$

$CL = 1 - \alpha = 0.99$

$$\alpha = 0.01$$
$$\alpha/2 = 0.005$$
$$1 - \alpha/2 = 0.995$$
$$r = 7$$

The estimates are

$$\text{MTBF}(l) = \frac{2(21{,}000)}{\chi^2_{0.016;\ 16}} = \frac{42{,}000}{32.000} \approx 1313 \text{ hours} \qquad (14\text{-}16)$$

and

$$\text{MTBF}(u) = \frac{2(21{,}000)}{\chi^2_{0.99;\ 14}} = \frac{42{,}000}{4.660} \approx 9013 \text{ hours} \qquad (14\text{-}17)$$

14.14 ESTIMATES OF MTBF WHEN NO FAILURES ARE OBSERVED

Failures are common early in the life of a product or product line. If we are good designers and good reliability engineers, we reduce these failures as the program matures, and it is common to conduct tests in which no items fail. Even when this happens, It is possible to estimate MTBF, and to place confidence limits on our estimates.

It is tempting to say that, if no failures occur in a test, the MTBF is infinite. In reality, however, we know that a failure might have occurred the instant after we terminated the test. To estimate the MTBF for this case, we use the values listed in Table 14-2. Obviously, this is a time-terminated test, and an estimate of the upper limit is meaningless. It is interesting to compare this case to that of the lower one-sided confidence limit for a failure-terminated test in which one failure occurs. From Table 14-2, the number of degrees of freedom for the latter case is $2r$, or 2 for $r = 1$. Thus it may be noted that MTBF(l) for a test in which no failures occur actually is estimated on the assumption that a failure would have occurred a moment after termination of the test.

14.15 CONFIDENCE LIMITS ON RELIABILITY ESTIMATES FOR EXPONENTIALLY-DISTRIBUTED FAILURES

Confidence limits on reliability can be obtained by using the methods of described above and the equation

$$R = e^{-t/\text{MTFB}} \qquad (14\text{-}18)$$

If, for example, we obtain lower and upper estimates for MTBF of 1550 and 2700 hours for a product, at the 90% confidence level, we can estimate its lower and upper reliability limits as

$$R(l) = e^{-t/1550} \tag{14-19}$$

and

$$R(u) = e^{-t/2700} \tag{14-20}$$

If we want to estimate the reliability of this product over the next 100 hours of service, with 90% confidence, we can obtain the limits of this estimate by

$$R(l) = e^{-100/1550} = e^{-0.0645} = 0.9375 \tag{14-21}$$

and

$$R(u) = e^{-100/2700} = e^{-0.0370} = 0.9636 \tag{14-22}$$

Thus, we can estimate with 90% confidence that the reliability of the product will be between 0.9375 and 0.9636 over the next 100 hours.

14.16 SYSTEM RELIABILITY CALCULATIONS

For systems with n components, each of which has a constant hazard rate, and each of which must operate in order for the system to operate, the system reliability is calculated using the equation

$$R_{system} = e^{-\lambda_1 t} e^{-\lambda_2 t} e^{-\lambda_3 t} \cdots e^{-\lambda_n t} = e^{-(\lambda_1 + \lambda_2 + \lambda_3 \cdots + \lambda_n)t} \tag{14-23}$$

Thus the failure rate of the system is the sum of the failure rates of its component parts. This is the basis of the probabilistic methods discussed in Chapter 15.

Consider the example of a system with 1000 components, all with equal failure rates, which has a system MTBF of 15,000 hours. We can calculate the required failure rate for each component as follows:

$$R_{system} = e^{-t/MTBF} = e^{-t/15,000} = e^{-\lambda_1 t} e^{-\lambda_2 t} \cdots e^{-\lambda_{1000} t} \tag{14-24}$$

$$-\frac{t}{15,000} = -\lambda_1 t - \lambda_2 t - \lambda_3 t - \cdots - \lambda_{1000} t \tag{14-25}$$

$$\frac{t}{15,000} = 1000 \lambda t \tag{14-26}$$

$$\lambda = \frac{1}{15,000,000} = 0.067 \text{ failures per million hours} \tag{14-27}$$

14.17 DETERMINING TEST TIMES FOR MTBF ESTIMATES

The equations presented so far in this section have been directed toward estimating MTBF from a given accumulated test time and failures occurring within that test time. In setting up a test, it is often necessary to go in the other direction. For example, given a requirement to demonstrate a reliability of 15,000 hours, what should the sample size be, and how long should the test last?

The quick and accurate answer is to work backward using the methods and equations presented above, and calculate the required t_a . t_a is the number of items tested multiplied by the total clock hours for each item. If a large number of items can be tested, and few failures are expected, then the accumulated device hours can be approximated by multiplying the number of items entering the test by the clock hours. If these assumptions cannot be made, care must be taken to calculate actual hours accumulated. This could get further complicated if failed items are replaced.

After a required t_a has been calculated, the actual number of samples and test duration can be determined by considering the cost and availability of samples, the schedule restrictions, and the level of confidence desired from the results.

Before leaving this subject, a word should be said about t_a and MTBF. Theoretically, 1 million hours of accumulated time could be attained by testing one item for 1 million hours, or 1 million items for 1 hour, or any other combination whose product equals 1 million. Although this is mathematically possible, we know that some such combinations are unrealistic. MTBF tells us nothing about the life of the product. For example, the MTBF of mosquitoes in the typical backyard in Minnesota during the summer is probably about 1 billion hours, and yet no individual mosquito lives more than a few days. In setting up a reliability test, care must be taken to test the items over a time range which truly represents the operating conditions, and which considers the lifetime of the individual items. This can only be determined from knowledge of the product, its likely causes of failure, the operating conditions, and the operating requirements.

14.18 CENSORING

For most of the data used in the examples so far in this chapter, the exact time to failure for each failure has been known. For reliability work, this is rarely the case, and the data are said to be *censored*. If failures occur before a certain time, but the exact time is not known, the data are *left censored*. Conversely, data for failures which are known to occur after a certain time, but whose exact times of failure are unknown, are *right censored*. Many reliability tests are *time censored*; that is, they are terminated after a certain time, regardless of how many of the items under test have failed. Less common are *failure-censored* tests, in which the test is terminated after a certain number of failures have occurred.

Interval data are collected from tests in which all items are tested at specified time intervals. In this case, it is known that failures occurred within a given interval, but the exact times within that interval are not known. Some experimenters assume that all failures occurred at the end of the interval, while others distribute them evenly throughout the interval. There is no rule for this, except that, whatever approach is used, it should be consistent. Interval data are very common in reliability testing.

In evaluating data from reliability tests or in-service performance, it is apparent that, even though an item has not failed, it contributes information about the reliability of the product. Such items are called *suspended items* or *suspensions*. It is therefore desirable to have analytical methods to account for data from both failed and suspended items. To analyze data from reliability testing or products in service, and which include both failed and suspended items, the following steps are followed:

1. List the data in order of running time; for example, list the items in the chronological order in which they are placed in service. For a repairable item, the first record would be the time at which it was placed in service when new; the second record would be the time it was placed in service after the first relevant repair, etc.
2. Identify the relevant failures.
3. Record the number of preceding items for each item.
4. Calculate the *mean order number i_t*, using

$$i_t = i_{t-1} + N_t \qquad\qquad (14\text{-}28)$$

where

$$N_t = \frac{(n + 1) - i_{t-1}}{1 + (n - \text{number of preceding items})} \qquad\qquad (14\text{-}29)$$

in which n is the sample size.

The above approach is illustrated by the example of a repairable industrial product, data for which are shown in Tables 14-3 through 14-6. Table 14-3 shows the first 30 records for a product that was placed into service in early 1991, representing serial numbers 9–25. (The first eight serial numbers are missing because those items were used for testing or other purposes, and were not placed into service.) As an example, serial number 9 was placed into service on April 11, 1991, and as of the analysis date, December 6, 1993, it had not yet experienced a failure. Serial number 10 was placed in service on March 8, 1991, and experienced its first failure (or was returned to the manufacturer) on September 28,

TABLE **14-3**   Censored Field Return Data (Analysis Date: 12/6/93)

Record no.	S/N	Ship date	Return date	Time to return (days)	Failed or unfailed?
1	9	4/11/91		970	U
2	10	3/8/91	9/28/91	204	F
3	10	10/28/91		770	U
4	11	4/26/91	10/26/92	549	F
5	11	11/26/92		375	U
6	12	4/5/91		976	U
7	13	4/5/91	8/12/91	129	F
8	13	9/12/91		816	U
9	14	4/5/91	10/16/91	194	F
10	14	11/16/91	2/4/93	446	F
11	14	3/4/93	5/18/93	75	F
12	14	6/18/93		171	U
13	15	4/5/91	6/16/91	72	F
14	15	7/16/91	7/14/92	364	F
15	15	8/14/92		479	U
16	16	4/5/91	5/4/91	29	F
17	16	6/4/91		916	U
18	17	4/5/91	8/9/91	126	F
19	17	9/9/91		819	U
20	18	4/5/91		976	U
21	19	4/5/91		976	U
22	20	4/5/91		976	U
23	21	4/5/91		976	U
24	22	4/11/91	7/29/91	109	F
25	22	8/29/91		830	U
26	23	4/11/91	2/18/92	313	F
27	23	3/18/92		628	U
28	24	4/11/91		970	U
29	25	4/12/91	4/5/92	359	F
30	25	5/5/92		580	U

1991. It was repaired and returned to service on October 28, 1991. Because it was placed into service twice, serial number 10 contributes two records to this analysis.

The data are sorted according to time to return in Table 14-4. The time to return is converted to hours, and the data needed for calculating $F(t)$ and $R(t)$ also are shown in Table 14-4. To illustrate the calculations, consider the examples of failures 1 and 8.

TABLE 14-4 Field Return Data from Table 14-3, Sorted According to Length of Service, and for Which $F(t)$ and $R(t)$ Are Calculated

Days to return	F/U?	Failure	Hours to return	No. preceding	N_i	i_i	$F(t)$	$R(t) = 1 - F(t)$
29	F	1	166	0	1.00	1.00	2.30%	97.70%
72	F	2	411	1	1.00	2.00	5.59%	94.41%
75	F	3	429	2	1.00	3.00	8.88%	91.12%
109	F	4	623	3	1.00	4.00	12.17%	87.83%
126	F	5	720	4	1.00	5.00	15.46%	84.54%
129	F	6	737	5	1.00	6.00	18.75%	81.25%
171	U			6				
194	F	7	1109	7	1.04	7.04	22.18%	77.82%
204	F	8	1166	8	1.04	8.04	25.47%	74.53%
313	F	9	1788	9	1.05	9.05	28.77%	71.23%
359	F	10	2051	10	1.05	10.05	32.06%	67.94%
364	F	11	2080	11	1.05	11.05	35.36%	64.64%
375	U			12				
446	F	12	2548	13	1.11	12.11	38.85%	61.15%
479	U			14				
549	F	13	3137	15	1.19	13.19	42.39%	57.61%
580	U			16				
628	U			17				
770	U			18				
816	U			19				
819	U			20				
830	U			21				
916	U			22				
970	U			23				
970	U			24				
976	U			25				
976	U			26				
976	U			27				
976	U			28				
976	U			29				

TABLE **14-5** Censored Field Return Data (Analysis Date: 12/6/93)

Record no.	S/N	Ship date	Return date	Time to return (days)	Failed or unfailed?
1	400	12/18/91	2/12/93	422	F
2	400	5/12/93		208	U
3	401	12/4/91		733	U
4	402	12/4/91		733	U
5	403	12/18/91	1/5/93	384	F
6	403	4/5/93		245	U
7	404	12/18/91		719	U
8	405	12/18/91		719	U
9	406	12/18/91		719	U
10	407	12/18/91		719	U
11	408	12/18/91		719	U
12	409	12/18/91		719	U
13	410	12/18/91		719	U
14	411	12/18/91		719	U
15	412	12/18/91	9/11/92	268	F
16	412	12/11/92		360	U
17	413	12/18/91	10/21/92	308	F
18	413	1/21/93		319	U
19	414	12/18/91		719	U
20	415	1/8/92		698	U
21	416	12/18/91		719	U
22	417	12/18/91		719	U
23	418	12/18/91		719	U
24	419	1/8/92		698	U
25	420	12/18/91		719	U
26	421	12/18/91		719	U
27	422	1/8/92		698	U
28	423	12/18/91		719	U
29	424	12/18/91		719	U
30	425	12/18/91		719	U

TABLE 14-6 Field Return Data from Table 14-5, Sorted According to Length of Service, and for Which $F(t)$ and $R(t)$ Are Calculated

Time to return	F/U?	Failure	Hours to return	No. preceding	N_t	it	$F(t)$	$R(t) = 1 - F(t)$
208	U			0				
245	U			1				
268	F	1	1531	2	1.07	1.07	2.53%	97.47%
308	F	2	1760	3	1.07	2.07	5.83%	94.17%
319	U			4				
360	U			5				
384	F	3	2194	6	1.16	3.16	9.41%	90.59%
422	F	4	2411	7	1.17	4.17	12.72%	87.28%
698	U			8				
698	U			9				
698	U			10				
719	U			11				
719	U			12				
719	U			13				
719	U			14				
719	U			15				
719	U			16				
719	U			17				
719	U			18				
719	U			19				
719	U			20				
719	U			21				
719	U			22				
719	U			23				
719	U			24				
719	U			25				
719	U			26				
719	U			27				
733	U			28				
733	U			29				

For failure number 1:

$$N_t = \frac{(30 + 1) - 0}{1 + (30 - 0)} = 1$$

and

$$i_t = 0 + 1 = 1$$

For failure number 8:

$$N_t = \frac{(30 + 1) - 6}{1 + (30 - 7)} = 1.04$$

and

$$i_t = 6 + 1.04 = 7.04$$

Using this approach, it is possible to calculate the cumulative percent failed $F(t)$ and the cumulative percent unfailed $R(t)$ for each failure, taking into consideration that there are some items in the sample that have not failed, even though they have accumulated more hours than some of the failed items. The $R(t)$ results are plotted as an exponential distribution in Figure 14-2.

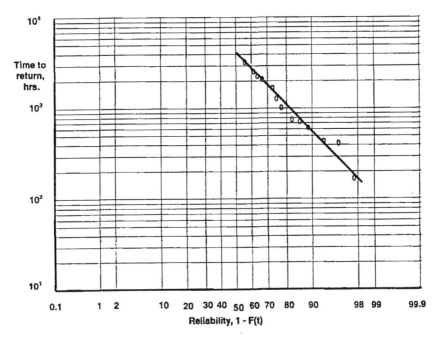

FIGURE 14-2 Reliability plot of the data in Table 14-4.

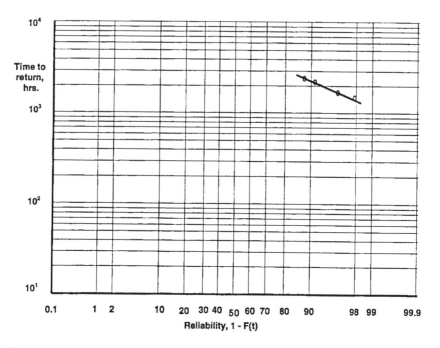

FIGURE 14-3   Reliability plot of the data in Table 14-6.

Data for 30 records of the same product, for serial numbers 400–425, are shown in Tables 14-5 and 14-6, and in Figure 14-3. Clearly, the product has improved in the time intervening between these two sets of serial numbers.

If the appropriate information regarding the costs of failures and repairs, the approach described above can be used in a variety of ways to estimate system repair costs. Although that subject is beyond the scope of this book, Refs. 2–5 provide a good basic discussion.

14.19   SUMMARY

In this chapter, a number of basic reliability calculations have been presented. In applying them, the most important concern is to know the type of distribution to be used. The calculations in this chapter are based on the exponential distribution, for which the failure rate is constant. Graphical and calculation methods for other distributions are presented in Chapter 15.

REFERENCES

1.   B. Epstein and M. Sobel, Life testing, Journal of the American Statistical Association 48(263):486–502, 1953.

2. W. Nelson, Graphical analysis of system repair data, Journal of Quality Technology 20(1):24–35, January 1988.
3. W. Nelson, Hazard plotting of left truncated life data, Journal of Quality Technology, 22(3):230–238, July 1990.
4. W. Nelson, Confidence limits for recurrence data–applied to cost or number of product repairs, Technometrics 37(2):147–157, May 1995.
5. W. Nelson, An application of graphical analysis to repair data, Quality and Reliability Engineering International 14:49–52, 1998.

Chapter 15

Reliability Calculations: Various Failure Rate

15.1 INTRODUCTION

The vast majority of reliability calculations are made with the assumption that the failure rate is constant, using many of the methods described in Chapter 14. This is due to several reasons, not the least of which is that constant failure rate calculations are easier to understand—not only by those who conduct them, but by those who must review them and make decisions based on them. Much of the system reliability software is based on the assumption of constant failure rate. For these reasons, the temptation is strong to assume the failure rate is constant, even when it is not appropriate to do so.

Usually, but not always, constant failure rates are coupled with a *probabilistic* approach to reliability. In this approach, the different failure causes, modes, and mechanisms are not distinguished, and they are all lumped together for reliability calculations. *Deterministic reliability,* on the other hand, usually includes the assumption that the failure rates are not constant. It also includes an attempt to analyze the individual failures, causes, and mechanisms, and to treat each of them separately in reliability analyses. Usually, deterministic methods are more difficult and time consuming than probabilistic ones, but they also reveal more and better information and lead to more accurate conclusions.

Deterministic reliability is more a management approach than a set of methods. It is based on a management decision to understand the product, its design,

198

its manufacturing methods, and the way it is used and to take responsibility for its reliability under all reasonable operating and environmental conditions. An outline of a recommended deterministic approach is presented in this chapter, along with some calculation methods using the Weibull and lognormal distributions.

15.2 DETERMINISTIC RELIABILITY

No set of reliability methods or calculations can be said to be exclusively probabilistic or deterministic. The probabilistic approach is usually associated with the exponential distribution, for which calculations are presented in Chapter 14. Calculations using the Weibull and lognormal distributions are usually associated with a more deterministic approach, and they are presented in this chapter. These calculations should be viewed within the context of the outline of the deterministic approach to reliability. That approach was presented initially in Chapter 3, Section 3.6 and is expanded upon here.

15.2.1 Define Realistic System Requirements

This seems obvious. It is surprising, therefore, to find that it is often done poorly or incompletely. Even if it is done properly, it is often communicated or understood poorly or incompletely. It is not extreme to state that the biggest single cause of product problems is poor definition, communication, and understanding of the product requirements. Requirements definition requires a high degree of correlation between the customer and the producer of the product. This correlation is attained differently in different markets. The two extremes are represented by the consumer market, on the one hand, and the military market on the other.

In the consumer market, it is rare to find a case in which a customer spells out the specifications for a product, which are then met by the producer. Instead, the producer gauges the needs of the market to the best of his ability, and then produces products to meet perceived needs. Although there is no formal transaction between customer and producer to define the product requirements, this is a very real process. The degree to which the producer is able to perceive and meet market needs is the degree to which the producer will be successful. The fact that many producers fail in the consumer marketplace is evidence that this is not always done well.

At the other extreme, the military market, product specifications are spelled out in detail, often with great attention to how the quality and reliability of the product will be demonstrated, in a request for proposal to a specific producer. Yet it is also obvious that many unsatisfactory products are produced in this marketplace. Too often in the military market, a "propose as requested" mentality prevails. Even if an RFP contains requirements that, if modified slightly, could result in a better, more cost-effective product, it is often impossible for the producer to suggest changes. Often the true requirements for the product are not known or not stated in a way that allows the producer to deliver the best product for the application. Furthermore, once a product goes into production, it is often impossible or prohibitively expensive to implement changes to improve cost, quality, or reliability.*

Every system requirement that impacts reliability must be defined. The definition should include all functional outputs, and the range of environmental conditions over which the product must operate. It should include reliability requirements in the form of number of missions, required operating life, MTBF for repairable systems, and some way of assessing cost tradeoffs among maintenance, repair, replacement, and failure options. It also must also include ranges of variation for all important factors.

There is room for considerably more interaction than currently exists between producer and customer at the proposal and design stage, to improve products. There is also a need to produce auditable and archivable records of the results of this interaction to put all markets into a "continuous improvement" mode.

15.2.2 Define the Design Usage Environment

All operating and storage conditions which may impact reliability must be known and communicated to the product designer and manufacturer. This includes not only ambient environmental conditions but also those generated by the electrical and mechanical loading of the product. Examples of such conditions include temperature, temperature changes, operating voltage, humidity, and mechanical stresses.

Often, the critical system requirements are not all known by either the customer or the producer. In this case, conscious assumptions must be made, and communicated. The biggest mistakes are not made when the numerical value of a condition is misjudged, but when the condition is neglected entirely; it is therefore better to make an incorrect, but conscious assumption about a condition than to ignore it because it cannot be known exactly.

If the customer cannot or will not define a critical usage parameter, then the manufacturer must define it, and communicate what has been decided to the customer. Furthermore, this information must be stored in an archivable manner for future reference. It is no longer acceptable for the producer of a product to abdicate responsibility for understanding and communicating the reliability of a product in reasonably expected operating environments, whether specified by the customer or not.

---

* Fortunately, as this is written, the situation described here is changing; and there is reason to hope the relationship between military suppliers and customers will advance to the point that realistic product requirements are discussed thoroughly before being hardened into unrealistic ones.

To be useful in designing for reliability, the ambient and operating conditions must be known in terms of specific temperatures, humidities, vibration limits, voltages, etc. Recently, much progress has been made in defining the important environmental conditions, quantitatively, for various products. Chapter 17 contains information about specific environmental conditions for various products.

Increasing attention is being paid to random events which temporarily overstress certain parts of a product and cause failure. In the past, it was thought that these events were truly random, that is, they were beyond the capability of either the producer or the user to specify or anticipate. As we learn more about such events, it is becoming increasingly realistic to design products that are robust with respect to them. This is not easily addressed, but if it causes a significant proportion of failures, it must be addressed.

15.2.3 Identify Potential Failure Sites and Failure Mechanisms

It is no easy task to identify potential failure sites and mechanisms, but fortunately it is not as difficult as it seems. Most failure mechanisms can be identified by those familiar with the product by considering the possible effects of various operating and environmental stress conditions.

A *failure mechanism* is a structural change in the material of a product which causes failure. Examples are the initiation and growth of a crack, the diffusion atoms of one metal into those of another, and the removal of atoms or molecules from the surface of a material due to friction or wear. A *failure mode* is the evidence by which a failure is observed. Examples are the lowering of the yield strength of a material due to the presence of a crack, the embrittlement of a material due to intermetallics formed by solid state diffusion, and the decrease in thickness of a material due to wear.

Specific failure acceleration models and their mathematical statements are presented in later chapters. In this chapter, we discuss failure mechanisms in more general terms. Dasgupta and Pecht [1] identified four conceptual models for failure which must be considered in the deterministic approach. They are listed below.

15.2.3.1 Stress-Strength Failures

A stress-strength failure occurs if and only if the applied stress (either operating or environmental) exceeds the strength of the item. As long as the level of the applied stress remains below the item's strength, there is no damaging effect on the item; therefore, an unfailed item is as good as new up until the moment of failure. Usually, this type of failure is ascribed to random events in operation, and to the extent that these events have constant probability of occurrence in time, the stress-strength failure distribution is exponential.

Examples of stress-strength failures are the failure of an electrical fuse due to a power surge (although this would indicate the fuse is working properly, and might not fit some definitions of failure), failure due to lightning strikes, or to human-caused damage by misuse.

15.2.3.2 Damage-Endurance Failure

Damage-endurance failures are caused by the accumulation of damage during use. Although the accumulated damage may be evident if the item is inspected prior to failure, the performance of the item may not be degraded until the point at which enough damage is accumulated to cause catastrophic failure. During service before the item fails, the damage accumulates, and does not disappear when the stresses are removed. Holden et al. [2] showed that failures of this type in microcircuits could be ascribed to a series of microscopic events, and that the failure distribution could be described by a Weibull function.

Examples of damage-endurance failures are mechanical fatigue of metals, corrosion, and time-dependent dielectric breakdown. Accumulated damage failures are those which can be modeled and measured by accelerated testing.

15.2.3.3 Challenge-Response Failures

For a challenge-response failure, a defect can exist undetected in an item until a specific event or, more likely a combination of events, "challenges" the item and causes failure. An obvious and frustrating example of this is a computer software bug. This type of failure also occurs in a product that has been operating successfully in one application, but experiences failure in another. Electronic microcircuits with many logic gates or memory cells might operate for a long time with defects, until they are operated in a certain manner which causes failure. Challenge-response failures are extremely difficult to deal with.

15.2.3.4 Tolerance-Requirement Failures

In some cases, an item may operates within its specification limits, but with degraded performance. This type of failure is familiar to most of us as consumers, who are sometimes willing to accept marginal performance rather than incur the cost to correct it. An example of this type of failure is electrostatic discharge (ESD) damage to a microcircuit which causes leakage, but still allows the device to operate satisfactorily in other respects.

15.2.3.5 Failure-Causing Stresses

Stresses that cause or precipitate failures, and the ways in which they operate, are dealt with in detail in later chapters. Here we merely wish to point out that the deterministic approach to reliability is based on understanding both the applied stresses and the product's capability to withstand them. Often, it is unexpected combinations of stresses, rather than single stresses acting alone, which cause failure.

15.2.4 Purchase Reliable Materials and Components

Quality and reliability do not end when a material or component is selected and the supplier is chosen. Processes must be in place to guarantee that the continuing supply of such components and materials is consistent, that changes in the supplier's processes will not adversely affect the product, and that methods of evaluating new suppliers and designs are in place. These processes must be documented and auditable. SPC is an excellent tool to assure that the product is consistent, and DoE is effective in evaluating proposed changes.

15.2.5 Design Reliable Products, Within the Capabilities of the Materials and Manufacturing Processes

It is ironic that many industries have elaborate systems in place to assure that purchased components are properly qualified and controlled—but almost nothing to guarantee the compatibility of the design of the product with manufacturing processes and materials.

Many electronics manufacturers have a parts-failure mentality. Their approach to product quality, reliability, failure is based upon the premise that bad parts are the major, if not the only, cause of failure. Many of the U.S. military specifications have promoted this type of thinking. It is common to find failure analysis reporting systems that consider little except the parts replaced.

Table 15-1 shows that many other features of a product besides the parts contribute to failure. The design process must assure that all materials and processes are applied properly. Variations in the manufacturing processes and materials must be understood and controlled, as well as accounted for in the product design. Materials must be tested and understood for the proposed application prior to being designed into the product.

The design review can include a process capability analysis of key product features to quantify the ability of the manufacturing operation to produce the product. If a new material or process must be introduced, DoE is a quick and efficient way to optimize it before production begins.

15.2.6 Qualify the Manufacturing and Assembly Processes

Some progressive manufacturers require the quality and reliability of all new manufacturing processes and materials to be verified by a properly designed ex-

TABLE **15-1** Causes of Field Removal of Aerospace Electronic Equipment

Cause of removal	%
From Ref. 3:	
Unjustified	30
Assembly errors, handling damage, etc.	60
Design, improper installation, etc.	9
Parts manufacturing problems, etc.	1
Unexplained	<1
From Ref. 4	
Not verified	36
Apparent electrical	22
Part application	15
Bad lot	6
Process problem	12
Design	6
Miscellaneous	2
Workmanship	1
From Ref. 5:	
Retest okay	31
Manufacturing/quality/design defects	18
Thermal mechanical stress and switching	21
Electrically damaged	30
From Ref. 6	
Electrically related aircraft incidents	
Interconnections	36
Instruments	26
Power systems	25
Electromagnetic devices	10
Passive components	3
PWB part replacement data	
Integrated circuits	27
Transistors	14
Hybrid circuits	12
Capacitors	10
Resistors	10
Diodes	10
Solder joints	3
Other	10
Causes of aircraft equipment failures:	
Connectors	40
Parts on PWBs	30
Connections on PWBs	20
Other	10

periment prior to implementation. Some electronics manufacturers have test circuits, coupons, and cards designed exclusively to evaluate new materials and processes.

15.2.7 Control the Manufacturing and Assembly Processes

All manufacturing processes must operate within their control limits; that is, they must not be subject to special causes which introduce unnatural and unexpected variation. After the process control limits have been established by statistical process control (SPC), a continuous improvement mentality must be pervasive.

SPC can provide valuable information about the capability and level of control of a process, but other tools are necessary to improve them. Design of Experiments is the most effective and efficient of these tools. DoE used in conjunction with accelerated testing is a quick and efficient way to determine the reliability of a process, or a proposed process improvement.

15.2.8 Manage the Life Cycle Usage of the Product

The most difficult part of a comprehensive, deterministic reliability plan is usually the management of the product after it is delivered to the customer. By definition, a deterministic approach must be based on credible data. Data collection for the previous seven steps is within the control of the manufacturer and, although it may not be easy, quick, or inexpensive, it is relatively straightforward. Even some life cycle data may be available without involving the customer. Life test data, design reliability evaluation data, e.g., from step stress testing, and environmental stress screening data can be collected by the manufacturer and used as life cycle data. Effective life cycle data collection, however, depends at least in part on the customer.

Collecting useful data from products in service requires a systems approach. In most cases, existing systems must be changed substantially to provide useful information. This is expensive and time consuming, and is sometimes impossible without a change in attitude regarding field data. The manufacturer can help by providing detailed explanations of exactly what information is required, how it will be used, and how the resulting analyses will help both the manufacturer and the customer become more efficient and cost effective.

Often, in-service data collection is viewed as just one more task that will be used to produce one more report that no one will read. Suppliers and customers must identify what data are necessary, how they will be used, and how each party will benefit. This is almost always an iterative process, since the full potential of any data base cannot be known until some experience has been gained with it.

15.3 DETERMINISTIC LIFE DISTRIBUTIONS

In Chapters 12–14, several life distributions are mentioned, and four are discussed in some detail. No distribution can be said to be exclusively probabilistic, nor can any be called exclusively deterministic. Calculations using the exponential distribution are included in Chapter 14 because of the constant failure rate of that distribution, and hence its wide use in probabilistic methods.

There is no rule stating which failure distribution should be used for a given failure mechanism; this must be determined empirically. Perhaps the most common, however, are the Weibull and lognormal, and we shall concentrate on those two distributions in deterministic reliability calculations.

Generally, the Weibull distribution is used to describe failures that are mechanical in nature. Among the failures that can be described by the Weibull distribution are fatigue tests at constant alternating stress, early failures in environmental stress screening (with a decreasing failure rate), and any set of items with defects with a range of severity.

The lognormal distribution can be used to describe failures in any system in which the susceptibility to failure of an item depends on its age. The lognormal distribution is commonly applied to electronic products, and it usually is assumed that most semiconductor failures follow a lognormal distribution.

The exponential distribution also can be deterministic, especially if a deterministic approach is taken toward external conditions and events that have heretofore been considered random, and if the failure mechanisms are determined.

15.4 DETERMINISTIC RELIABILITY CALCULATIONS

Reliability calculations using the Weibull and lognormal distributions are quite complicated, and most of them are beyond the scope of this book. As a first choice, it is highly recommended that the capabilities of graphical analysis be exploited before trying the more complicated methods.

15.4.1 A Weibull Example

Consider the case of a sample consisting of 10 power supplies, of which the first 6 fail at the following times: 200, 500, 1000, 1300, 2400, and 3000 hours. The Weibull cumulative failure distribution for this sample is plotted in Figure 15-1, using the method of median ranks. If this distribution represents the true distribution of failures, we can say several things about the expected reliability of the power supplies.

The Weibull slope β is approximately 0. 9. Since it is slightly negative, we can say that the failure rate is decreasing slightly, and the product is probably

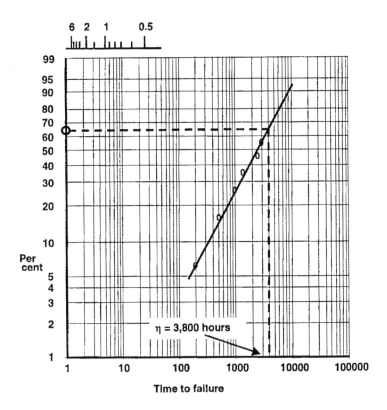

6 2 1 0.5

Per cent

Time to failure

FIGURE 15-1   A Weibull plot of power supply data.

in its useful life stage. Although traditional thinking is that the failure rate is constant during useful life for electronic products, there is increasing evidence that it decreases throughout the useful life of the product (see for example, Ref. 7 and 8).

By extrapolation, we can estimate that about 1% of the power supplies will fail before 23 hours, that t_{50} is about 2600 hours, and that 90% will have failed by 10,000 hours.

If we want to know the reliability of these items for a mission of 100 hours, starting at time zero, we can note that the cumulative percent failed at 100 hours is 3.7%, or

$$F(100) = 0.037 \qquad (15\text{-}1)$$

Thus

$$R(t) = 1 - F(t) \qquad (15\text{-}2)$$

and

$$R(100) = 1 - F(100) = 1 - 0.037 = 0.967 \tag{15-3}$$

For a mission of the same length, 100 hours, starting at 1000 hours, the reliability is estimated by

$$R(t_0, t_1) = \frac{R(t_0 + t_1)}{R(t_0)} \tag{15-4}$$

where t_0 is the time at the beginning of the mission and t_1 is the time at the end of the mission.

From Figure 15-1, we see that

$$R(t_0) = 1 - F(t_0) = 1 - F(1000) = 1 - 0.26 = 0.74 \tag{15-5}$$

and

$$R(t_0 + t_1) = 1 - F(t_0 + t_1) = 1 - F(1100) = 1 - 0.28 = 0.72 \tag{15-6}$$

Then

$$R(10000,1100) = \frac{0.72}{0.74} = 0.973 \tag{15-7}$$

$R(t_0, t_1)$ is called the *conditional reliability* for a mission of 100 hours, given successful operation to 1000 hours.

It may be noted that the reliability for a mission of 100 hours, starting at 1000 hours, is higher than the reliability for a mission of 100 hours, starting at zero hours. This is because of the decreasing failure rate, where $\beta < 1$. The reader is encouraged to construct an example in which the failure rate is increasing ($\beta > 1$), and show that the conditional reliability is lower for a mission starting at 1000 hours.

The failure rate for the Weibull distribution is not constant unless $\beta = 1$. Thus it must be calculated for a specific time, according to

$$\lambda(t) = \frac{\beta}{n}\left(\frac{t - \gamma}{\eta}\right)^{\beta-1} \tag{12-17}$$

In this case, since the distribution falls on a straight line, $\gamma = 0$ and the equation reduces to a two-parameter Weibull. Using the graphical method described in Chapter 13, η is determined to be 3800 hours. The failure rate at $t = 1000$ hours is thus

$$\lambda(1000) = \frac{0.9}{3800}\left(\frac{1000}{3800}\right)^{0.9-1} = 0.000039 \text{ failures/hour} \tag{15-8}$$

Recently developed methods of reliability calculation using the Weibull curve are reported by Nash [8], Wang [9] and Wang and Lu [10].

15.4.2 A Lognormal Example

The method presented here is actually another graphical method, based on the work of Goldthwaite [11], Peck [12], and Peck and Trapp [13].

Goldthwaite calculated the lognormal failure distribution in terms of t/t_m and λt_m, as shown in Figure 15–2 (Goldthwaite plotting paper is available from Technology Associates, Portola Valley, CA). In this figure, which applies to operations or tests in which failures are not replaced, t is the time for which the failure rate is to be calculated, t_m is the median, or t_{50}, of the lognormal life distribution, λ is the failure rate, and σ is the standard deviation, which for the lognormal distribution is $\ln(t_{50}/t_{16})$.

If we have a population of parts whose life distribution is lognormal with $t_m = 2$ million hours and $\sigma = 2.0$, we can calculate the failure rate at $t = 1000$ hours as follows:

$$\frac{t}{t_m} = \frac{1000}{2,000,000} = \frac{1 \times 10^3}{2 \times 10^6} = 5 \times 10^{-4} \tag{15-9}$$

In Figure 15-2, we enter the chart at 5×10^{-4} on the ordinate, as indicated by the arrow. We then go upward along the bold line until we reach the $\sigma = 2.0$ curve. We proceed horizontally along the bold line to the abscissa, and read

$$\lambda t_m \approx 2.8 \times 10^4 \tag{15-10}$$

Then

$$\lambda \approx \frac{2.8 \times 10^4}{2.0 \times 10^6} = 0.14 \text{ failures/million hours} \tag{15-11}$$

15.5 DETERMINISTIC RELIABILITY MANAGEMENT

In this chapter, considerable space is devoted to an outline of the deterministic method, and only a few calculation techniques are illustrated. This is appropriate, since deterministic reliability is first of all a management discipline.

The primary management responsibility in producing a reliable product is to decide that the product will be reliable because its design, its manufacturing materials and processes, and its operating conditions are understood by the manufacturer. Reliability must be understood as more than just a set of tasks to be completed, or a set of forms to be filled out. Manufacturers must be pro-active in reliability assurance, instead of relying on external standards.

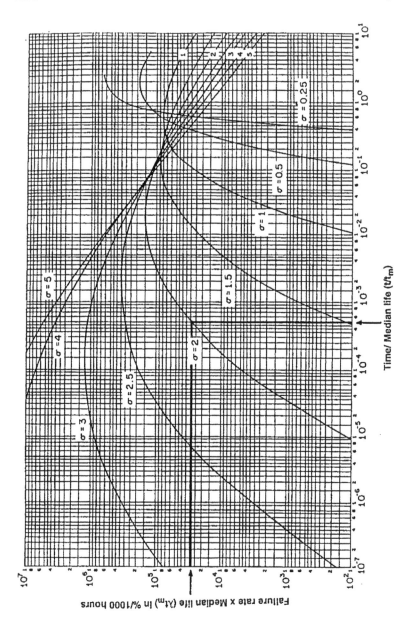

FIGURE 15-2 Lognormal failure rate estimate using Goldthwaite plotting paper.

The most important task of deterministic reliability is to collect and analyze credible data. This is no easy task, since by definition reliability data must be collected over a long period of time. Such data may be collected by field experience, or by accelerated reliability testing. Some methods and models for data collection by accelerated testing are presented in a later section of this book.

REFERENCES

1. A. Dasgupta and M. Pecht, Material failure mechanisms and damage models, IEEE Transactions on Reliability 40(5):531–536, 1991.
2. A. J. Holden, R. W. Allen, K. Beasley, and D. R. Parker, Death by a thousand cuts: The physics of device failure through a series of activated, microscopic events, Quality and Reliability Engineering International 4:247–254, 1998.
3. C. T. Leonard, Improved techniques for cost-effective electronics, Proceedings of the Annual Reliability and Maintainability Symposium, IEEE, 1990.
4. Rockwell International, presentation to CALCE EPRC, University of Maryland, 1990.
5. Private communication from Westinghouse, 1992.
6. D. Galler and G. Slenski, Causes of aircraft electrical failures, IEEE AES Systems Magazine, 1991.
7. K. L. Wong, The Exponential Law, IEEE Reliability Society Newsletter 38(1):3, 1992.
8. F. R. Nash, Making reliability estimates when zero failures are seen in laboratory aging, Materials Research Society Symposium Proceedings, vol. 184 pp. 3–21, 1990.
9. C. J. Wang, Sample size determination of Bogey tests without failures, Quality and Reliability Engineering International, vol. 7, pp. 35–38, 1991.
10. C. J. Wang and M-W. Lu, A Two-stage Sampling Plan for Bogey Tests, Quality and Reliability Engineering International, 7:29–35, 1992.
11. L. R. Goldthwaite, Failure rate study for the lognormal distribution, Proceedings of the Symposium on Reliability and Quality Control, vol. 7, pp. 208–213, 1961.
12. D. S. Peck, The analysis of data from accelerated stress tests, Proceedings of the International Reliability Physics Symposium, IEEE 1972, pp. 69–77.
13. D. S Peck and O. D. Trapp, Accelerated Testing Handbook, Technology Associates and D. S. Peck Consulting Group, Portola Valley, CA, 1987.

Chapter 16

System Reliability Methods

16.1 INTRODUCTION

Most of the reliability methods presented in this book are associated with *components* of a system; that is, the methods deal with failures of the basic elements of a system. Such failures usually result in total failure of the element, but not necessarily failure of the system. Usually, component failures are not repairable. Furthermore, component reliability methods usually are aimed at *preventing*, rather than *accommodating*, failures. As a result, only a segment of the total discipline of reliability engineering is presented in this book.

Many reliability engineers devote considerable attention to the reliability of relatively large systems that are made up of a number of subsystems or components. Failure of these sub systems or components usually does not result in complete failure of the entire system; furthermore, the failures usually are repairable, at least at the system level. We do not want to minimize the importance of system reliability; however, many other excellent texts deal with this field, leaving us free to concentrate on component reliability. Many of the component reliability methods presented in this book are used to provide data inputs to system reliability calculations and analyses.

This chapter (1) provides historical insight into the development of reliability engineering; (2) describes the relationship of component reliability methods to system reliability engineering; and (3) provides brief descriptions of some of the more important system reliability methods.

16.2 DEVELOPMENT OF RELIABILITY METHODS

Reliability has a relatively short history as a distinct discipline. Table 16-1 shows a brief (and oversimplified) summary of its development, along with trends in other areas. Prior to World War II, little attention was given to product reliability, or to the need for it. Manufactured products were mostly mechanical or electro-mechanical; product designs were not complex; and the number of components in a given product was small by the standards of today. Two major "reliability methods" were the location of power lines underground to avoid disruptions due to weather, and designing electronic equipment with sockets so that electron tubes could be replaced easily when they failed [1].

With the beginning of World War II, the need for more complex products became apparent. The fact that these products were unreliable also became apparent. It is estimated that 50% of all military spare equipment became unserviceable before use; that 60–75% of radio vacuum tubes failed prematurely; and that electronic gear on bombers gave no more than 20 hours of trouble-free operation [2].

Various government and industrial agencies, committees, and other groups were formed over the next ten years to study problems associated with reliability, especially in military products [3]. One of these groups concluded that [4]:

1. There needs to be better reliability data collected from the field.
2. Better components need to be developed.
3. Quantitative reliability requirements need to be established.
4. Reliability needs to be verified by test before full-scale production.
5. A permanent committee needs to be established to guide the reliability discipline.

Probabilistic methods dominated early reliability work because the available tools would support no other approach. When the need for reliability disciplines attained widespread notice in the industrial and scientific communities in the 1940s and 1950s, the primary computational tool was the slide rule, and analytical capability was limited to simple chemical analyses and mechanical testing. Although Design of Experiments had been developed in the West, it was not widely known or used. Reliability methods were therefore limited to broad measurements of field events (which included failures and removals for causes other than component or system failure) and statements of their statistical probabilities of occurrence.

TABLE 16-1 Development of Reliability Engineering

Decade	Electronic products	Computational capabilities	Analytical capabilities	TQM methods	Reliability methods
1940s	Electrical and electro-mechanical; vacuum tubes	Slide rules	Wet chemistry; mechanical testing; microscopes		Poor reliability observed; reliability methods needed
1950s	Discrete solid state components	Early main frames			AGREE; reliability notebook
1960s	Integrated circuits	Solid state main frames	Early electronc analysis		Handbook methods
1970s	Microprocessors	Hand-held calculators			Probabilistic methods dictated by military and other external standards
1980s	Large logic programmable and memory devices; software	Minicomputers and PCs	Advanced electronic analysis	SPC	
1990s	Advanced logic programmable and memory devices; software dominates	PCs and work stations		Design of experiments	Accelerated testing; design of experiments
2000s	Shrinking geometries, voltages, and temperatures; short development times, software dominates				Improved service data collection and analysis; combined DoE and accelerated testing

In the 1960s and 1970s, the hand-held calculator and mainframe computers increased our computational ability, and electronic analysis equipment began to mature as a tool for failure analysis. New mathematical analytical methods, such as finite element modeling, were under development. Chemists, materials scientists, and physicists were able to apply their models to reliability. These capabilities were new to us, however, and we made many mistakes, unwarranted assumptions, and false starts. Probabilistic methods were further developed and refined, and continued to be the methods of choice.

With the advent of the personal computer in the 1980s, and user-friendly software in the 1990s, our computational capability now enables us to handle reliability data in a deterministic manner. Finite element modeling and physical testing have progressed to the point where we can reasonably expect to determine causes of failures, model their likelihood in use, and describe their mechanisms. Also in the 1980s, we rediscovered quality tools, such as SPC and Design of Experiments; and we began to appreciate the use of accelerated testing. All of these capabilities allow us to collect, analyze and organize data, analyze failures, and conduct tests in a timely fashion. More details of the historical development of reliability engineering may be found in Ref. 5.

16.3   THE PROBABILISTIC APPROACH

In viewing the parallel history of reliability engineering with other technical capabilities, it is not surprising that a probabilistic viewpoint was prominent in its development. The constant failure rate distribution was assumed, and the emphasis was on *accommodating* failures at the system level, rather than on *preventing* them at the component level. In addition to the use of statistical distributions, as described in earlier chapters of this book, some system reliability methods were developed, including reliability block diagrams, fault tree analysis, failure modes and effects analysis, and Markov modeling. Some of these methods are described briefly at the end of this chapter. They are important for us, in that the results of many of the methods presented in more detail in this book are used as inputs to them.

Most of the formal reliability methods were developed and implemented by military and aerospace organizations; and they have been used in less formal ways by other industries, such as telecommunications, medical instruments, and automotive. In military programs, the major program reliability document was Military Standard 785, Reliability Program for Systems and Equipment Development and Production, which defined a series of tasks to be performed by contractors for military programs. The tasks were grouped into program surveillance and control, design and evaluation, and development and production testing. A host of other government documents described, in detail, the tasks to be performed in each group.

The intent of the military reliability standards was admirable, but they led to a system in which military contractors and military program managers concentrated more on the accomplishment of the tasks and the filing of reports than on producing reliable products. Although some improvements in reliability were observed, the costs grew to prohibitive levels, and it was observed that the reliability of products in other industries (which did not use the military approach) was superior to that of military products. As a result, in 1994, the U.S. Secretary of Defense mandated that the Department of Defense reduce its dependence on military standards, especially those that dictate how the product must be designed and built rather than what it should do.*

MIL-STD-785 and many other standards were scheduled for cancellation. As of August 1998, the U.S. Navy reported that, of 3658 military standards under consideration, 403 had been canceled, 979 had been inactivated, 566 had been transferred to the preparing agency, 983 had been replaced by commercial specifications, 105 had been converted to performance specifications, and 622 had been retained and updated [6].

The alternative to the task-based approach to reliability is the objective-based approach [7], in which the customer states the objectives to be accomplished by the supplier's reliability program, and the supplier documents the processes to be used in accomplishing the objectives. The objectives stated in Ref. 7 are:

1. Understand the customer's reliability requirements.
2. Determine a process to satisfy the requirements.
3. Assure the customer that the requirements have been met.

Although the above objectives appear to be little more than common sense, it is surprising to see how easily and often they are forgotten, especially when a "conformance to standards" mentality takes over. In the objective-based approach, the organization that designs and produces the product takes the lead in describing the processes that will be used to accomplish the objectives. This is appropriate, since the designer and manufacturer should know the most about how to make the product reliable. The objective-based approach to reliability has been described in the IEEE Standard Reliability Program for the Development and Production of Electronic Systems and Equipment (1998), and in a similar document prepared by the Society of Automotive Engineers. It has also been adopted by the Ministry of Defence in the United Kingdom.

The reliability standards mentality had some beneficial results, but it also produced two rather serious problems: (1) It distracted expensive and valuable

---

* It should be noted that the Secretary of Defense did not state that all standards are bad; merely that the Department of Defense would no longer promote the use of prescriptive military standards that built in cost with little added value.

resources into paperwork exercises which added little value to products, and (2) it got us to thinking of reliability as a bureaucratic function, instead of a critical product feature. In some cases, if certain procedures and requirements were followed, the resulting product was defined as reliable by the manufacturer, the customer, and the government. If such a product proved to be less than reliable in service, no one was held responsible, because the proper procedures were followed.

Thankfully, we are getting past the standards mentality in reliability, but we still have far to go. Because of it, reliability as a discipline is not always held in high regard; the skills of its practitioners are not always considered relevant; and its positive influence on the product design is not great. Some reliability practitioners have added to these dismal perceptions by defining their contributions in bureaucratic terms. We must continue to work to change these perceptions.

16.4 RELIABILITY PREDICTION

The development of the reliability prediction process illustrates the recent trend from probabilistic to deterministic reliability. Electronic equipment reliability prediction began in the 1950s, and was based on a "bottom-up" approach, in which it was assumed that system failure rates were simply a combination of the failure rates of the individual components of the system. Furthermore, the individual component failure rates were assumed constant; and systems were assumed to be in the flat portion, or useful life, region of the bathtub curve. The constant failure rate assumes an exponential distribution, with

$$R(t) = e^{-\lambda t} \tag{16-1}$$

The failure rate λ is the important variable in the exponential distribution, and it is the basic unit in the probabilistic methods of Refs. 8–14. Failure rates are calculated for all components in a system, and combined to produce the system failure rate.

The general forms of the failure rates for several probabilistic methods are quite similar as shown in Table 16-4. They usually consist of a base failure rate λ_B, which is modified by the π factors, to account for environmental stresses, operating conditions, and component features. Although the exact definitions and numerical values of these terms vary from method to method, the approach can be illustrated by [8]

$$\lambda = \lambda_B(C_1\pi_T + C_2\pi_E)\pi_Q\pi_L \tag{16-2}$$

λ_B, the base failure rate, usually is obtained from government or industrial data bases, and based on field maintenance and repair operations, or other sources.

C_1, the circuit complexity factor, varies according to the number of gates and the type of technology of the device. C_2, the package complexity factor, varies according to the package type and number of leads.

π_T, the temperature acceleration factor, is based on the Arrhenius relationship, in which steady-state temperature is assumed, with no consideration for temperature cycling or other temperature conditions.

π_E is the environmental factor. The descriptions of the environments are qualitative descriptions of the locations and conditions in which the equipment is operated, with no quantitative descriptions of conditions such as temperature, vibration, humidity, etc.

π_Q, the quality factor, depends on the amount of screening and testing at the part level.

π_L, the learning factor, varies according to the production life of the device.

The result of a calculation from one of the probabilistic methods is a list of component failure rates, which are then combined to provide the failure rate of a system. Since the failure distribution is assumed exponential, the inverse of the failure rate is the MTBF.

A second assumption is that all system failures are due to part failures, which is not supported by the data of Table 15-1 and other sources.

Because of continued criticisms [15–26], the handbook approach to reliability prediction has become less popular in recent years. New methods have been proposed [27, 28], which are more comprehensive, in that they try to account for all relevant factors, including design and manufacturing processes, variations among users, etc. The new approaches also illustrate the trend away from the use of government standards in general, and in reliability standards in particular. They are embodied in industry standards for reliability prediction, published by the Institute of Electrical and Electronic Engineers [29], the International Electrotechnical Commission [30], and the Reliability Analysis Center [31].

16.5 SYSTEM RELIABILITY METHODS

This chapter is not intended to describe all the system reliability methods available today; instead, we list here some of the most widely used methods. Most of the methods listed here use inputs from component reliability determinations to calculate the expected reliability, dependability, or availability of a system.

16.5.1 Reliability Block Diagrams

Reliability block diagrams represent a system's reliability, and facilitate system reliability calculations, in much the same way that an electronic circuit is represented schematically to calculate its outputs. A series reliability block diagram

FIGURE 16-1 A reliability block diagram showing two components, with reliabilities R_1 and R_2, in series.

is shown in Figure 16-1. The components represented by the two boxes are independent of each other and have constant failure rates λ_1 and λ_2. The reliabilities of the two components, R_1 and R_2, are $\exp[-\lambda_1 t]$ and $\exp[-\lambda_2 t]$. The reliability of the system is $R_1 R_2 = \exp[-(\lambda_1 + \lambda_2)t]$. The general rules for the reliability of a series of n statistically independent components are

$$R = \prod_{i=1}^{n} R_i \tag{16-3}$$

and

$$\lambda = \sum_{i=1}^{n} \lambda_i \tag{16-4}$$

where i indicates the ith component.

If a system consists of two components in parallel, as shown in Figure 16-2, and if the system operates if either of the components is operable, we have an *active parallel redundant system*. The general equation for active parallel redundancy is

$$R = 1 - \prod_{i=1}^{n} (1 - R_i) \tag{16-5}$$

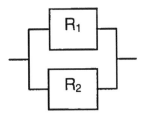

FIGURE 16-2 A reliability block diagram showing two components, with reliabilities R_1 and R_2, in parallel.

For the two components in parallel with constant hazard rates,

$$R = e^{-\lambda_1 t} + e^{-\lambda_2 t} - e^{-(\lambda_1 + \lambda_2)t} \tag{16-6}$$

The general case of active parallel redundancy is one containing n components, of which only k need be operable at any given time. With constant failure rate, the general expression for k-out-of-n redundancy is

$$R = 1 - \frac{1}{(\lambda t + 1)^n} \sum_{i=0}^{k-1} \binom{n}{i} (\lambda t)^{n-i} \tag{16-7}$$

Some systems have standby components, which are active only when switched on by a failure in a primary component. The general expression for a system with n equal units in standby redundancy is

$$R = \sum_{i=0}^{n-1} \frac{(\lambda t)^i}{i!} e^{-\lambda t}. \tag{16-8}$$

From this extremely brief discussion it should be apparent that reliability block diagrams can become quite complex, and that much more space than is available here would be required to treat it properly.

16.5.2 Markov Analysis

Markov analysis is also called *state-space analysis*. Reliability of a system is modeled over a series of intervals in time, during which the component has a given probability of going either from the unfailed state to the failed state, or of going from failed to unfailed. It is often used to describe availability of a system that is subject to failure and repair. The probabilities are constant throughout the operation of the component, and independent of past history.

As an example of Markov analysis, consider a system with the following probabilitites:

$$P_{u \to f} = 0.1 \qquad \text{and} \qquad P_{f \to u} = 0.3$$

Figure 16-3 shows a tree diagram for the first three time intervals. If the system starts in the unfailed state, then at the end of the first time interval, it will have a probability of 0.9 of being in the unfailed state, and a probability of 0.1 of being failed. (The probabilities of being failed and unfailed must add to 1.) For each interval thereafter the probabilities of all the U's are added. The probability of each U is determined by multiplying the probabilities of all its legs from the beginning of the first interval. Thus, at the end of the first interval, the probability of the component being in the unfailed state is 0. 9. At the end of the second

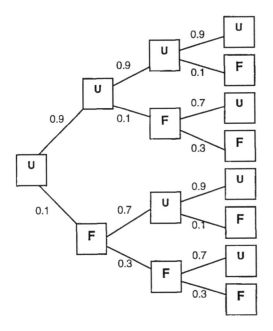

FIGURE 16-3   A tree diagram for a Markov analysis.

interval, it is

$$(0.9 \times 0.9) + (0.1 \times 0.3) = 0.84$$

At the end of the third interval, it is

$$(0.9 \times 0.9 \times 0.9) + (0.9 \times 0.1 \times 0.3) + (0.1 \times 0.3 \times 0.9)$$
$$+ (0.1 \times 0.7 \times 0.3) = 0.\ 80$$

It is easy to see that Markov analysis can get very complicated very quickly, if a number of time intervals and several components are involved. Matrix methods are used for complex systems. These and other extensions of Markov methods are beyond the scope of this discussion.

16.5.3.   Risk Analysis

Risk analysis is a method to quantify both the likelihood of a given type of failure, the severity of its effect on the entire system, and in some cases, a measure of its detectability. One category of risk assessment is the bottom-up approach, in which the reliability of a system is calculated on the basis of failures of its individual components. In this approach, each component is considered separately, and all the possible effects of its failure on system performance are considered. If these effects are traced to the highest level of the system, the types and likelihood of system failure may be determined. Commonly used bottom-up risk analysis methods are failure modes and effects analysis (FMEA) and failure modes, effects, and criticality analysis (FMECA). An example of an FMECA is shown in Table 16-2 [32].

At the opposite end of the spectrum is the top-down approach, in which every type of system failure is considered, and the potential causes of each are traced downward to the component level. The most common top-down method is fault tree analysis. Table 16-3 shows some of the symbols used in fault tree analysis, and Figure 16-4 shows an example fault tree anlysis for an alarm system.

16.5.4 Reliability Growth Analysis

Most manufacturers of complex electronic, mechanical, or electromechanical systems have observed that the reliability of their products improves during the product development stage, and that this improvement often continues well into production. This is logical, because most organizations experience "learning curve" effects, in which improvement over time is observed, even though no conscious efforts are made. Duane [33] quantified this effect with the relationship

$$MTBF_{cum} = \frac{1}{K}t^{\alpha} \qquad (16-9)$$

where

$MTBF_{cum}$ = cumulative MTBF
K = constant determined by the initial MTBF
α = growth rate
t = cumulative test time.

The growth rate α is the slope of a log-log plot of cumulative MTBF vs. time.
A second reliability growth model is the AMSAA model [34]:

$$r_c(t) = \lambda t^{\beta-1} \qquad (16-10)$$

where

$r_c(t)$ = the cumulative failure rate at time t
t = total test time
β = estimate of the time value of the growth parameter
λ = scale parameter of the Weibull distribution

Equations (16-9) and (16-10) are empirical, and the growth rates may vary from system to system. Other reliability growth models may be found in Refs. 35 and 36.

TABLE 16-2 Failure Modes and Effects and Criticality Analysis (FMECA) for the Main Assembly Superstructure of an Excavator

Component (1)	Component function (2)	Failure mode (3)	Effects of failure (4)	Cause of failure (5)	Occurrence (6)	Severity (7)	Detection (8)	Risk no. $(6 \times 7 \times 8)$ (9)	Corrective action (10)
Shutdown valve	Allows fuel to pass into injector	Coil does not magnetize	Fuel supply is cut off. Engine does not function	a. Improper operation b. Defective material	6	9	3	162	Improve quality
Hydraulic pump	Delivers fluid	a-does not supply oil b-pressure drop c-pressure does not build up	Hydraulic system does not function. Component damage	a. Foreign material in pump b. Fluid does not go into system	3	8	7	168	a-check gate valve b-ensure fluid is clean
Zako rings	Pipe joint	Cracks during operation or assembly	Hydraulic fluid leaks	a. High material hardness b. Improper assembly	8	8	6	384	Improve quality
O rings	Seals joints	Fails during testing	Excessive fluid leakage	a. Low strength b. Material defects	8	7	6	336	Improve quality

Source: Adapted from Ref. 32. © 1995 John Wiley & Sons, used by permission.

TABLE 16-3 Fault Tree Symbols

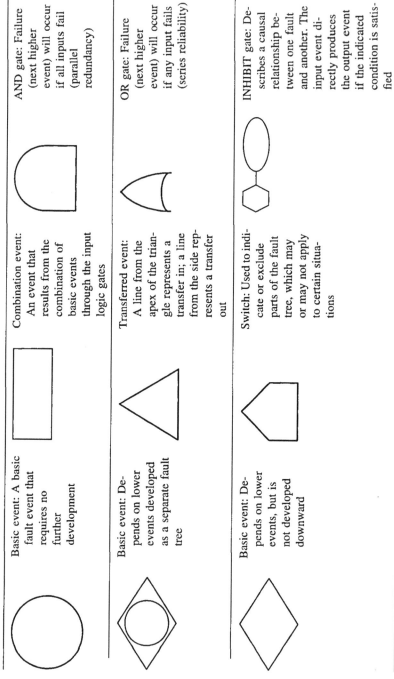

Symbol	Description
(circle)	Basic event: A basic fault event that requires no further development
(rectangle)	Combination event: An event that results from the combination of basic events through the input logic gates
(AND gate)	AND gate: Failure (next higher event) will occur if all inputs fail (parallel redundancy)
(diamond with circle)	Basic event: Depends on lower events developed as a separate fault tree
(triangle)	Transferred event: A line from the apex of the triangle represents a transfer in; a line from the side represents a transfer out
(OR gate)	OR gate: Failure (next higher event) will occur if any input fails (series reliability)
(diamond)	Basic event: Depends on lower events, but is not developed downward
(house/switch)	Switch: Used to indicate or exclude parts of the fault tree, which may or may not apply to certain situations
(INHIBIT gate)	INHIBIT gate: Describes a causal relationship between one fault and another. The input event directly produces the output event if the indicated condition is satisfied

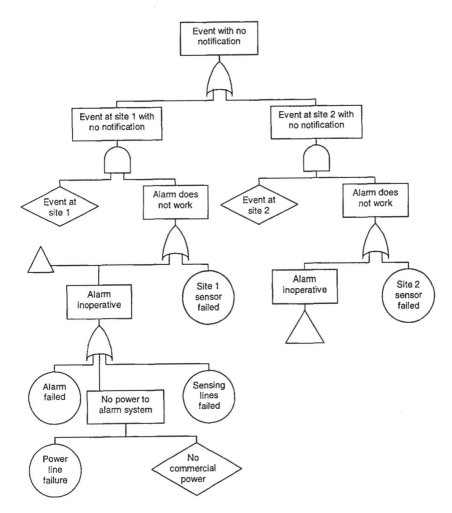

FIGURE 16-4 Example fault tree.

The use of this and similar equations to predict and monitor the reliability of products under development has come to be called reliability growth monitoring. In military terms, the monitoring effort is called reliability growth testing (RGT), and the improvement effort is called test-analyze-and-fix (TAAF).

The usual approach to reliability growth monitoring is to begin plotting the cumulative MTBF early in the program, and to continue plotting it throughout the process. The growth rate α can be determined from prior experience with

TABLE **16-4** Equations for Calculating Failure Rates for Microelectronic Devices According to Seven Different Probabilistic Methods

Method	Model
MIL-HDBK-217	$\lambda = (C_1\pi_T + C_2\pi_E)\pi_Q\pi_L$
Bellcore	$\lambda = \lambda_B\pi_Q\pi_V\pi_T$
British Telecom	$\lambda = \lambda_B\pi_T\pi_Q\pi_E$
NTT	$\lambda = \lambda_B\pi_Q(\pi_E + \pi_T\pi_V)$
CNET	$\lambda = (C_1\pi_T\pi_T\pi_V + C_2\pi_\beta\pi_E\pi_\sigma)\pi_L\pi_Q$
Siemens	$\lambda = \lambda_B\pi_V\pi_T$
SAE 870050	$\lambda = \lambda_B\pi_T\pi_Q\pi_T C_2$

similar products, or from early data on the product being monitored. This means that the early, high-failure-rate data will always be included. The cumulative and instantaneous MTBF lines are usually parallel. Using this method, it is possible both to monitor the reliability, and to predict when it will reach a given level. It is also possible to predict the amount and cost of testing and analysis required to reach the stated goal.

Reliability growth methods allow a manufacturer to begin introducing a product to service before its ultimate reliability is reached. The initial reliability, the ultimate reliability, and the growth rate are sometimes contractually specified. A proactive reliability improvement plan, such as TAAF, is usually part of the contract.

If properly managed, reliability growth methods can be effective in planning and improving product reliability. There are many rules to be obeyed in its use, and effective technical and management judgments are critical to its success.

REFERENCES

1. R. A. Evans, Electronics Reliability: A Personal View, IEEE Transactions on Reliability 4(3-SP):SP-329–SP-332, September 1998.
2. D. Kececioglu, Reliability Engineering Handbook, Prentice Hall, Englewood Cliffs, NJ, 1991.
3. A. Coppola, Reliability engineering of electronic equipment: A historical perspective, IEEE Transactions on Reliability R-33:29–35, April 1984.
4. W. Denson, The history of reliability prediction, IEEE Transactions on Reliability 47(3-SP):SP-321–SP-328, September 1998.
5. 50th Anniversary Special Publication, IEEE Transactions on Reliability 47(3-SP), September 1998.
6. Report of a Survey Conducted at the Naval Air Warfare Center, Lakehurst, NJ, Best Manufacturing Practices Center of Excellence, College Park, MD, August 1998.

7. A. Malhotra, A. Strange, L. Condra, I. Knowles, T. Stadterman, I. Boivin, A. Walton, and M. Jackson, Framework for an objective and process-based reliability program standard, Communications in RMSL, January 1996.

8. MIL-HDBK-217F, Reliability Prediction of Electronic Equipment, Rome Laboratory, Griffiss AFB, NY, 1991.

9. TR-TSY-000332, Reliability Prediction Procedure for Electronic Equipment, Issue 2, Bellcore, 1988.

10. Handbook of Reliability Data for Components Used in Telecommunications Systems, Issue 4, British Telecom, 1987.

11. Standard Reliability Table for Semiconductor Devices, Nippon Telegraph and Telephone Corporation, 1985.

12. Recueil De Donnees De Fiabilite Du CNET, Centre National D'Etudes des Telecommunications, 1983.

13. SN 29500, Reliability and Quality Specification Failure Rates of Components, Siemens Standard, 1986.

14. SAE 870050.

15. P. F. Manno, RADC Failure Rate Prediction Methodology—Today and Tomorrow, Image Sequence Processing and Dynamic Scene Analysis, NATO ASI Series vol. F3, edited by J. K. Skwirzinsky, Springer-Verlag, Berlin, Heidelberg, 1983.

16. J. B. Bowles, A survey of reliability—prediction procedures for microelectronic devices, IEEE Transactions on Reliability 41(1), 1992.

17. J. L. Spencer, The Highs and Lows of Reliability Predictions, Proceedings of the Annual Reliability and Maintainability Symposium, pp. 156–162, 1986.

18. H. S. Blanks, Reliability prediction: A constructive critique of MIL-HDBK-217E. Quality and Reliability Engineering International 4:227–234, 1988.

19. L. R. Webster, Field vs. predicted for commercial SATCOM terminal, Proceedings of the Annual Reliability and Maintainability Symposium, pp. 88–91, 1986.

20. S. J. Flint and J. J. Steinkirchner, Reliability models—Practical constraints, in Image Sequence Processing and Dynamic Scene Analysis, NATO ASI Series, vol. F3, edited by J.K. Skwirzinsky, Springer-Verlag, Berlin, Heidelberg, pp. 201–212, 1983.

21. L. Condra and M. Pecht, Commercial microcircuit options in military avionics systems demand reliability, Defense Electronics, pp. 43–46, 1991.

22. K. L. Wong, The bathtub does not hold water any more, Quality and Reliability Engineering International 4:279–282, 1988.

23. J. A. McLinn, Constant failure rate—A paradigm in transition? Quality and Reliability Engineering International 6:237–242, 1990.

24. C. T. Leonard and M. Pecht, How failure rate methodology affects electronic equipment design, Quality and Reliability Engineering International 6:243–250, 1990.

25. K. L. Wong, What is wrong with the existing reliability prediction methods? Quality and Reliability Engineering International 6:251–258, 1990.

26. C. T. Leonard, Improved techniques for cost-effective electronics, Proceedings of the Annual Reliability and Maintainability Symposium, 1990.

27. W. K. Denson and S. Keene, A new reliability prediction tool, Proceedings of the Reliability and Maintainability Symposium, IEEE, pp. 15–29, 1998.

28. L. Condra, C. Bosco, R. Deppe, L. Gullo, J. Treacy, and C. Wilkinson, Reliability

assessment of aerospace electronic equipment, Quality and Reliability Engineering International 15:253–60.

29. IEEE Standard Method for Reliability Predictions and Assessment for Electronic Systems Equipment, 1998.

30. IEC QC 001007–1-3, Guide for Reliability Assessment of Electronic Equipment, draft revision F, International Electrotechnical Commission Quality Assessment System for Electronic Components (IECQ).

31. New System Reliability Assessment Method, Reliability Analysis Center, final report for IITRI project no. A06830, Rome, NY, June 1, 1998.

32. S. K. Majumdar, Study on reliability modelling of a hydraulic excavator system, Quality and Reliability Engineering International 11:49–63, 1995.

33. J. T. Duane, Learning curve approach to reliability monitoring, IEEE Transactions on Aerospace 2(2):563–566, 1964.

34. Reliability Growth Analysis Using the Duane and AMSAA Models, RAC Technical Brief, Rome, NY, 1988.

35. J. Quigley and L. Walls, Measuring the effectiveness of reliability growth testing, Quality and Reliability Engineering International 15:87–93, 1999.

36. J. Donovan and E. Murphy, Reliability growth—A new graphical model, Quality and Reliability Engineering International 15:167–174, 1999.

Chapter 17

Accelerated Testing

17.1 INTRODUCTION

One of the most critical requirements in reliability work is to know early in the life of a product (preferably in the design stage) how that product will perform at some time in the future. This means that reliability information must be obtainable in a short time, and that it must be predictive. Most product managers have available to them at least some data from prior performance of similar products, or from earlier tests, or from component suppliers, or other sources. These are usually the least expensive data available, and should be used as extensively as possible. In some cases, however, these sources are insufficient, and experimental data must be collected. Design of Experiments (DoE) is a means of obtaining quick, efficient, and accurate experimental data. DoE in combination with accelerated testing can facilitate reliability prediction in a relatively short time.

About the only noncontroversial thing that can be said about accelerated testing is that it can be controversial. Although it is widely used in almost all industries in some form or another, there are many differences of opinion about how to set up accelerated tests and interpret data collected from them. Properly understood and applied, accelerated testing can add much value to a product. Misunderstood and improperly applied, it can lead to serious mistakes.

With this chapter, a section on accelerated testing is introduced. This section (Chapters 17–20) includes descriptions of accelerated tests and models

229

(Chapter 17), discussion of typical product environmental and operating stresses (Chapter 18), presentation of some examples of testing and data analysis (Chapter 19), and a comprehensive program of accelerated testing (Chapter 20).

Several test methods and models are discussed in these chapters. They are not presented in the expectation that they will be considered the "correct" ones, but to illustrate the various approaches to accelerated testing. To be effective, any accelerated test must be developed and applied with a knowledge of the product, its operating requirements, and of how it behaves in its operating environment. The examples presented in this chapter have been idealized to illustrate the methods. In practice, accelerated testing is rarely ideal, and some more realistic examples from actual accelerated tests are presented in Chapter 19.

17.2 THE FUNDAMENTAL PRINCIPLE OF ACCELERATED TESTING

Accelerated testing is based on the fundamental principle that the unit under test (UUT) will exhibit the same behavior in a short time at high stresses, that it will exhibit in a longer time at lower stresses.

Generally speaking, accelerated testing is useful for obtaining data for damage-endurance and tolerance-requirement failure mechanisms. It is not generally effective for stress-strength and challenge-response failures. (See Chap. 15 for definitions of these terms.)

Figure 17-1 shows a plot of stress vs. time, with higher stresses being applied at short times, and lower stresses being applied for longer times. Time-to-failure distributions are shown for four different constant stress levels. These

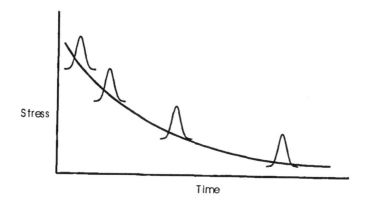

FIGURE 17-1   Model of accelerated stress testing showing time to failure at constant stress levels.

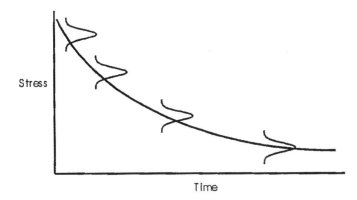

FIGURE 17-2 Model of accelerated stress testing showing stress to failure at constant time.

distributions could be of the types discussed in earlier chapters. In this case, the samples are monitored throughout the test, and their exact times of failure are recorded. An example of this type of test is leakage monitoring of a sample set of capacitors operated at various applied voltages, for which failure is defined as leakage above a specified value.

Figure 17-2 also shows a plot of stress vs. time. In this case, the distributions are of stress to failure at constant times. The samples are not monitored during testing, but are tested at intervals, and the stress at which they fail is measured. An example is testing of the mechanical strength of a metal product weakened by stress corrosion, in which samples are held at various temperatures in a corrosive atmosphere, and removed at intervals for measurement of their tensile strength.

17.3 ACCELERATED TESTING MODELS

Mathematical models are used to relate the behavior of items at one stress level to their behavior at another level. They are the equations that describe the curves in Figures 17-1 and 17-2. The most rigorous models are those which describe the failure mechanisms at the structural, or atomic, level. They are called *structural, closed-form, constitutive,* or physics-of-failure models. An example of such a model is Fick's work in diffusion [1].

Another type of model is the *empirical model.* Empirical models are not based on descriptions of structural changes, but describe mathematically the data collected from testing or use. They can be viewed as curve-fitting, although a good knowledge of the physics-of-failure mechanisms is often applied to the exercise. Examples of this type of model are some of those developed for humidity testing.

The models do not tell us anything about the shapes of the small distributions shown in Figures 17-1 and 17-2, but only about their locations.

Models range from the very simple to the very complex. Usually, the simpler models can be said to apply to a wider range of cases, while the more complex models are specific to a rather narrow set of applications. Also, some of the more complex models can be quite difficult to use. Engineering judgment is required to select the simplest model which gives satisfactory results. Perhaps the best advice in this regard is that given by Weibull [2]:

> ... there may exist two or more true relationships of different shapes. Facing this abundance, the only reasonable way to act seems to be to choose the one which most easily gives answers to posed questions.

A variety of model forms is available to the accelerated test engineer, and all reasonable models should be considered. In this chapter, three general forms are presented: (1) the Arrhenius model, (2) the inverse power law, and (3) the Eyring model. Most of the popular models in use today are variations of one of these three models. They, and other models, are described in many publications, and Refs. 3–7 are listed as examples.

17.3.1   The Arrhenius Model

Svante Arrhenius [8] presented this model in 1889 to describe the inversion of sucrose. It is used to describe thermally activated mechanisms such as solid state diffusion, chemical reactions, many semiconductor failure mechanisms, battery life, etc. The underlying life distributions for these mechanisms may be lognormal, Weibull, or exponential. The Arrhenius model describes the temperature dependence of some feature of the distribution (e.g., mean t_{16}) as a rate equation:

$$r = r_o e^{-E_a/kT} \tag{17-1}$$

where

r = reaction rate
r_o = a constant
E_a = activation energy, in electron volts*
k = Boltzmann's constant (8.617×10^{-5} eV/K)
T = reaction temperature, in K

---

* Usually, the activation energy is reported in electron volts, but sometimes it is reported in calories per mole. 1 eV = 23,000 cal/mole.

If this relationship holds, then the product of the reaction rate and the time for it to occur is constant over its temperature range of applicability or

$$r_1 t_1 = r_2 t_2 \tag{17-2}$$

for two different reaction temperatures T_1 and T_2. Thus for a given mechanism, with time to failure expressed as t_f, rt_f is a constant, and

$$t_f \propto \frac{1}{r} \propto e^{E_a/kT} \tag{17-3}$$

The acceleration factor for accelerated testing with the Arrhenius equation is

$$\text{AF} = \frac{t_u}{t_t} = \exp\left[\frac{E_a}{k}\left(\frac{1}{T_u} - \frac{1}{T_t}\right)\right] \tag{17-4}$$

where the subscripts u and t indicate use and test, respectively.

To illustrate the use of the Arrhenius equation, consider the example of a thermocompression bond between two dissimilar metals. The strength of this bond is reduced in time by the formation of voids or brittle intermetallics by solid-state diffusion, which has an activation energy of 0.9 eV. The use temperature is 25°C, and an accelerated test is conducted at 100°C. The times to failure for 10 samples are 130, 140, 160, 180, 185, 195, 205, 205, 240, and 260 hours.

These data are lognormally distributed, and the distribution is plotted in Figure 17-3. The mean is 185 hours.

We can estimate the life of this bond in service by using Eq. (17-4), with

t_u = life of the bonds in use
t_t = mean life of the bonds in test = 185 hours
$E_a = 0.9$
$k = 8.617 \times 10^{-5}$ eV/K
$T_u = 25°C = 298K$
$T_t = 100°C = 373K$

Then

$$t_u = t_t \exp\left[\frac{E_a}{k}\left(\frac{1}{T_u} - \frac{1}{T_t}\right)\right]$$

$$= 185 \exp\left[\frac{0.9}{8.617 \times 10^{-5}}\left(\frac{1}{298} - \frac{1}{373}\right)\right] \tag{17-5}$$

$$= 268{,}435 \text{ hours} \approx 30 \text{ years}$$

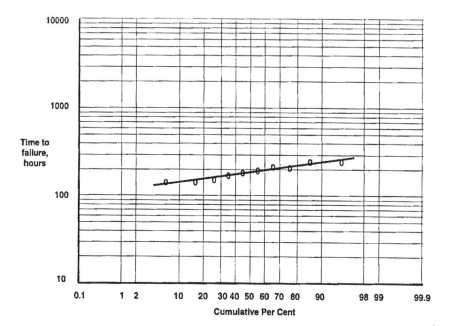

FIGURE 17-3   Lognormal plot of time to failure for thermocompression bonds in accelerated testing.

Although calculations using the Arrhenius equation are not difficult, it is often easier to use graphical methods. Figure 17-4 shows an example of an Arrhenius plot (paper is available from Technology Associates, Portola Valley, CA). The Arrhenius equation is linear if the log of time is plotted vs. $1/T$, and the slope of such a plot is equal to the activation energy. In Figure 17-4, the mean time to failure of the above data, 185 hours, is plotted at the test temperature, 100°C, and a line is drawn through it with a slope equal to the E_a, or 0.9 eV, using the nomograph. This line intersects the 25°C line at approximately 268,000 hours.

The choice of the mean time to failure, 185 hours, was somewhat arbitrary. Any other point on the distribution could have been chosen, and a conservative choice would have been the lowest value, 130 hours. This example was used merely to illustrate the process. As an exercise, the reader is encouraged to estimate the life of the bond using the 130 hour figure (time to first failure), both mathematically and graphically.

The Arrhenius equation and plotting methods can also be used in the reverse direction. Accelerated tests can be conducted at several different temperatures, and E_a can be determined from the slope of the line drawn between them.

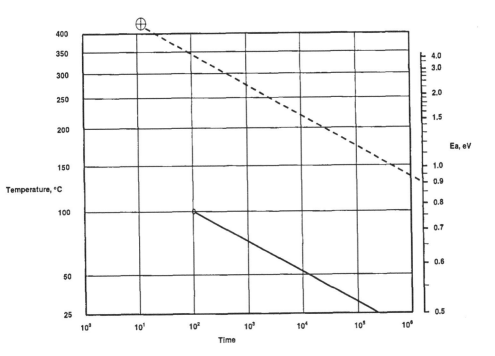

FIGURE 17-4   Arrhenius plot of mean time to failure for thermocompression bonds.

It is apparent that the important factor in the Arrhenius model is the activation energy E_a. If E_a is known, then data from tests at a higher temperature can be used to estimate times for certain events to occur at lower temperatures. The published literature contains many references that report values for various E_a mechanisms. In general, E_a for semiconductor failure mechanisms range from 0.3–0.6 eV; for intermetallic diffusion, it is in the 0.9–1.1 eV range.

17.3.2   The Inverse Power Law

The inverse power law is used when the life of a system is inversely proportional to an applied stress. The underlying failure rate distribution can be lognormal, exponential, or Weibull, but the Weibull distribution seems to be used more often than the others.

Two common applications of the inverse power law are for voltage stress, and for fatigue due to alternating stress. Fatigue may be low cycle, such as that resulting from temperature cycling, or high cycle, resulting from mechanical vibration.

The general form of the inverse power law is

$$\tau = \frac{A}{S^n} \tag{17-6}$$

where

τ = time for an event (such as failure) to occur
A = a constant characteristic of the product
S = applied stress
n = an exponent characteristic of the product

Many different forms of the inverse power law have been developed for various applications. One of the most common is the Coffin-Manson law for fatigue testing [9–10]. It is stated as

$$N_f = A \left(\frac{1}{\Delta\varepsilon_p} \right)^B \tag{17-7}$$

where

N_f = number of cycles to failure
A = constant related to the material
$\Delta\varepsilon_p$ = plastic strain range
B = constant related to the material

This equation has been modified by many different investigators to fit a variety of situations. Some examples are given in Refs. 11–16. These modifications are too numerous to mention here, but some of them are used in later examples. The Coffin-Manson law applies to both isothermal mechanical fatigue cycling and fatigue due to mechanical stresses resulting from thermal cycling.

If the total applied stress is much higher than the elastic strain range for a fatigue test, a simplified acceleration factor for isothermal fatigue testing is

$$AF = \frac{N_{fu}}{N_{ft}} = \left(\frac{\Delta\varepsilon_t}{\Delta\varepsilon_u} \right)^B \tag{17-8}$$

where u and t denote use and test. The $\Delta\varepsilon$'s could be due to displacement in bending, elongation in tension, or other mechanical strains. Similarly, a simplified acceleration factor for fatigue testing in temperature cycling is

$$AF = \frac{N_{fu}}{N_{ft}} = \left(\frac{\Delta T_t}{\Delta T_u} \right)^B \tag{17-9}$$

where the ΔT's are the applied temperature cycling ranges. Some qualifications on the use of Eq. (17-8) are noted in Ref. 14, and for Eq. (17-9) in Ref. 16.

Based on testing experience, the following approximate values are commonly used for B in Eqs. (17-8) and (17-9):

Metals	2–3
Electronic solder joints	2–3
Microelectronic plastic encapsulants	4–8
Microelectronic passivation layers	12
Cratering of microcircuits	7
Al-Au intermetallic fatigue failures	4–7

The inverse power law plots as a straight line on log-log paper, with a slope equal to B. This line is called the S-N curve. Such a curve is shown in Figure 17-5, for a set of electronic solder joints which have been fatigue-tested at a displacement $\Delta \varepsilon_t$ of 0.0008 in, with a mean time to failure of 10 cycles. If these solder joints are subjected to a displacement in use $\Delta \varepsilon_u$ of 0.00005 in, and the exponent B is known to be 2.5, we can calculate the service life using Eq. (17-8), with

$$N_{ft} = 10$$
$$\Delta \varepsilon_t = 8 \times 10^{-4}$$
$$\Delta \varepsilon_u = 5 \times 10^{-5}$$

The number of cycles to failure in use is then

$$N_{fu} = N_{ft} \left(\frac{\varepsilon_t}{\varepsilon_u} \right)^B = 10 \left(\frac{8 \times 10^{-4}}{5 \times 10^{-5}} \right)^{2.5} = 10{,}240 \text{ cycles} \qquad (17\text{-}10)$$

This estimate is shown graphically in Figure 17-5.

Another common form of the inverse power law is the Paris equation for crack propagation [17]:

$$\frac{da}{dN} = A(\Delta K)^n \qquad (17\text{-}11)$$

where da/dN is the crack growth rate, A and n are material constants, and K is the stress intensity factor.

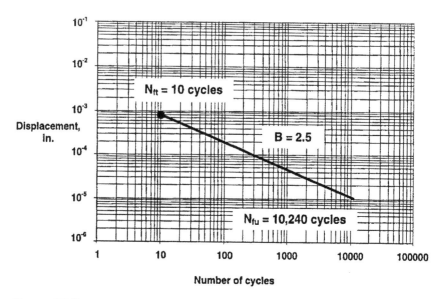

FIGURE 17-5   Graphical determination of fatigue life using the inverse power law.

17.3.3   The Eyring Model

The Arrhenius model and the inverse power law each contain one term for stress. The Eyring model [18] contains two stress terms, one of which is temperature. Its general form is

$$\tau = A\left(\frac{1}{S}\right)Be^{E_a/kT} \tag{17-12}$$

where

τ = measure of product life
A, B = constants
S = an applied stress
E_a/kT = Arrhenius exponent

The applied stress S can be almost any stress that exists in combination with temperature. It can be used in a variety of transforms, such as $1/S$, $\ln S$, etc. Two commonly used stresses are humidity and voltage. S can also be an additional temperature term, such as temperature cycling range or rate. An example of the Eyring equation is Peck's temperature-humidity relationship for electronic microcircuits [19, 20]:

$$t_f = A(RH)^{-n}e^{E_a/kT} \tag{17-13}$$

where

t_f = time to failure
A = a constant
RH = percent relative humidity
n = a constant
E_a/kT = Arrhenius exponent

n and E_a are determined empirically, and may vary from one testing situation to another; generally, for electronics applications, n is equal to 3.0 and E_a is equal to 0.9 eV.

The acceleration factor for this equation is

$$AF = \left(\frac{RH_u}{RH_t}\right)^{-3.0} \exp\left[\frac{E_a}{k}\left(\frac{1}{T_u} - \frac{1}{T_t}\right)\right] \tag{17-14}$$

It may be noted that, in this case, the Eyring model is the product of the inverse power law for humidity and the Arrhenius equation for temperature.

As an example of the use of Eqs. (17-13) and (17-14), consider the results of a life test of a set of microelectronic circuit samples, conducted at 85°C and 85% relative humidity, under operating bias. (This is called a temperature-humidity-bias, or THB, test.) If the mean time to failure is 800 hours, we can predict the mean time to failure at operating conditions of 40°C and 60% RH.

$$AF = \left(\frac{60}{85}\right)^{-3.0} \exp\left[\frac{0.9}{8.617 \times 10^{-5}}\left(\frac{1}{313} - \frac{1}{358}\right)\right] \approx 182 \tag{17-15}$$

The predicted mean life at operating conditions is then

$$182 \times 800 \text{ hours} = 145,600 \text{ hours} \approx 16.6 \text{ years} \tag{17-16}$$

This estimate may also be made graphically, by combining the methods illustrated in Figures 17-4 and 17-5.

17.4 SELECTING THE APPROPRIATE MODEL

The purpose of a mathematical model is to aid in understanding and predicting the behavior of the unit under test. Models, by their very nature, are not exact. The most effective ones usually represent a compromise between the two extremes of (1) attempting to describe the situation so completely that they become so complex and data-hungry that they are unusable, and (2) being so simple that they are inaccurate. Jensen [21] lists three rules in selecting and using models:

1. The assumptions that underly the model should be clearly stated and realistic and recognizable by practitioners.
2. Any data required by the model must be both practical to gather and represent the real world.
3. The end result, your mathematical model, must be presented in uncluttered terms that clearly represent a solution to the practical problem.

Although the three general model forms presented above are the most common ones used in accelerated testing, they are certainly not the only ones. Often, a transform of the applied stress S, must be used to accurately describe the failure mechanism. Some commonly-used transforms are

$$A\left(\frac{1}{S}\right), A + B \ln S, AS^B, A + \frac{B}{S}, \frac{1}{A + BS}, \text{ and } A + BS$$

The task of the test engineer is to understand the product, its requirements, and its operating conditions well enough to know which form of which model to apply. In many cases, this task is an iterative one, in which a model is empirically fit to the data, and then the physics-of-failure implications are assessed. Based on this assessment, the model is modified, and the process is repeated.

17.5 COMBINATIONS OF STRESSES

In the above discussions, it was assumed that the applied stresses are well defined and constant. Many models describe only a single stress, and are silent about the effects of other stresses that are applied simultaneously. In reality, the situation is always more complicated. Every product operating environment consists of many stresses, which vary in intensity and range during use. They also vary in relative importance. An excellent example of this variation is an aircraft power supply.

Power supplies generate heat during operation, so an important failure acceleration model for an aircraft power supply during aircraft operation is the Arrhenius model for steady-state temperature. When the power supply is switched on and off, the temperature rises and falls; and when the power supply is not operating, it is subject to temperature changes based on ambient climate conditions, in which case a form of the inverse power law for low-cycle fatigue is appropriate. Vibration intensity varies according to the flight conditions, and in this case the inverse power law for high-cycle fatigue should be applied.

Various methods of handling the variation in stress type and level have been developed and used in accelerated testing. Options include testing sequentially under a variety of conditions, and testing under a set of conditions applied simultaneously. A danger, of course, is that the test conditions will be so complicated as to make realistic analysis impossible. A commonly used approach to analysis of data from combined stresses is Miner's rule [22], which is used to estimate life when a range of applied stresses is likely. For fatigue testing, it is stated as

$$R = \frac{n_1}{N_1} + \frac{n_2}{N_2} + \frac{n_3}{N_3} + \cdots + \frac{n_j}{N_j} \quad \text{or} \quad R = \sum_{i=1}^{j} \frac{n_i}{N_i} \quad (17\text{-}17)$$

where the ratio R is the fraction of life that is exhausted at a given point, n_i is the number of cycles at a specific applied stress level, and N_i is the number of cycles to failure at that level. n_i/N_i is the fraction of a product's useful life that is used up at each level of each applied stress. Although Eq. (17-15) is usually applied to fatigue testing, it has also been applied to other failure mechanisms.

Barker et al. [23] derived R for combined vibrational and thermal fatigue testing of solder joints as

$$R = R_v + R_{th} = n_{th}\left(\frac{f_v/f_{th}}{N_v} + \frac{1}{N_{th}}\right) \quad (17\text{-}18)$$

where

n_{th} = number of applied thermal cycles
f_v/f_{th} = ratio of the frequency of vibration cycles to that of thermal cycles
N_v = number of cycles to failure in vibration
N_{th} = number of cycles to failure in thermal cycling

The estimated solder joint fatigue life, for samples subjected to combined thermal and vibration testing, is

$$N_f = \frac{1}{R} \quad (17\text{-}19)$$

Recently, considerable work has been done in the area of combined stresses in accelerated testing, and commonly accepted practices are updated frequently. Newer approaches, such as finite element modeling, seem to be better at describing test results than some of the older empirical closed-form models. Upadhyay-ala and Dasgupta [24] describe ways to design and conduct accelerated tests to yield realistic and trustworthy results.

17.6   SELECTING THE TEST CONDITIONS

Accelerated test conditions should be selected on the basis of expected failure mechanisms and anticipated operating and environmental stresses. Previous experience with similar products, physics-of-failure analyses, preliminary testing, and good engineering judgment must be used. For example, it might be known that a product will be exposed to elevated temperatures, and that several thermally activated failure mechanisms are possible. Each of the possible mechanisms might have a different activation energy, and it might not be known which of them will occur in the given product. It might also be possible that unanticipated thermally activated failures will occur. In this case, the test conditions (time and temperature) must be selected such that, whichever ones are present, the relevant failure mechanisms will be observed in the test.

Finite element analysis is a useful tool for locating the points of maximum applied stress, and for calculating the level of stress at those points.

Experience has shown that the best accelerated tests are those that expose the samples to the environmental conditions that may cause failures, based on a general knowledge of the possible effects of the environment. When this is done properly, any failures that may occur in use will be detected, even if the specific mechanism was not anticipated when the test was designed. Over the years, specific failure mechanisms, and the test conditions which expose them, have emerged as critical for various products. Following is a short discussion of some of these conditions and their possible effects.

17.6.1 Thermally Activated Failures

Thermally activated failures are caused by chemical reactions according to the Arrhenius equation. They include diffusion of solids, liquids, and gases, and the effects of such diffusion. Solid-state diffusion can form brittle intermetallic compounds, weaken a local area due to Kirkendall voiding, or cause high electrical impedance. Diffusion can also weaken a strain-hardened alloy by precipitating the strengthening component. The chemical reaction of oxygen or other chemicals with a metal can cause corrosion. Plastics age more rapidly at high temperatures.

A general assumption is that the higher the temperature, the more a failure mechanism is accelerated. This is not always true, and the opposite is true for some mechanisms, such as the embrittlement of tin, the effects of condensation, and the effects of hot electrons in microelectronic devices.

The important term in the Arrhenius equation is the activation energy E_a. Because it is in the exponent, small variations in E_a can cause enormous changes in extrapolated values. Experimentally determined values of activation energies for various semiconductor device failure mechanisms are shown in Table 17-1.

In using the Arrhenius equation, the tests must be conducted within its temperature range of applicability. If an accelerated test is conducted at a temperature above which the properties of the test samples are different from those of the product at the use temperature, the test results cannot be extrapolated to the use temperature. If the sample to be tested is an alloy, the phase diagram should

TABLE **17-1** Activation Energies for Some Semiconductor
Device Failure Mechanisms

Failure mechanism	Activation energy, eV
Metal migration	1.8
Charge injection (slow trapping)	1.3
Ionic contamination	1.0–1.1
Gold-aluminum intermetallic growth	1.0
Corrosion in humidity	0.8–1.0
Electromigration in aluminum	0.5
Electromigration of silicon in aluminum	0.9
Time dependent dielectric breakdown	0.3–0.6
Electrolytic corrosion	0.3–0.6
Hot carrier trapping	−0.1
Surface charge accumulation	1.0
Gate oxide defects	0.3
Silicon junction defects	0.8
Metallization defects	0.5
Charge loss	0.6
Dark line and dark spot defects in laser diodes	0.6

be consulted for phase transformation temperatures. It is also risky to extrapolate through the glass transition temperature of a polymer.

Even when no phase transformation occurs, the properties of some materials change with temperature. An important property in this regard is the *homologous temperature* of a metal. The homologous temperature T_h is defined as

$$T_h = \frac{T_t}{T_m} \tag{17-20}$$

where T_t is the test temperature, in K, and T_m is the melting temperature, also in K. As a rule of thumb, the properties of a given metal begin to change significantly when T_m exceeds 0.5.

Another risk of predicting failures with the Arrhenius equation is that of *hidden failure mechanisms*. Figure 17-6 illustrates such a case. If the accelerated test is conducted at higher temperatures, shown as line A, the activation energy is estimated at 0.9 eV, and an extrapolated service life of over 100,000 hours is obtained for an operating temperature of 100°C. What is not observed in these tests is the hidden failure mechanism, shown as line B, with an activation energy of 0.5 eV. If this product is placed in service at 100°C, it will fail after a little over 10,000 hours due to the mechanism of line B. The mechanism of line B is undetectable above the crossover point, 175°C.

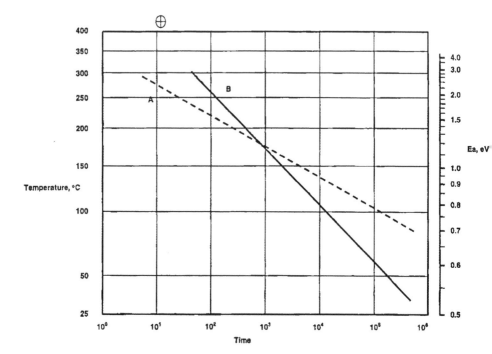

FIGURE 17-6 Arrhenius plot showing a hidden failure mechanism. If tests were conducted above 175°C (line A, the failure mechanism at lower temperatures (line B) would be hidden.

Because of its familiarity and ease of use, the Arrhenius equation has been applied widely. Unfortunately, it is often applied to failure mechanisms indiscriminately and incorrectly. Table 17-2 shows the role of temperature for some microelectronic circuit failure mechanisms [25]. Only those mechanisms for which the dominant temperature dependence is listed as T can be accurately modeled with the Arrhenius equation. An extensive bibliography (75 references) for temperature dependence of microelectronic failures can be found in Ref. 25.

17.6.2 Temperature Cycling Failures

Temperature cycling is a prominent cause of failure, due to mechanical stresses resulting from differential thermal expansion. These stresses may arise in a product with several materials having different coefficients of thermal expansion, or in a product with a single material having spatial thermal gradients.

Temperature cycling failures are detected by exposure to accelerated tests at ΔT ranges wider than the operating range of the product. Obviously, the ΔT

TABLE 17-2   Temperature Dependencies for Some Microelectronic Circuit Failure Mechanisms

Failure site	Failure mechanism	Dominant temperature dependence	Nature of absolute temperature dependence under normal operation
Wire	Flexure fatigue	ΔT	Independent of absolute temperature function.
	Shear fatigue	ΔT	Independent of absolute temperature.
Wire bond	Kirkendall voiding	T	Independent of absolute temperature below 150°C; independent of absolute temperature above lower temperatures ($t < 150$°C) in presence of halogenated compounds.
Die	Fracture	$\Delta T, \nabla T$	Primarily dependent on temperature cycling.
Die adhesive	Fatigue	ΔT	Independent of absolute temperature.
Encapsulant	Reversion	T	Independent of absolute temperature below 300°C (glass transition for typical epoxy molding resin for plastic packages).
	Cracking	$\Delta T, \nabla T$	Independent of absolute temperature below the glass transition temperature of the encapsulant.
Package	Stress corrosion	dT/dt	Mildly absolute temperature dependent.
Die metallization	Corrosion	dT/dt	Only occurs above dew point temperature. Mildly absolute temperature dependent under normal operation.
	Electromigration	∇T	Absolute temperature dependent above 150°C.
	Hillock formation	T	Hillocks in die metallization can form as a result of electromigration, or due to extended periods under temperature cycling conditions (thermal aging). Extended periods in the neighborhood of 400°C produce hillocks.
	Metal migration	T	Independent of absolute temperature below 500°C.
	Contact spiking	T	Independent of absolute temperature below 400°C.
	Constraint cavitation	T	Absolute temperature dependent above 25°C.
Die	Electrical overstress	T	Independent of absolute temperature below 160°C (the temperature at which the thermal coefficient of resistance changes signs).
Device oxide	Electrostatic discharge	T	ESD voltage (resistance to ESD) reduces with temperature increase (from 25 to 125°C). Not a dominant mechanism in properly protected devices.
	TDDB	$1/T$	Absolute temperature dependence is very weak. TDDB is a dominant function of voltage.

Source: From Ref. 25. © 1990 John Wiley & Sons, used by permission.

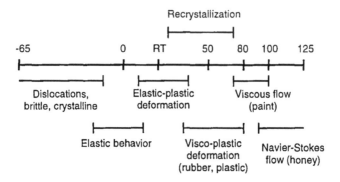

FIGURE 17-7 Properties of soft solder in the temperature range −65 to +125°C. (From Ref. 26. © 1987 IPC, used by permission.)

ranges in test must not extend to temperatures where the properties of the materials being tested are different from those in use. Electronic assemblies containing soft solders are often tested in accelerated temperature cycling tests. Figure 17-7 shows how the properties of soft solders can vary over some of the common test temperature ranges. The potential for erroneous conclusions is apparent.

Another danger in extrapolating high-strain data to low strain conditions for both thermal and mechanical tests is that some materials have an endurance limit at lower strains, as illustrated in Figure 17-8. Strains below the endurance limit will induce no accumulated damage, and will not affect the life of the product. Failure to take this into account may result in overly conservative reliability estimates. Figure 17-9 shows that the S-N curve for solder is composed of three different types of strain. The material will behave differently, depending upon which type is operating at any given time.

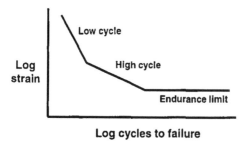

FIGURE 17-8 The regions of the S-N curve.

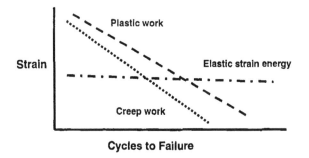

FIGURE 17-9 Components of the S-N curve for soft solder.

If the failure mechanism to be accelerated is crack initiation and growth, then misleading results can be obtained from temperature cycling tests if the test range initiates cracks that would not be initiated in use.

17.6.3 Vibration Failures

Vibration failures are commonly called high-cycle fatigue failures. The most common failure mechanism in high cycle fatigue is crack initiation and growth. Vibration failures are modeled by the inverse power law, using one of the following equations [27]

$$T_1 G_1^b = T_2 G_2^b$$
$$N_1 Z_1^b = N_2 Z_2^b \qquad\qquad (17\text{-}21)$$
$$N_1 G_1^b = N_2 G_2^b$$

where

N = number of cycles
T = total elapsed test time
G = acceleration level
Z = displacement amplitude
b = exponent equal to the slope of the S-N curve on a log-log plot

Vibration testing is probably the most widely used, and least understood, of all the accelerated tests in use today. The simplest vibration tests are sine wave inputs, at constant frequency and amplitude, to the unit under test. While easy to understand, sine wave testing is not very good at representing the actual vibration experience of products in use. Swept sine testing came next, but it was superseded by random vibration. Random vibration inputs usually are represented by plots of frequency vs. G^2/Hz, which are called *autospectral density* or *power spectral*

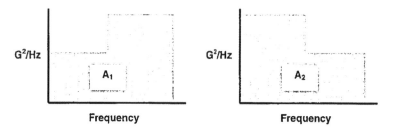

FIGURE 17-10 Two PSD vibration curves with equal areas and G_{rms} values, but with unequal inputs to the unit under test.

density (PSD) curves. The area under a PSD curve is the G_{rms}. Although G_{rms}, also called the *g* level, is often used as the measure of energy input in a vibration test, it is almost always naïve to do so. Figure 17-10 shows two PSD curves with equal areas, but they are quite different in their potential to damage the unit under test. In order for the G_{rms} value to have meaning, its value at the natural frequencies of the failure mechanisms being investigated must be known. Natural frequencies may be calculated, provided that sufficient information is available to do so, or may be determined experimentally, for example using the shock-response-spectrum approach [28].

Interpreting results from random vibration tests can be quite difficult. Several methods have been proposed, and some of them are described in Refs. 29–32.

Random vibration can be imposed by single-axis electrodynamic or servo-hydraulic shakers, which produce coherent vibration across the entire surface of the shaker table; or by multiaxis repetitive shock machines, in which pneumatic hammers produce noncoherent responses on the table. Reference 33 contains a good description of repetitive shock machines.

In conducting a vibration test, it is important to measure the vibration input at the site of the failure mechanism of interest. If measurements are made anywhere else, mechanical filtering can produce misleading results.

17.6.4 Humidity

Typical humidity failure mechanisms are corrosion, electrolytic corrosion, and metal migration. Humidity tests usually are conducted in conjunction with elevated temperatures, and a common test for electronic equipment is the temperature-humidity-bias (THB) test. A standard THB condition is 85°C-85% relative humidity, under operating electrical bias. It is possible to conduct THB tests above 100°C with the use of pressure vessels. This is called *highly accelerated stress testing* (HAST).

In humidity testing, care must be taken to ensure that the local conditions on the test sample are the same as the ambient conditions of the test chamber. If a test sample generates heat, such as is the case with a power semiconductor, the humidity in its immediate vicinity will be much lower than in other parts of the chamber. In such cases, it is a common practice to switch the power on and off at intervals to allow the moisture to contact the samples.

Recent publications [34–37] indicate that the actual mechanism for THB failures is the ingression of moisture into cracks caused by mechanical or thermal stresses. For this reason, some THB tests are now being conducted in sequence with fatigue tests.

There are many published acceleration models for humidity failures, but the most widely used is still that of Eqs. (17-13) and (17-14). Reference 38 contains an extensive bibliography of 56 references describing the role of humidity in microelectronic devices.

17.6.5 Electrical Stresses

Voltage and current stresses can cause a variety of failure mechanisms. High voltages can cause dielectric breakdown or arcing. Atomic or ionic movement due to electric fields can result in failures by electrolytic corrosion and shorting. Power cycling can cause both temperature and voltage gradients, which in turn can lead to a variety of failures depending on the structures and properties of the materials involved. Current flow can cause electromigration, which is the movement of metal atoms along a conductor, leaving voids at one end and hillocks at the other.

The general equation for time to failure for electromigration failure of a microcircuit conductor is Black's equation [39]:

$$t_f = AJ^{-n}e^{E_a/kT} \tag{17-22}$$

where

t_f = time to failure
A = a constant related to the geometry of the conductor
J = current density
n = parameter related to the current density which accounts for effects of current flow other than Joule heating of the conductor

A reported value for n is approximately 2, and for E_a approximately 0.5 eV [4].

A major cause of failures in integrated microcircuits is *time-dependent dielectric breakdown* (TDDB), which is the breakdown of an oxide dielectric during exposure to a constant electric field. The acceleration factor for TDDB is

$$AF = \exp\left(\frac{E_t - E_u}{E_{EF}}\right) \qquad (17\text{-}23)$$

where

E_t = electric field in test, in MV/cm
E_u = electric field in use, also in MV/cm
E_{EF} = empirical constant for a given acceleration rate

The acceleration rate used for E_{EF} is given by

$$E_{EF} = \frac{1}{\ln 10^\gamma} \qquad (17\text{-}24)$$

γ varies with electric field levels, temperature, and oxide thickness. Reported values for γ vary from 1.7 to 7 [4, 40–44]. Berman [44] reported γ as

$$\gamma = 0.4e^{(0.7/kT)} \qquad (17\text{-}25)$$

17.6.6 Other Stresses

Other environmental stresses can cause failures in some products. Examples include radiation, ultraviolet light, biological contaminants, and electrostatic discharge. While these and other stresses should be considered for comprehensive reliability testing, detailed discussions are beyond the scope of this text.

17.7 SUMMARY

Properly understood and applied, acceleration models can help to understand, predict, and assure the reliability of a product during the design stage. Too often, however, they are not understood or applied properly, and serious mistakes can be made. Pecht, Shukla, Kelkar, and Pecht [45] list six factors that limit the usefulness of acceleration models, and Meeker and Escobar [46] list 10 pifalls to be avoided in accelerated testing.

A wide array of accelerated tests, models, and acceleration factors has been discussed in this chapter. It is an indication of the scope and complexity of the subject that, even in this relatively long chapter, we have been able to do little more than introduce the subject of accelerated testing. This is a subject that cannot be learned from a textbook. Models and theories of the type presented here are necessary, but the knowledge gained from actually conducting accelerated tests and interpreting the results will both require and promote learning far beyond what is presented here. Accelerated testing is discussed in more depth in the following chapters, where some examples of real-life accelerated tests are presented and discussed.

REFERENCES

1. A. Fick, Uber Diffusion, Poggendorff's Annalen, vol. 94, p. 59, 1855.
2. W. Weibull, Statistical design of fatigue experiments, Journal of Applied Mechanics, pp. 109–113, March 1952.
3. D. Kececioglu and J. Jacks, The Arrhenius, Eyring, inverse power law and combination models in accelerated life testing, Reliability Engineering 8:1–9, 1984.
4. D. S. Peck and O. D. Trapp, Accelerated Testing Handbook, Technology Associates, Portola Valley, CA, 1987.
5. W. Nelson, Accelerated Testing, John Wiley & Sons, New York, 1990.
6. D. J. Klinger, On the notion of activation energy in reliability: Arrhenius, Eyring, and thermodynamics, Proceedings of the Reliability and Maintainability Symposium, pp. 295–300, 1991.
7. J. M. Hu, D. Barker, A. Dasgupta, and A. Arora, Role of failure mechanism identification in accelerated testing, Proceedings of the Reliability and Maintainability Symposium, pp. 181–188, 1992.
8. S. Arrhenius, Z. Physik. Chem., vol. 4, 1889.
9. L. F. Coffin, Jr., A study of the effects of cyclic thermal stresses on a ductile metal, Transactions of the ASME 76 (5):931–950, 1954.
10. S. S. Manson, Fatigue: A complex subject—some simple approximations, Experimental Mechanics 5(7):193–226, 1965.
11. L. F. Coffin, Jr., The effect of frequency on the cyclic strain and low cycle fatigue behavior of Cast Udimet 500 at elevated temperature, Metallurgical Transactions 2: 3105–3113, 1971.
12. W. Engelmaier, Fatigue life of leadless chip carrier solder joints during power cycling, IEEE Transactions on Components, Hybrids, and Manufacturing Technology, vol. CHMT-6, no. 3, 1985.
13. D. R. Olsen and H. M. Berg, Properties of bond alloys relating to thermal fatigue, IEEE Transactions on Components, Hybrids, and Manufacturing Technology, vol. CHMT-2, 1979.
14. H. D. Solomon, Low cycle fatigue of surface mounted chip carrier/printed wiring board joints, Proceedings of the 39th Electronic Components Conference, IEEE, pp. 277–292, 1989.
15. J. K. Hagge, Predicting fatigue life of leadless chip carriers using Manson-Coffin equations, Proceedings of the IEPS, pp. 199–208, 1982.
16. C. F. Dunn and J. W. McPherson, Temperature-cycling acceleration factors for aluminum metallization failure in VLSI applications, Proceedings of the 28th International Reliability Physics Symposium, IEEE, pp. 252–258, 1990.
17. P. C. Paris, The growth of fatigue cracks due to variations in load, Ph.D. Thesis, Lehigh University, 1962.
18. S. Glasstone, K. J. Laidler, and H. E. Eyring, The Theory of Rate Processes, McGraw-Hill, New York, 1941.
19. D. S. Peck, Comprehensive model of humidity testing correlation, Proceedings of the 24th International Reliability Physics Symposium, IEEE, pp. 44–50, 1986
20. D. S. Peck and W. R. Thorpe, Highly accelerated stress test history: Some problems and solutions, Tutorial Notes, 28th Reliability Physics Symposium, IEEE, pp. 4.1–4.27, 1990.

21. F. Jensen, How to succeed in modeling, Quality and Reliability Engineering International 15:159, 1999.

22. M. A. Miner, Cumulative damage in fatigue, Transactions of the American Society of Mechanical Engineers 67:A-159–A-164, 1945.

23. D. B. Barker, A. Dasgupta, and M. G. Pecht, PWB solder joint life calculations under thermal and vibrational loading, Proceedings of the Reliability and Maintainability Symposium, IEEE, pp. 451–459, 1991.

24. K. Upadhyayala and A. Dasgupta, Physics-of-failure guidelines for accelerated qualification of electronic systems, Quality and Reliability Engineering International 14: 433–447, 1998.

25. M. Pecht, P. Lall, and S. Whelan, Temperature dependence of microelectronic device failures, Quality and Reliability Engineering International 6:275–284, 1990.

26. W. Engelmaier, Is Present-Day Accelerated Cycling Adequate for Surface Mount Attachment Reliability Attachment? Institute for Interconnecting and Packaging Electronic Circuits (IPC), Lincolnwood, IL, 1987.

27. D. S. Steinberg, Vibration Analysis for Electronic Equipment, John Wiley & Sons, New York, 1988.

28. S. A. Smithson, Shock response spectrum analysis for ESS and STRIFE/HALT measurement, Proceedings of the Institute for Environmental Sciences, 1991.

29. N. E. Dowling, Fatigue failure predictions for complicated stress-strain histories, Journal of Materials, JMLSA 7(2):71–87, March 1972.

30. R. G. Lambert, Case histories of selection criteria for random vibration screening, The Journal of Environmental Sciences, pp. 19–24, January/February 1985.

31. R. G. Lambert, Fatigue damage prediction for combined random and static mean stresses, Journal of the IES, pp. 25–32, May/June 1993.

32. R. G. Lambert, Fatigue life prediction for various random stress peak distributions, Shock and Vibration Bulletin, no. 52, part 4, pp. 1–10, 1982

33. G. R. Henderson, Dynamic characteristics of repetitive shock machines, Proceedings of the Institute for Environmental Sciences, May 1993.

34. S. LeRose, G. DeLeuze, and M. Brizou, Evaluation of standard plastic IC's reliability after accelerated sequential testing, Proceedings of the 7th International Conference on Reliability and Maintainability, Brest, France, pp. 419–427, 1990.

35. R. C. Blish and P. R. Vaney, Failure rate model for thin film cracking in plastic IC's, Proceedings of the International Reliability Physics Symposium, IEEE, 1991.

36. L. W. Condra, S. O'Rear, T. Freedman, L. Flancia, M. Pechr, D. Barker, Comparison of plastic and hermetic microcircuits under temperature cycling and temperature humidity bias, IEEE Transactions on Components, Hybrids, and Manufacturing Technology 15(5), October 1992.

37. E. B. Hakim, Microelectronic reliability/temperature independence, quality and reliability engineering international 7(4):215–220, 1991.

38. D. J. Klinger, Humidity acceleration factor for plastic packaged electronic devices, Quality and Reliability Engineering International 7(5):365–370, 1991.

39. J. R. Black, Electromigration—A brief survey and some results, IEEE Transactions on Electron Devices, vol. ED-16, 1969.

40. D. L. Crook, Method of determining reliability screens for time dependent dielectric

breakdown, Proceedings of the 17th International Reliability Physics Symposium, IEEE, pp. 1–7, 1979.

41. E. S. Anolick and G. R. Nelson, Low field time dependent integrity, Proceedings of the 17th International Reliability Physics Symposium, IEEE, pp. 8–12, 1979.

42. J. W. McPherson and D. A. Baglee, Acceleration factors for thin gate oxide stressing, Proceedings of the 23rd International Reliability Physics Symposium, IEEE, 1985.

43. Y. Hokari, T. Baba, and N. Kawamura, Reliability of thin SiO_2 films showing intrinsic dielectric integrity, IEDM Technology Digest, 1982.

44. A. Berman, Time zero dielectric reliability test by a ramp method, Proceedings of the 19th International Reliability Physics Symposium, IEEE, pp. 204–209, 1981.

45. M. Pecht, A. Shukla, N. Kelkar and J. Pecht, Criterian for the assessment of reliability models, IEEE Transactions on Components, Packaging, and Manufacturing Technology—Part B, 20(3):229–234, August 1997.

46. W. Q. Meeker and L. A. Escobar, Pitfalls of accelerated testing, IEEE Transactions on Reliability 47(2):114–118, June 1998.

Chapter 18

Environmental and Operating Stresses

18.1 INTRODUCTION

The accelerated test models introduced in Chapter 17 are used to extrapolate from a set of accelerated test conditions to a set of unaccelerated use conditions. The accelerated test conditions are usually quite well known and controlled. To complete the process, we need to know the use conditions. In reality, the use conditions are never known as well as the test conditions, but we need to know enough about them to make reasonable extrapolations. In this chapter, we look at the use conditions, and we also consider some requirements for reliability for products operating in those conditions.

Product reliability specifications usually contain two parts: (1) a statement of required reliability, such as useful life, mean time between failures (MTBF), etc., and (2) a description of the environmental and operating conditions to which the product will be exposed. Both types of requirements must be considered in the design of the product. Although it might seem axiomatic that both requirements must be stated, it is surprising how often they are unknown, or partially known. Therefore, it is worth some time and effort to discuss them.

It is not always as easy as one might expect to understand the environmental conditions. Some products, such as household appliances and office equipment, operate in stable, well-controlled, and well-defined environments. Others, such as outdoor telecommunications equipment, operate in well-defined and relatively stable environments, but they cannot be called well controlled. Still others, such as automobiles, aircraft, and ships, operate in environments that are highly variable, uncontrolled, and unstable.

Often, the local conditions for a component within a system are quite different from those of the overall system environment. For example, a power semiconductor device might operate in an electronics system whose ambient conditions are well known, but are quite different from the local conditions at the site of the device. It is therefore critical for the product manufacturer to determine the local conditions, and that can be expensive and time consuming.

Reliability requirements are commonly stated as (1) some form of useful life, such as time to first-failure, and (2) failure rate, such as MTBF. The design and manufacturing steps taken to ensure a given useful life are not necessarily the same as those taken to ensure that an acceptable failure rate within the useful life period will be attained. Three types of reliability testing to evaluate both useful life and failure rate are (1) life testing, (2) design testing, and (3) production testing. They are discussed in some detail in Chapter 20.

Generally, useful life implies operating life. Products such as bullets, flashbulbs, and missiles have a very short useful operating life, but a relatively long dormant storage period. Each product has either an explicit useful life requirement spelled out in a specification or contract, or an implied useful life which both the customer and the producer have accepted through experience.

18.2 STANDARDS FOR ENVIRONMENTAL CONDITIONS

A number of "standard" or "generic" descriptions of environments and operating conditions are available for some products, but most of them are too general to provide the information necessary to assess and predict reliability of a given product. An example of this type of description is shown in Table 18-1, from Mil-Hdbk-217 [1].

MIL-STD-810 [2] and UK Defence Standard 07-55 [3] list test methods for a variety of conditions. These tests are of the pass-fail type. This type of test is sometimes called an *elephant test* [4]. The idea behind an elephant test is that if a product still works after an elephant steps on it, it must be reliable. It is obvious that this type of test is not satisfactory to the deterministic reliability professional, since it provides no quantitative information on reliability. MIL-HDBK-781 [5] contains information on how to assess environmental conditions and design tests, and MIL-STD-210 [6] contains climatic information, for use in MIL-STD-810.

TABLE 18-1 Descriptions of Operating Requirements from MIL-HDBK-217F

Environment	π_E symbol	Description
Ground, benign	G_B	Nonmobile, temperature- and humidity-controlled environments readily accessible to maintenance; includes laboratory instruments and test equipment, medical electronic equipment, business and scientific computer complexes, and missiles and support equipment in ground silos.
Ground, mobile	G_M	Equipment installed on wheeled or tracked vehicles and equipment manually transported; includes tactical missile ground support equipment, mobile communication equipment, tactical fire direction systems, handheld communications equipment, laser designations, and range finders.
Airborne, inhabited, cargo	A_{IC}	Typical conditions in cargo compartments which can be occupied by an air crew. Environment extremes of pressure, temperature, shock and vibration are minimal. Examples include long mission aircraft such as the C130, C5, B52, and C141. This category also applies to inhabited areas in lower performance smaller aircraft such as the T38.
Airborne, inhabited, fighter	A_{IF}	Same as A_{IC} but installed on high-performance aircraft such as fighters and interceptors. Examples include the F15, F16, F111, F/A 18, and A10 aircraft.
Space, flight	S_F	Earth orbital. Approaches benign ground conditions. Vehicle neither under powered flight nor in atmospheric reentry; includes satellites and shuttles.
Missile, launch	M_L	Severe conditions related to missile launch (air, ground, and sea), space vehicle boost into orbit, and vehicle reentry and landing by parachute. Also applies to solid rocket motor propulsion powered flight, and torpedo and missile launch from submarines.

18.3 ENVIRONMENTAL CONDITIONS FOR SPECIFIC PRODUCTS

This section includes some published information about typical environmental conditions for specific product groups.

Although there are no "standard" environments and operating conditions for computers, the environments may be grouped into four broad categories. The first is the controlled server room, which includes rack-mounted servers and storage units. The humidity and temperature are controlled continuously within relatively narrow ranges, and the air may be filtered to remove dust and other airborne contamination. Mechanical shock and vibration stresses are essentially nonexistent. Typically, the temperature range for server rooms is 25 ± 5°C, the humidity ranges from 40 to 60 %RH, and the barometric pressure ranges from 70 to 105 kPa (0–2500 ft).

Controlled office environments for desktop and rack-mounted computers include temperature ranges similar to those for server rooms; and the humidity may range from 20 to 70% RH. The equipment is seldom moved from one location to another, but there is some exposure to mechanical shock and, less so, to vibration.

The general-purpose computer environment includes the extremes of temperature, humidity, pressure, mechanical shock, and vibration expected for desktop equipment in offices, laboratories, manufacturing areas, and other workplaces

TABLE 18-2. Typical Automobile Environmental Conditions

Location	Temperature range, °C	Relative humidity (% at 40°C)	Salt spray	Vibration and shock
Under hood				
Above exhaust	−40 to +650	80	Yes	50 g to 1 kHz
Intake manifold	−40 to +125	95	Yes	Over 100 g
Fireproof wall	−40 to +140	80	Yes	1 g to 600 Hz
Vehicle frontal zone	−40 to +85	98	Yes	1 g to 600 Hz
On chassis				
Inside a wing	−40 to +85	98	Yes	2 g to 2 kHz
Near the exhaust system	−40 to +125	98	Yes	2 g to 2 kHz
Extreme conditions	−40 to +175	98	Yes	Over 100 g
Car interior				
Dashboard	−40 to +120	98	No	1 g to 20 kHz
Rear window	−40 to +100	98	No	1 g to 20 kHz

Source: Ref. 8.

TABLE **18-3** Thermal Conditions During a 24-hour Time Period for Under-the-Hood Automotive Electronics

Cycle	Winter cycle		Summer cycle	
	°C	Hours	°C	Hours
Cold	−55	12.0	+55	12.0
Warmup	−55 to +5	0.5	+55 to +105	0.5
Warm-hot	+5 to +50	1.0	+105 to +150	1.0
Hot-hot	+50	8.0	+150	8.0
Hot-cool	+50 to +55	2.5	+150 to +55	2.5
Average	−9.1		+91.2	

Source: Ref. 8.

in all geographic locations. The temperature may range from below zero to +40°C or higher, and the humidity range may be from nearly zero to nearly 100%. Exposure to dust and other airborne contamination should be expected. Equipment may be moved frequently, increasing the exposure to mechanical shock and vibration. Equipment also may be subjected to liquid, such as beverage spills, occasional rain, etc.

The fourth computer environment is that of hand-held equipment, including keyboards, mice, other small input devices, and laptop and palmtop computers.

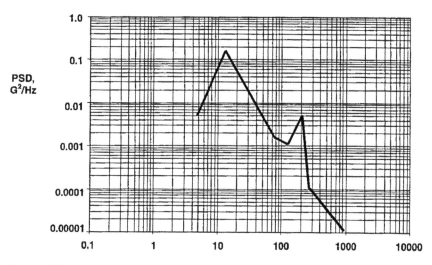

FIGURE **18-1** Typical PSD (vertical axis) for a small truck.

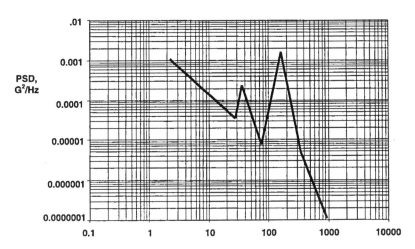

FIGURE 18-2 Vibration input to the payload of a 10-ton truck. (Adapted from Ref. 9.)

TABLE 18-4 Worst-Case Use Environments for Electronic Products

Use category		T_{min}, °C	T_{max}, °C	ΔT, °C	Dwell time, hours	Cycles per year	Years of service
Consumer		0	+60	35	12	365	1–3
Computers		+15	+60	20	2	1460	~5
Telecommunications		−40	+85	35	12	365	7–20
Commercial aircraft		−55	+95	20	2	3000	~10
Industrial and automotive-		−55	+65	20	12	185	~10
passenger compartment				& 40	12	100	
				& 60	12	60	
				& 80	12	20	
Military ground and ship		−55	+95	40	12	100	~5
				& 60	12	265	
Space:							
Low earth orbit		−40	+85	35	1	8760	5–20
Geosynchronous orbit		−40	+85	35	12	365	
Military avionics	*a*	−55	+95	40	2	500	~5
	b	−55	+95	60	2	500	
	c	−55	+95	80	2	500	
				& 20	1	1000	
Automotive, under hood		−55	+125	60	1	1000	~5
				& 100	1	3000	
				& 140	2	40	

Source: Ref. 10.

This environment includes the entire range of environmental conditions in which humans may have to work.

Table 18-2 shows typical environmental conditions for various parts of an automobile [7], and Table 18-3 lists thermal conditions for under-the-hood electronics [8]. Figure 18-1 shows a typical vibration PSD for a small truck. Richards and Hibbert [9] measured the vibration inputs to the payloads of 22 different 10-ton trucks traveling over the same 40 km roundtrip, and a PSD adapted from their results is shown in Figure 18-2.

Table 18-4 [10] contains information about environmental conditions for electronics used in consumer products, computers, telecommunications equipment, commercial aircraft, automobiles, military ground and ship applications, space equipment, and military avionics equipment. It should be noted that the numbers shown are only typical, and that wide ranges of conditions apply to all the products listed. Except for products that operate in well-controlled environments, environmental conditions vary throughout the day and year.

18.4 PRODUCT SPECIFICATIONS

Product specifications range from the very informal, such as those for some commercial products, to the very formal, such as those for custom military and aerospace products. In this section, we look at formal specifications, using some aerospace examples.

Commercial airliners have useful life expectancies of 20 years or more, and their failure rates are made up of the individual failure rates of their component parts, such as structures, engines, and avionics. The operating life requirements are deduced from the major mission requirements. For example, a typical utilization scenario for a commercial airliner capable of intercontinental flight is shown in Table 18-5. Aircraft built to this specification must be capable of any combination of long, nominal, and short missions; however, larger airplanes generally fly the longer missions, and smaller airplanes generally fly the shorter ones.

TABLE 18-5 Typical Utilization Scenarios for a Commercial Airplane

Mission type	Flight hours per mission	Operating hours per mission	Missions per year	Flight hours per year	Operating hours per year
Long mission	12.5	14.5	350	4375	5075
Nominal mission	4.9	6.0	850	4165	5100
Short mission	1.5	2.5	2000	3000	5000

TABLE 18-6 Nominal Flight Profile for a Commercial Airplane

Flight segment	Time (hours)
1. Taxi	0.167
2. Takeoff (to 1500 ft)	0.033
3. Climb (to cruising altitude)	0.333
4. Cruise	4.000
5. Descent	0.417
6. Approach and landing	0.083
7. Taxi	0.167
Flight time, hours	4.863
Total time, hours	6.000

A nominal flight profile is shown in Table 18-6. The major variable between long and short missions is the cruise time. Local conditions for the various components of an airplane during each of the segments of a flight vary according to location and duty cycle.

The ambient conditions to which a product is exposed also vary throughout each day and from day to day. Figure 18-3 shows some measured temperature

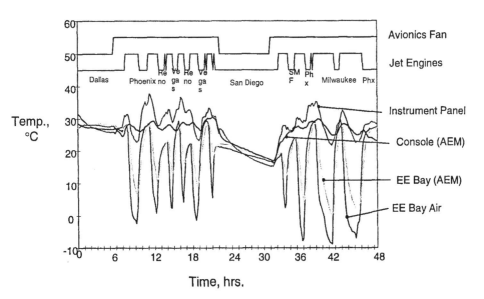

FIGURE 18-3 Temperature variations in a commercial airplane. (From Ref. 11.)

TABLE **18-7** Typical Specification for Internal Operating
Temperature and Pressure for Avionics in a Commercial Aircraft

Operating temperature, °C				Transient and shock	Minimum pressure at operating altitude
Ground		Flight			
Min	Max	Min	Max		
−40	+45	+25	+45	+50	5 PSIG

TABLE **18-8** Hot, Standard, and Cold Days in a Year (in °C)

MIL-STD-210C 1% hot		Standard		MIL-STD-210C 10% cold	
Min	Max	Min	Max	Min	Max
32.2	72.2	9.4	46	−54	−45.6

TABLE **18-9** Percent of Time Spent
in Various Temperature Ranges for an
Avionics Product

Temperature range	Percent of time
−54 to −20°C	5
−20 to −5°C	10
−5 to +85°C	70
+85 to +110°C	10
+110 to +121°C	5

variations for various locations in a commercial airplane during flight, takeoff, and landing at a number of airports [11]. Table 18-7 shows a typical specification for internal operating temperature and pressure for avionics in a commercial airplane. The number of hot, standard, and cold days per year is shown in Table 18-8, and Table 18-9 lists the percent of time spent in various parts of the temperature specification range for an avionics product.

18.5 SUMMARY

In this chapter, several examples of product operating conditions have been listed. Obviously, there are many more such conditions for specific products. The job of the reliability and design engineers is to develop meaningful tests to be conducted during the design stage, which will predict reliability during use. This is quite a challenge, given that so many different combinations of stress and their levels are operable, and that these stresses also vary greatly with both location and time.

In Chapters 19 and 20 we consider some approaches to analyzing, testing, and ensuring reliability over the ranges of stresses encountered by products.

REFERENCES

1. MIL-HDBK-217F, Reliability Prediction of Electronic Equipment, available from the National Technical Information Service, Springfield, VA.
2. MIL-STD-810, Environmental Test Methods, available from the National Technical Information Service, Springfield, VA.
3. UK Defence Standard 07–55, Environmental Testing.
4. W. Nelson, Accelerated Testing, John Wiley & Sons, New York, 1990.
5. MIL-HDBK-781, Reliability Testing for Engineering Development, Qualification and Production, available from the National Technical Information Service, Springfield, VA.
6. MIL-STD-210, Climatic Information to Determine Design and Test Requirements for Military Systems and Equipment, available from the National Technical Information Service, Springfield, VA.
7. M. G. Priore and J. P. Farrell, Plastic Microcircuit Packages: A Technology Review, Reliability Analysis Center, 210 Mill St., Rome, NY, 1992.
8. T. R. Poulin and Y. Belopolsky, Automotive Environment and Reliability of Materials Systems, Materials Development in Microelectronic Packaging Conference Proceedings, Montreal, Quebec, pp. 317–321, 1991.
9. D. P. Richards and B. E. Hibbert, A round robin exercise on road transportation data, Journal of the IES, pp. 19–27, July/August 1993.
10. IPC-SM-785, Guidelines for Accelerated Reliability Testing of Surface Mounted Solder Attachments, Institute for Interconnecting and Packaging Electronic Circuits (IPC), 1989.
11. K. D. Cluff, Characterizing the humdity and thermal environments of commercial avionics for accelerated test tailoring, Ph. D. dissertation, University of Maryland, 1996.

Chapter 19

Interpreting Data from Accelerated Tests

19.1 INTRODUCTION

Traditionally, the best reliability data have been considered those which come from the operation of a product in service. For many of the reasons spelled out in Chapter 2, Section 2.4.1, however, credible service data are difficult to obtain. Furthermore, when such information is available, it is often in a form that is difficult to use. This difficulty, coupled with continuing progress in the development of accurate, efficient, and cost-effective accelerated reliability testing, have made accelerated testing an increasingly viable source of reliability data.

The important features of an effective accelerated test plan are (1) a knowledge of the functional and reliability requirements of the product; (2) an understanding of the product's operating and environmental conditions; (3) the expected failure modes and mechanisms; (4) acceleration models for each of the failure mechanism in order to design the test and analyze results; (5) a technique for designing and conducting the test, such as Design of Experiments; (6) the capability to analyze physical failures; and (7) the capability to interpret the results and draw accurate conclusions from them.

The first five of these features have been discussed at length in previous chapters. The sixth, physical failure analysis, is vital and should not be discounted. Physical failure analysis is product specific, and therefore beyond the scope of this text. In this chapter, we consider examples of the seventh feature:

interpretation of results and drawing conclusions. It should never be forgotten, though, that this type of analysis is successful only if conducted in parallel with effective physical failure analysis.

Test data analysis and drawing conclusions therefrom are emphasized differently in different markets, and are used most extensively on products which must be highly reliable and are produced in small volumes, such as military and aerospace products.

Each of the examples in this chapter illustrates at least one important technique for interpreting results from accelerated testing. The examples include:

- Accelerated life testing of microcircuits using the inverse power law and a constant failure distribution
- Accelerated life testing of microcircuits using the inverse power law and a Weibull distribution
- Failure distribution parameters at different stress levels
- Accelerated life testing with the Arrhenius relationship over a range of stress levels
- Accelerated testing of transistors with lognormal distributions over a range of stress levels
- Accelerated testing at different stress levels when more than one failure mechanism occurs

19.2 ACCELERATED LIFE TESTING OF MICROCIRCUITS USING THE INVERSE POWER LAW AND A CONSTANT FAILURE DISTRIBUTION

This example involves analysis of data from accelerated life testing of plastic-encapsulated microcircuits for use in a commercial avionics product [1]. In the product design, 1200 microcircuits of the type to be tested are assembled into an electronics box with a required MTBF of 15,000 hours, and a useful life of 20 years. (Some technical expertise and judgment are required to determine whether the samples being tested are sufficiently similar to those being used in the product.) The test is conducted to determine if the reliability of the microcircuits is sufficient to meet that goal. The most likely cause of failure is mechanical stresses resulting from temperature cycling. Since the package type is a leaded dual in-line package (DIP), the most likely failure mechanisms are those which occur inside the package, and not in the interconnection of the package to the printed circuit card.

The samples were tested after being soldered onto circuit cards: 1600 samples were divided into four groups of 400 each, and the temperature cycling ranges for the four groups were: -55 to $+125°C$; 0 to $+75°C$; $+50$ to $+125°C$;

and +75 to +150°C. The acceleration model for temperature cycling failures is the inverse power law; therefore test points were equispaced on a logarithmic scale at 0, 10, 30, 100, 300, and 1000 cycles.

No failures were observed through 1000 cycles. This means that no failure rate distribution is yet evident, and a constant failure rate distribution must be assumed. For this type of distribution, the failure rate is usually expressed as mean time between failures (MTBF). Strictly speaking, for nonrepairable systems such as microcircuits, the correct term is mean time to failure (MTTF). In this example, however, a large number of such devices are included in a repairable system, and MTBF may be used. Since the unit of stress is a temperature cycle, the actual reliability term is mean cycles between failures (MCBF). The lower limit of the MCBF is estimated using (see Sec. 14.13).

$$\text{MCBF}(l) = \frac{2t_a}{X^2_{\Upsilon;dF}} \tag{19-1}$$

where t_a is the accumulated device-cycles, which in this case is 1000×1600, or 1,600,000 cycles; and the X^2 subscripts are obtained from Table 14-2 and Appendix D. We can estimate the lower limit on the MCBF, with 95% confidence, as

$$\text{MCBF}(l) = \frac{2t_a}{X^2_{0.05;2}} = \frac{(2)(1.6 \times 10^6)}{5.991} = 5.34 \times 10^5 \text{ cycles} \tag{19-2}$$

These are, of course, accelerated cycles. Before we can estimate the operating MCBF, we must determine the acceleration factors for the test conditions. Since no failures have yet occurred, we will assume that failures will occur, if we continue testing, according to the inverse power law, for which the acceleration factor is (see Chapter 17)

$$\text{AF} = \frac{N_{fu}}{N_{ft}} = \left(\frac{\Delta T_t}{\Delta T_u}\right)^B \tag{19-3}$$

The critical term in Eq. (19-3) is the exponent B, which may range from 2 to 12 or higher depending on the failure mechanism, as reported in Section 17.3.2. (The abstract, introduction, and references sections of Ref. 2 contain an excellent list of values of this exponent for various semiconductor failure mechanisms.) The lower the value of B, the lower the acceleration factor and the more conservative will be the estimate of MCBF. Since we want to be conservative, we assume the failure mode to be cracking of the plastic encapsulation, which has a reported value of $B = 4$. To be even more conservative, however, we assume $B = 3$.

ΔT_u for commercial aircraft is shown as 20°C in Table 18-4. To be conservative, we use 30°C for ΔT_u. Two different ΔT_t's were used in the this experiment, so there are two acceleration factors. Four hundred samples were tested at a ΔT_t of 180°C, for an acceleration factor of

$$(AF)_1 = \left(\frac{\Delta T_t}{\Delta T_u}\right)^B = \left(\frac{180}{30}\right)^3 = 216 \tag{19-4}$$

The remaining 1200 were tested at a ΔT_t of 75°C, and the acceleration factor is

$$(AF)_2 = \left(\frac{75}{30}\right)^3 = 15.625 \tag{19-5}$$

To calculate an overall acceleration factor, we use a form of Miner's rule, which in this case is just the weighted average of the two acceleration factors from Eqs. (19-4) and (19-5):

$$AF = \frac{(AF)_1 n_1 + (AF)_2 n_2}{n_1 + n_2} = \frac{(216)(400) + (15.625)(1200)}{400 + 1200} \approx 66 \tag{19-6}$$

Thus, the overall acceleration factor is applied to the MCBF estimate of 534,000 cycles from Eq. (19-2), and the lowest estimate of MCBF, with 95% confidence, is

$$MCBF = (5.34 \times 10^5)(66) = 3.5 \times 10^7 \text{ cycles} \tag{19-7}$$

This number can be converted to MTBF in hours by using the information from Table 18-5, which shows that, for a nominal mission, each flight (or each temperature cycle) is equivalent to 6.0 operating hours. Thus the 3.4×10^7 MCBF is equivalent to an MTBF of approximately 2.0×10^8 hours.

The failure rate for the microcircuits is

$$\lambda = \frac{1}{MTBF} = \frac{1}{2 \times 10^8} = 5.0 \times 10^{-9} \text{ failures per hour} \tag{19-8}$$

This is sometimes expressed as 5.0 FITs. (FIT is short for *failure unit*, which is one failure in a billion device hours.) The 1200 microcircuits in each box thus have an estimated aggregate failure rate of

$$(1.2 \times 10^3)(5.0 \times 10^{-9}) = 6.0 \text{ failures per million hours} \tag{19-9}$$

If we assume that half of the total failures in the box will be due to microcircuit failures (the rest will be due to other components, connectors, assembly defects, materials etc.), we can estimate an overall box failure rate of 12.0 failures per million hours.

The required failure rate for the electronics box is

$$\lambda = \frac{1}{MTBF} = \frac{1}{15,000} = 67 \text{ failures per million hours} \tag{19-10}$$

Thus, we can estimate with at least 95% confidence that the microcircuits will exhibit the required MTBF in the avionics application.

We must now address the question of useful life. To do so, we use only the data from the samples with the highest acceleration factor. To this point, we have achieved 1000 temperature cycles of 180°C; and with an acceleration factor of 216, this equates to 216,000 cycles of 30°C. From Table 18-5, we see that the aircraft will have a nominal 850 flights per year, and that there is one temperature cycle per flight. Thus our 1000 test cycles are equivalent to

$$\frac{(1000)(216)}{850} \approx 254 \text{ years} \tag{19-11}$$

and the useful life goal is met. (Note that the number of device hours does not enter into the useful life calculation.)

Our extrapolated failure rate is less than one fifth of the allowable rate, and the extrapolated useful life is over 10 times the required useful life. It appears that the expected reliability of the microcircuits is well beyond that required for the product application, especially since we made conservative assumptions regarding the operating temperature and the exponent B. While this is indeed the case, we should remember that this type of extrapolation is based on acceleration factors that are quite general, and which have not been clearly demonstrated in this experiment. Furthermore, the ΔT_t of 180°C is quite large for the materials involved, and there is reason to be cautious.

In general, if we do not obtain reliability estimates far better than those of the requirements, there is room for some uneasiness in assuming that the product is reliable. In this case, however, our extrapolated reliability is so much greater than the requirement that we can be confident that the product will be reliable.

19.3 ACCELERATED TESTING OF MICROCIRCUITS USING THE INVERSE POWER LAW AND A WEIBULL FAILURE DISTRIBUTION

Engelmaier [3–7] developed forms of the inverse power law to describe the reliability of surface mounted electronic microcircuit solder joints in accelerated fatigue testing.* His equations are

---

* Recent research has indicated that closed-form models may be inadequate to describe electronic solder joints in detail, and that numerical methods such as finite element analysis are required. The Engelmaier model, however, is still used as one of the best closed-form approximations available.

$$N_{fu} = \frac{1}{2}\left[(2N_{ft})^{c_t}\left(\frac{\Delta T_u}{\Delta T_t}\right)\right]^{1/c_u} \qquad (19\text{-}12)$$

for leadless packages, and

$$N_{fu} = \frac{1}{2}\left[(2N_{ft})^{c_t}\left(\frac{\Delta T_u}{\Delta T_t}\right)^2\right]^{1/c_u} \qquad (19\text{-}13)$$

for leaded packages, where the subscripts t and u represent "test" and "use," N_f is the number of cycles to failure, ΔT is the temperature cycling range, and

$$c = -0.442 - 0.0006T_\delta + 0.0174 \ln\left(1 + \frac{360}{t_D}\right) \qquad (19\text{-}14)$$

where T_δ is the mean cyclic solder joint temperature in °C, and t_D is the half cycle dwell time in minutes.

In Ref. 3, Engelmaier reports test results for microcircuits with two types of leads, which we shall call "type B" and "type N." Twenty-four samples of each type were temperature-cycled between 25°C and 82°C (solder joint temperature), with dwell times of about 225 seconds at the high and low temperatures. The test was terminated at 100,000 cycles, by which time all of the type B samples had failed, and none of the type N samples had failed.

Results of this experiment are shown on Weibull plots in Figures 19-1 and 19-2. (It is believed by many that the Weibull distribution is always the best one for fatigue failures, but this cannot be proven and should be confirmed in all cases.) The first type B failure occurred at approximately 6800 cycles, and the last had occurred by 48,000 cycles. These data show a Weibull slope β of 2.0, with a median of about 15,000 cycles, which is consistent with previous work on mechanical fatigue testing, as reported in Ref. 3.

These results may be extrapolated to use conditions with Eq. (19-12). For type B samples, the following values are used:

$N_{ft} = 15,000$ cycles
$\Delta T_u = 30°C$ (from the example in section 19.2)
$\Delta T_t = 57°C$
T_δ (test) $= 53.5°C$
T_δ (use) $= 20°C$
t_D (test) $= 3.75$ min
t_D (use) $= 360$ min (from Table 18-5)

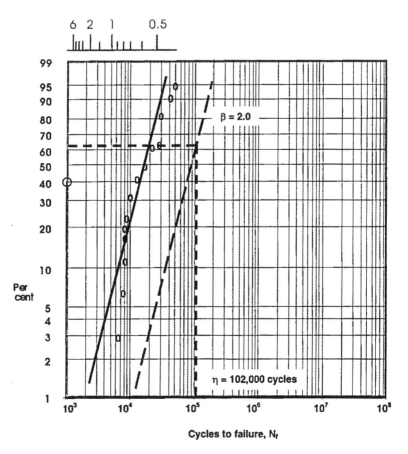

FIGURE 19-1 Weibull plot of the results of an accelerated life test of surface mount solder joints, type *B* leads. (From Ref. 3. © 1989 IEEE, used by permission.)

c_t for this example is then

$$c_t = -0.442 - 0.0006(53.5) + 0.0174 \ln\left(1 + \frac{360}{3.75}\right) = -0.394 \quad (19\text{-}15)$$

and c_u is

$$c_u = -0.442 - 0.0006(20) + 0.0174 \ln\left(1 + \frac{360}{360}\right) = -0.442 \quad (19\text{-}16)$$

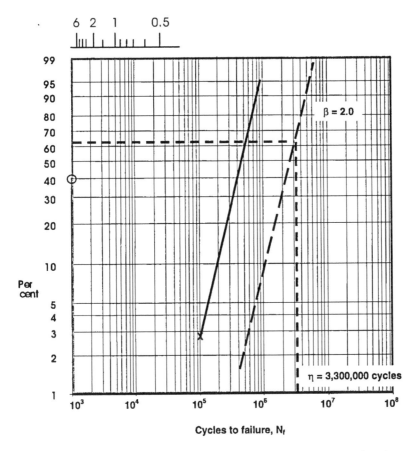

FIGURE 19-2 Weibull plot of the results of an accelerated life test of surface mount solder joints, type N leads. (From Ref. 3. © 1989 IEEE, used by permission.)

The median, or t_{50}, of cycles to failure in use is then

$$N_{fu} = \frac{1}{2}\left[([2][15{,}000])^{-0.394}\left(\frac{30}{57}\right)^2\right]^{-1/0.442} = 86{,}317 \text{ cycles} \qquad (19\text{-}17)$$

and the acceleration factor is

$$AF = \frac{86{,}317}{15{,}000} \approx 5.75 \qquad (19\text{-}18)$$

(This extrapolation is made to a set of use conditions different from that of Ref. 3; thus the acceleration factor is also different from that obtained in Ref. 3.)

In order to estimate the distribution at use conditions, we must make an assumption regarding the Weibull slope, β. The most obvious assumption is that it will remain unchanged, and that is the one made here. The scale parameter η is the number of cycles corresponding to 63.2% cumulative failures. As seen in Figure 19-1, this is approximately 102,000 cycles. Since a straight line was drawn through the plotted test data points, the location parameter γ is equal to zero, and the distribution is a two-parameter Weibull. (This decision might be considered subjective by some observers.)

It is now possible to estimate the useful life of the product. We shall define the useful life to be the time required for the first 5% of the product to fail, or t_5. From Figure 19-1, t_5 is about 45,000 cycles. From Table 18-5, a commercial aircraft will undergo about 850 cycles per year. Thus the useful life, as we have defined it, is equal to 45,000/850, or approximately 53 years.

The failure rate at the required useful life of 20 years may be determined by

$$\lambda(t) = \frac{\beta}{\eta}\left(\frac{t-\gamma}{\eta}\right)^{\beta-1} \tag{19-19}$$

From Table 18-5, each year is the equivalent of 850 temperature cycles, and

$$\lambda(17,000) = \left(\frac{2}{102,000}\right)\left(\frac{17,000}{102,000}\right)^{2-1} \tag{19-20}$$

$$= 3.3 \text{ failures per million cycles}$$

The type N leads present an interesting challenge, in that no failures were observed before the test was terminated at 100,000 cycles. (This is not an unusual situation in accelerated testing.) This problem is approached by assuming that the first failure will occur on the very next cycle after the test is terminated, and that the failure distribution is identical to that for type B samples, although displaced in time. This approach is illustrated in Figure 19-2, where the "test" data distribution is represented as a solid line with the "first failure" indicated by an X. The acceleration factor is calculated with Eq. 19-13, and the extrapolated distribution for use conditions is shown as the dashed line in Figure 19-2.

19.4 FAILURE DISTRIBUTION PARAMETERS AT DIFFERENT STRESS LEVELS

In the previous examples, failure data were obtained at only one accelerated stress level. It was assumed that the distribution parameters, such as the Weibull slope or the lognormal standard deviation, are the same at all stress levels, including operating stresses. In fact, this may not be a good assumption. The variation in distribution parameters at different stress levels is illustrated in this section by two examples.

Figure 19-3 shows Weibull plots of cumulative failures vs. temperature cycles for microcircuit accelerated life testing [8]. Individual data points are not shown. Each distribution represents a different combination of cycling range and microcircuit molding compound. (The specific conditions are not relevant to this discussion). The lines representing each of the distributions in Figure 19-3 are approximately parallel, indicating that the Weibull slopes are approximately equal. Figure 19-4 shows the results of plotting the same data as lognormal distributions. The lines are not parallel, indicating that if it is desired to represent these data as lognormal distributions, it must be assumed that their standard distributions are not equal. The author of the paper used this fact to conclude, correctly, that the data are best represented by Weibull distributions.

Figures 19-5 and 19-6 show lognormal plots of results for temperature-humidity testing of microcircuits from two different manufacturers at identical stress levels [9]. Figure 19-5 shows distributions obtained for various temperature-humidity conditions for manufacturer A. Since these lines are nearly parallel, it may be concluded that the standard deviations are nearly equal. Similar

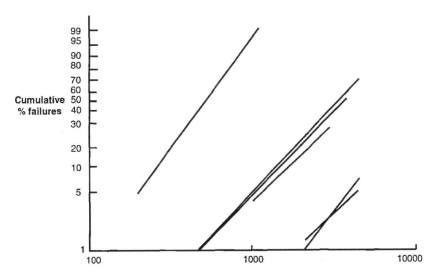

FIGURE 19-3 Weibull plot of the cumulative failure distributions of microcircuits with different molding compounds and temperature cycling ranges. They all have approximately the same Weibull slope. (From Ref. 8. © 1991 IEEE, used by permission.)

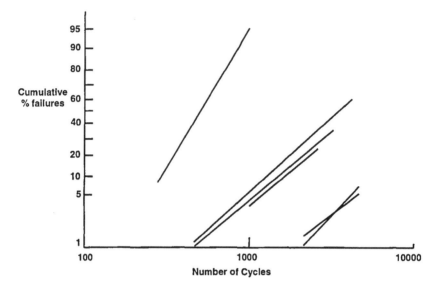

FIGURE 19-4 Lognormal plot of the same data shown in Figure 19-3. The different slopes indicate different standard deviations. (From Ref. 8. © 1991 IEEE, used by permission.)

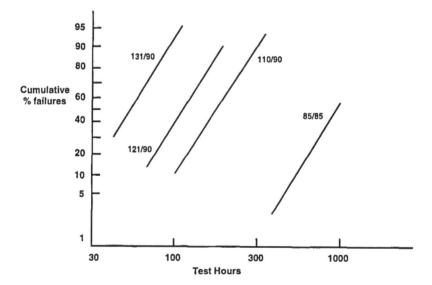

FIGURE 19-5 Lognormal plot temperature-humidity test data for microcircuits from manufacturer *A*. The standard deviations are equal for all test conditions. (From Ref. 9. © 1991 IEEE, used by permission.)

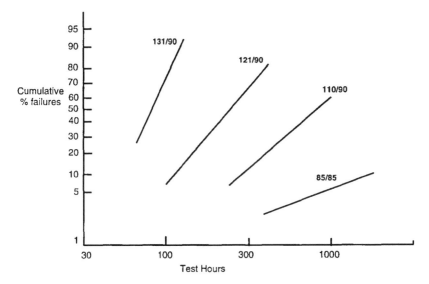

FIGURE 19-6   Lognormal plot of temperature-humidity test data for microcircuits from manufacturer *B*. Although the test conditions are the same as those for microcircuits from manufacturer *A* (Figure 19-5), the standard deviations are not the same for all test conditions. (From Ref. 9. © 1991 IEEE, used by permission.)

results for manufacturer *B*, shown in Figure 19-6, indicate that the lines are not parallel, and the standard deviations are not equal.

Discussion of the physical reasons for variations in the distribution parameters at different stress levels is beyond the scope of this text. These examples should, however, illustrate that there is no inherent failure distribution for a given set of samples under a given set of stress conditions. Furthermore, the parameters of a failure distribution under one set of stress conditions may not be the same for other levels of the same stresses. The choice of distribution and resulting parameters, therefore, requires judgment and engineering knowledge. The statistician can help in the manipulation of the data, but the engineer must be responsible for choosing models and drawing conclusions.

When possible, it is best to conduct accelerated tests at several different stress levels. This provides data about the shape of the distributions over a range of stresses and allows the experimenter to plot several points on the stress vs. time curve. In turn, this leads to more believable extrapolations to use conditions. The next example illustrates this point.

19.5 ACCELERATED LIFE TESTING OVER A RANGE OF STRESSES

Keller [10] investigated the reliability of external leads soldered to thin film hybrid circuits, in which the expected failure mechanism was bond strength degradation due to solid state diffusion of one metal into another. Tests were conducted by steady-state temperature aging of solder joint samples at 125, 150, and 175°C for times of up to 10,000 hours. The goal was to determine if the bonds would be reliable after 40 years of operation in telecommunications equipment at an operating temperature of 50°C.

The tests results yielded a series of distributions of bond strengths at times of 0, 0.3, 1, 3, 10, 30, 100, 300, 1000, 3000, and 10000 hours, for each of the three test temperatures. Means and standard deviations for each of the distributions are shown in Table 19-1. Each of the distributions represents approximately 60 data points, and some example normal distrubutions are shown in Figure 19-7. For the sake of clarity, the individual data points are not shown in this figure, although the original reference shows that they fit very well on the plotted lines. It should also be noted that these are distributions of mechanical bond strengths, not times to failure.

It is apparent from Table 19-1 and Figure 19-7 that the standard deviation of bond strength is higher at higher mean strengths than at lower mean strengths. Based on a least-squares curve fit of mean strength vs. standard deviation, the standard deviation is about 2.0 when the mean strength is 8 lb, and about 0.5 when the mean strength is 4 lb.

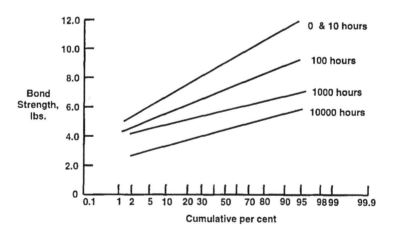

FIGURE 19-7   Bond strength distributions after aging at 150°C. (From Ref. 10. © 1979 IEEE, used by permission.)

TABLE **19-1** Mean and Standard Deviation Values of Bond Strength Distribution in High-Temperature Accelerated Life Testing

Test time, hours	125°C		150°C		175°C	
	Mean	Std. Dev.	Mean	Std. Dev.	Mean	Std. Dev.
0	9.0	1.8	9.0	1.8	9.0	1.8
0.3	8.4	1.9	9.0	2.2	5.5	0.8
1	9.6	1.8	9.0	2.4	6.0	1.2
3	9.6	2.2	10.0	1.9	5.9	0.9
10	9.3	1.7	8.9	1.9	5.6	0.9
30	9.3	2.1	7.8	2.0	5.5	0.8
100	8.7	2.0	7.1	1.2	5.8	1.4
300	8.8	2.0	6.2	1.2	5.2	0.8
1000	7.9	2.1	5.7	0.9	4.1	0.4
3000	6.7	1.5	5.0	1.0		
10000	5.6	1.1	4.3	0.6		

Source: From Ref. 10. © 1979 IEEE, used by permission.

Using the data from Table 19-1, it is possible to plot bond strength vs. time, as shown in Figure 19-8. (Note that this is not time to failure, since we have not yet defined a failure.) Any feature of the distribution, such as the lowest strength value, the 1% value, the mean, mean minus one standard deviation, etc., could have been chosen for this plot. Following the author's example, Figure 19-8 is a plot of mean strength vs. aging time. Since this plot will be used to obtain times to failure at the three aging temperatures, we must now define a failure. This could be any strength below which the equipment might not operate successfully, and the author chose a conservative value of 5.0 lb. The times to failure at each temperature were 315 hours at 175°C. 3,000 hours at 150°C, and 30,000 hours at 125°C. The latter value was obtained by extrapolation, but it is a safe extrapolation, since the curve is straight near this point, and the distance is short.

Since the failure mechanism is solid state diffusion, which is thermally activated, the appropriate acceleration model is the Arrhenius equation. If this equation applies, and we are dealing with a single failure mechanism, a plot of the log of time to failure vs. $1/T$ will be a straight line with the slope equal to the activation energy. Such a plot is shown in Figure 19-9. In this figure, the three times to failure are plotted, and the straight line drawn through them is extrapolated to the use temperature of 50°C. The activation energy is estimated at approximately 1.4 eV. (The activation energy can be calculated, or estimated graphically with Arrhenius plotting paper.) The estimated time for the mean bond

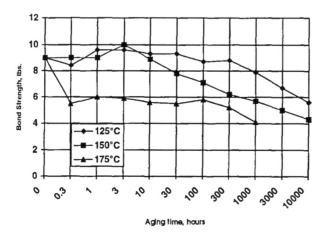

FIGURE 19-8 Plot of bond strength vs. time for three different aging temperatures. (From Ref. 10. © 1979 IEEE, used by permission.)

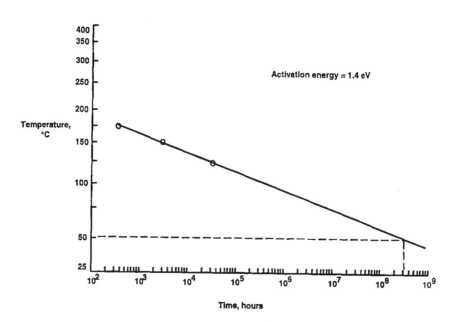

FIGURE 19-9 Arrhenius plot of bond strength aging data. (From Ref. 10. © 1979 IEEE, used by permission.)

strength to degrade to 5.0 lb. at 50°C is approximately 300 million hours, which equates to over 30,000 years.

The estimated life is well beyond that required by the product. However, the estimate is made by extrapolating results from a test of approximately 1 year's duration, and we are predicting what will happen far into the future. Obviously, no one will be able to check the accuracy of the lifetime estimate, but the potential for inaccuracy is great. A reasonable conclusion is that, even if the prediction is off by several orders of magnitude, the product will have an acceptable lifetime. In general, when reliability predictions like this are made, they should be made using very conservative assumptions. (Since many required lifetimes are in the 10–30 year range, it is easy to see that these predictions may be made with increasing boldness as the experimenter nears retirement age.)

19.6 ACCELERATED LIFE TESTING OF GALLIUM ARSENIDE TRANSISTORS WITH LOGNORMAL FAILURE DISTRIBUTIONS

Christou et al. [11] investigated the reliability of high-electron-mobility transistors (HEMTs) in high-temperature accelerated aging. HEMTs are relatively complicated devices, and a discussion of their functions and operating characteristics is beyond our scope.

The failure criteria for these transistors were considered to be a 20% degradation of transconductance and drain current. Often, the failure criteria for a given product are not well specified, and even may change from one application to another. In many cases, with a new product or technology, it is not possible to define a failure until after the reliability tests are complete, and an overview of the product capability is available. It is therefore always better to collect parametric data than simple pass-fail data from a reliability test, if possible.

In cases where failure criteria are not well established, the definition of a failure as a significant shift in a parameter is a good one, since it signals a major structural change in the unit under test.

The times-to-failure data obtained by Christou et al. for doped channel high-electron mobility transistors (DCHEMTs) are shown in Table 19-2 and plotted in Figure 19-10. The plots of Figure 19-10 show that the data can be fitted to a lognormal distribution, which is consistent with this type of failure mechanism. The lines drawn through the data points all have the same slope, indicating a constant standard deviation σ of approximately 0.6. The fit of the data to these lines is less than perfect, but it is a common practice in accelerated tests of this type to use a single slope which best fits all the individual distributions, rather than to fit curves with slightly different slopes to different distributions. Some engineering judgment is required in this area.

TABLE **19-2** Time-to-Failure Data for Accelerated
Life Testing DCHEMTs

Temperature	180°C	195°C	210°C
Time to failure, hours	250	190	105
	420	200	110
	1000	330	120
	1300	330	130
	1300	400	210
	1300	400	230
	1300	420	290
	1500	600	300
	1500	800	320
	1600		400

Source: From Ref. 11. © 1991 IEEE, used by permission.

The mean times to failure (MTTF) are shown on the Arrhenius plot of
Figure 19-11. The estimated activation energy is 0.85 eV, and the estimated
MTTF is 40,000 hours at an operating temperature of 110°C. Since the authors
assumed that the standard deviation was the same, 0.6, at all temperatures, then
the following MTTF could be predicted for various points on the lognormal failure distribution curve at 110°C:

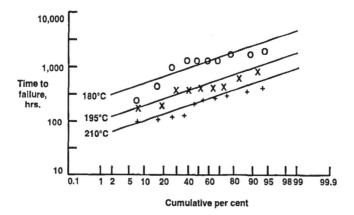

FIGURE **19-10** Lognormal plot of time to failure for accelerated life testing of
DCHEMTs at three different temperatures. (From Ref. 11. © 1991 IEEE, used by permission.)

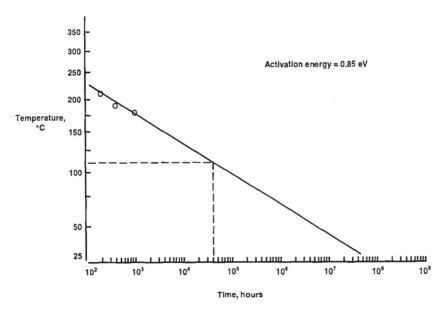

FIGURE 19-11   Arrhenius plot of MTTF vs. temperature for DCHEMTs in accelerated life testing. (From Ref. 11. © 1991 IEEE, used by permission.)

μ (t_{50}):	40,000 hours
$\mu - 1\sigma$ (t_{16}):	20,000 hours
$\mu - 2\sigma$ $(t_{2.3})$:	11,000 hours
$\mu - 3\sigma$ $(t_{0.135})$:	5500 hours

Thus, if a lifetime is defined as the point where 1.35 failures have occurred for every 1000 transistors (μ-3σ), then the lifetime is approximately 5500 hours at 110°C.

19.7   ACCELERATED LIFE TESTING WHEN MORE THAN ONE FAILURE MECHANISM OCCURS

In all the previous examples, only single-failure mechanisms have been encountered. In reality, many mechanisms are possible, and usually more than one occurs in any given test. Sometimes, one failure mechanism can occur, and the unit under test can remain operative with respect to other mechanisms of interest. More often, one failure mechanism causes the sample to fail completely, and it

TABLE 19-3 Failures in Temperature Cycling Between −40 and+130°C in the Beam Lead Reliability Test

Interval i	Cumulative cycles	Samples entering interval, n	Failure modes						Removals
			Total	A-B	B-1	B	B-2	LF-C	
1	105	107	20	17	0	0	6	0	0
2	205	87	12	10	0	0	3	1	0
3	305	75	7	7	0	0	1	0	0
4	405	68	2	0	2	0	0	0	0
5	510	66	5	3	1	0	3	2	0
6	1000	61	26	3	16	5	5	19	0
7	1490	35	25	1	25	8	6	20	0
8	1980	10	3	0	3	1	2	3	1
9	2470	6	4	4	4	1	1	2	2

Source: From Ref. 12. © 1978 IEEE, used by permission.

must be treated as a removal with respect to other mechanisms.[†] If, in the analysis of data for one failure mode, samples are removed because of failure by other failure modes, the results are said to be *multiply censored*.

In this section, we consider an example of accelerated life testing of beam-leaded microcircuits at two different temperature cycling ranges, using Ref. 12 as an example. In this test, each test unit had 16 leads, each of which could fail by any of 5 different failure mechanisms. Thus, there could be more than one failure per microcircuit sample.

The samples were divided into two groups. Group I had 107 microcircuits, and the temperature range was −40 to +130°C, for a ΔT of 170°C. Group II had 106 samples, the temperature range was +25 to +125°C, and the ΔT was 100°C. The samples were not monitored continuously during the test, but were removed for testing at various intervals. At a given test point, those microcircuits which had at least one failed lead (by whatever mechanism) were removed from the experiment, and the remainder were put back on test. At some test points, more than one lead on a given microcircuit had failed, and the failure mechanisms were not necessarily the same. The results of this experiment were a series of failure distributions representing each of the five failure mechanisms.

The raw data from this experiment are shown in Tables 19-3 and 19-4.

Analysis of data from this type of test is rather complicated. Nelson [13, 14] presents methods for this analysis when the exact time of each failure is

---

[†] See the discussion of censoring in Chapter 14.

TABLE 19-4 Failures in Temperature Cycling Between $+25$ and $+125°C$ in the Beam Lead Reliability Test

Interval i	Cumulative cycles	Samples entering interval n	Failure modes						Removals
			Total	A-B	B-1	B	B-2	LF-C	
1	100	106	1	0	0	0	1	0	0
2	200	105	2	2	0	0	0	0	0
3	300	103	0	0	0	0	0	0	0
4	400	103	0	0	0	0	0	0	0
5	500	103	1	1	0	0	0	0	0
6	1000	102	1	1	0	0	0	0	0
7	1500	101	2	1	0	0	2	0	0
8	2000	99	3	2	1	0	1	1	0
9	4000	96	20	0	19	2	13	0	11
10	8000	65	37	1	36	15	20	0	0
11	10,000	28	10	1	10	4	5	0	0
12	20,000	18	3	0	3	3	2	1	15

Source: From Ref. 12. © 1978 IEEE, used by permission.

known, but this treatment must be modified for the interval testing used in the present example [12, 15]. A detailed derivation of the analytical approach is beyond the scope of this book, but it can be summarized here. The cumulative failure distribution $F(t)$, can be calculated for each mechanism by

$$F(t) = 1 - \exp(-H(t)) \qquad (19\text{-}21)$$

where $H(t)$ is the cumulative hazard function. $H(t)$ is given by

$$H(t) = \sum_{i=1}^{j} h(t)\,\Delta t \qquad (19\text{-}22)$$

where $h(t)$ is the hazard rate for the ith time interval (out of a total of j intervals). $h(t)\,\Delta t$ is calculated by

$$h(t)\Delta t = \frac{1}{n} + \frac{1}{n-1} + \frac{1}{n-2} + \cdots + \frac{1}{n-(m-1)} \qquad (19\text{-}23)$$

where n is the number of samples entering the interval and m is the number of failures in that interval, due to the mechanism of interest. To illustrate this method, Table 19-5 shows the results of calculations for total and *A-B* type failures for the data in Table 19-3. In the first interval of 105 cumulative cycles, n is 107, and m for total failures is 20. Using Eq. (19-23), $h(t)\,\Delta t$ is calculated as

$$h(t)\,\Delta t = \frac{1}{107} + \frac{1}{106} + \frac{1}{105} + \cdots + \frac{1}{107 - (20 - 1)} = 0.206 \qquad (19\text{-}24)$$

For the second interval (205 cumulative cycles), n is 87 and m is 12. $h(t)\,\Delta t$ is then

$$h(t)\,\Delta t = \frac{1}{87} + \frac{1}{86} + \frac{1}{85} + \cdots + \frac{1}{87 - (12 - 1)} = 0.147 \qquad (19\text{-}25)$$

Using Eq. (19-22), $H(t)$ at the end of the second interval is equal to the sum of 0.206 and 0.147, or 0. 353. $F(t)$ is calculated for each interval using Eq. 19-21.

For *A-B* type failures, the n's are the same as for the total failures, but the m's are as shown in Table 19-5. The reader is encouraged to perform a few of these calculations for other intervals.

The resulting cumulative failure distribution data for both temperature cycling ranges are shown in Tables 19-6 and 19-7. These data are plotted as lognormal cumulative distributions in Figures 19-12 through 19-17. The plots for total failures in Figure 19-12 are actually the sums of plots for all types of failures, and it is not surprising that they have some inflection points. The lines drawn through the points in Figures 19-13 through 19-17 are parallel for the two temperature ranges, indicating that the standard deviations of the distributions are equal at different temperatures for a given failure mode. For *B*-2 type failures (Figure 19-16), there appears to be a point of inflection for both curves. The author of the paper attributed this to the presence of two different failure mechanisms causing the same failure type.

Acceleration factors can be calculated from the distributions shown in Figures 19-14 through 19-17, by comparing the times to achieve a given percent failure for the two ΔT's (170°C and 100°C). These calculated acceleration factors for the various failure types are:

Failure type	Acceleration factor
A-B	580
B-1	6
B	6
B-2	5 (upper portion)
LF-C	38

TABLE 19-5 Calculation of Cumulative Failure Distributions for Total and A-B Type Failures for the Data in Table 19-3

Interval i	Cumulative cycles	Samples entering interval n	Total failures				A-B type failures			
			m	$h(t)\Delta t$	$H(t)$	$F(t)$	m	$h(t)\Delta$	$tH(t)$	$F(t)$
1	105	107	20	0.206	0.206	0.186	17	0.172	0.172	0.158
2	205	87	12	0.147	0.353	0.298	10	0.121	0.293	0.254
3	305	75	7	0.097	0.451	0.363	7	0.097	0.391	0.323
4	405	68	2	0.030	0.480	0.381	0	0	0.391	
5	510	66	5	0.078	0.558	0.428	3	0.046	0.437	0.354
6	1000	61	26	0.549	1.108	0.670	3	0.050	0.487	0.386
7	1490	35	25	1.218	2.326	0.902	1	0.029	0.516	0.402
8	1980	10	3	0.336	2.662	0.930	0	0	0.516	
9	2470	6	4	0.950	3.612	0.973	4	0.095	1.466	0.769

TABLE 19-6 Failure Distributions by Failure Mode for Temperature Cycling Between −40 and +130°C

Interval i	Cumulative cycles	Cumulative percent failed by failure mode					
		Total	A-B	B-1	B	B-2	LF-C
1	105	18.6	15.8			5.6	
2	205	29.8	25.4			8.8	1.1
3	305	36.3	32.3			10.0	
4	405	38.1		2.9			
5	510	42.8	35.4	4.4		14.1	4.1
6	1000	67.0	38.6	29.3	8.1	21.1	33.7
7	1490	90.2	40.3	79.1	28.8	34.4	71.1
8	1980	93.0		85.0	35.6	46.9	79.3
9	2470	97.3	76.9	94.2	46.9	55.0	85.7

Source: From Ref. 12. © 1978 IEEE, used by permission.

TABLE 19-7 Failure Distributions by Failure Mode for Temperature Cycling Between +25 and +125°C

Interval i	Cumulative cycles	Cumulative percent failed by failure mode					
		Total	A-B	B-1	B	B-2	LF-C
1	100	0.9			0.9		
2	200	2.8	1.9				
3	300						
4	400						
5	500	3.8	2.8				
6	1000	4.7	3.8				
7	1500	6.6	4.7			2.9	
8	2000	9.4	6.7	1.0		3.9	1.0
9	4000	28.2		20.5	2.1	16.8	
10	8000	68.7	8.1	64.2	24.5	42.2	
11	10000	79.7	11.3	76.8	35.1	52.4	
12	20000	83.0		80.5	45.6	57.5	6.4

Source: From Ref. 12. © 1978 IEEE, used by permission.

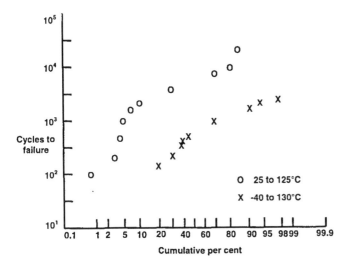

FIGURE 19-12   Lognormal plot of cycles to failure for total failures in accelerated life testing of beam leads at two different temperature cycling ranges. (From Ref. 12. © 1978 IEEE, used by permission.)

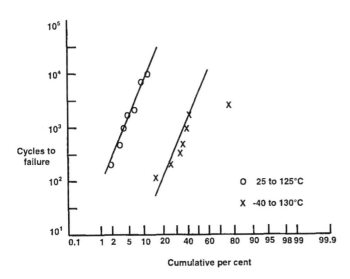

FIGURE 19-13   Lognormal plot of cycles to failure for *A-B* type failures in accelerated reliability testing of beam leads at two different temperature cycling ranges. (From Ref. 12. © 1978 IEEE, used by permission.)

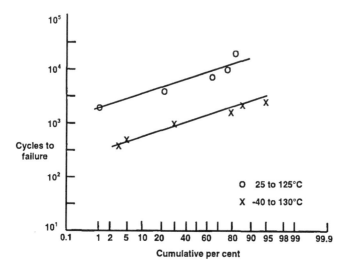

FIGURE 19-14 Lognormal plot of cycles to failure for B-1 type failures in accelerated reliability testing of beam leads at two different temperature cycling ranges. (From Ref. 12. © 1978 IEEE, used by permission.)

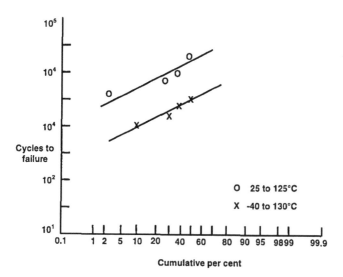

FIGURE 19-15 Lognormal plot of cycles to failure for *B* type failures in accelerated reliability testing of beam leads at two different temperature cycling ranges. (From Ref. 12. © 1978 IEEE, used by permission.)

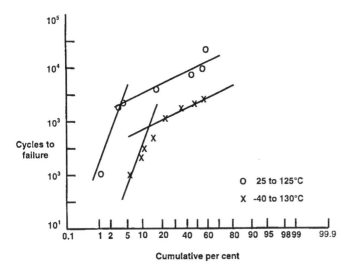

FIGURE 19-16 Lognormal plot of cycles to failure for *B*-2 type failures in accelerated reliability testing of beam leads at two different temperature cycling ranges. (From Ref. 12. © 1978 IEEE, used by permission.)

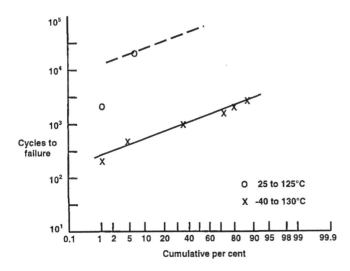

FIGURE 19-17 Lognormal plot of cycles to failure for LF-C type failures in accelerated reliability testing of beam leads at two different temperature cycling ranges. (From Ref. 12. © 1978 IEEE, used by permission.)

There is obviously a great difference in acceleration factors among the different possible failure mechanisms. This fact must be considered when setting up an accelerated test, because it is possible to simulate many years of life with respect to one failure mechanism while barely exercising the samples with respect to another failure mechanism which is equally likely in use conditions.

19.8 DEGRADATION TESTING

In the examples discussed above, the definition of a failure is somewhat subjective. This is due to the fact that many failure mechanisms in accelerated life testing occur gradually; that is, the unit under test experiences structural changes that cause it to degrade gradually over time, rather than to fail catastrophically at a given instant. This type of testing is called *degradation testing*, and in such cases, it is necessary for the experimenter to have sufficient knowledge of the product being tested, and of its expected performance in service, to be able to define a level of degradation that is equivalent to failure in service. In some cases, the degradation proceeds so slowly that none of the samples experience "failure" during the time allotted for the test. In these cases, it is necessary to measure the degradation of a given property or output of the unit under test, and to be able to extrapolate the time for it to achieve "failure" status. This is not always easily done. Lu and Meeker [16] provide a theoretical basis for degradation testing, and illustrate it with examples. The use of Design of Experiments in degradation testing is illustrated by Tseng et al. [17].

19.9 SUMMARY

Interpreting accelerated reliability test data is one of the most difficult tasks of reliability work, because it requires knowledge of all the major reliability tasks: reliability theory, accelerated test design, failure acceleration modeling, and data interpretation. It also requires a thorough understanding of the product or process involved. It is also one of the most powerful tasks of reliability work, since the results can be applied across a wide range of products, materials, processes, and operating conditions.

Accelerated reliability testing often is considered too expensive or time consuming, and is therefore rejected as an option. The reason for this rejection is often that the engineer or company does not have a thorough understanding of the purpose of testing, or of how to realize benefits from the testing that will justify the cost. The result is that testing is either not done at all, or some handbook test is applied in a cookbook fashion to satisfy contractual requirements.

To be most useful, accelerated testing must be conducted with a firm understanding of its purpose, and with a knowledge of how to apply the results. Various types of accelerated tests are routinely conducted on some products, but there is seldom a comprehensive view of the reasons for such tests, and of how they can be integrated into an overall program of reliability testing and reliability improvement. That is the subject of Chapter 20.

REFERENCES

1. L. Condra, G. Wenzel, and M. Pecht, Reliability evaluation of simple logic microcircuits in surface mount plastic packages, Proceedings of the ASME Winter Meeting, New Orleans, November 1993.

2. R. C. Blish II and P. R. Vaney, Failure rate model for thin film cracking in plastic IC's, Proceedings of the International Reliability Physics Symposium, IEEE, pp. 22–29, 1991.

3. W. Engelmaier and A. I. Attarwala, Surface mount attachment reliability of clip leaded ceramic chip carriers on FR-4 circuit boards, IEEE Transactions on Components, Hybrids, and Manufacturing Technology 12(2):284–296, 1989.

4. W. Engelmaier, Surface mount solder joint long term reliability: Design, Testing, Prediction, Soldering and Surface Mount Technology, no. 1, pp. 14–22, 1989.

5. W. Engelmaier, Is present day accelerated cycling adequate for surface mount attachment reliability evaluation? IPC-TP-653, Institute for Interconnecting and Packaging Electronic Circuits, 1986.

6. J. P. Clech, W. Engelmaier, R. W. Kotlowitz, and J. A. Augis, Surface mount solder attachment reliability figures of merit "Design for Reliability" Tools, Proceedings of the Surface Mounting and Advanced Related Technology (SMART) V Conference, 1989.

7. J. W. Evans, H. Chernikoff, and W. Engelmaier, SMT reliability for space flight applications, Surface Mount Technology, pp. 24–31, 1990.

8. R. L. Zelenka, A reliability model for interlayer dielectric cracking during temperature cycling, Proceedings of the 29th International Reliability Physics Symposium, IEEE, pp. 30–34, 1991.

9. R. P. Merrett, J. P. Bryant, and R. Studd, An appraisal of high temperature humidity stress tests for assessing plastic encapsulated semiconductor components, Proceedings of the 21st International Reliability Physics Symposium, IEEE, pp. 73–81, 1983.

10. H. N. Keller, Reliability of clip-on terminals soldered to Ta-Ta$_2$N-NiCr-Pd-Au thin films, Proceedings of the 29th Electronic Components Conference, IEEE, pp. 99–105, 1979.

11. A. Christou, J. M. Hu, and W. T. Anderson, Reliability of InGaAs HEMTs on GaAs substrates, Proceedings of the 29th International Reliability Physics Symposium, IEEE, pp. 200–205, 1991.

12. G. A. Dodson, Analysis of accelerated temperature cycle test data containing different failure modes, Proceedings of the 16th International Reliability Physics Symposium, IEEE, pp. 238–246, 1979.

13. W. Nelson, Theory and application of hazard plotting for censored failure data, Technometrics 14(4):945–966, 1972.

14. W. Nelson, How to Analyze Reliability Data, American Society for Quality Control, Milwaukee, pp. 39–45, 1983.

15. A. F. Siegel, University of Washington, Seattle, private communication, 1992.

16. C. J. Lu and W. G. Meeker, Using degradation measures to estimate a time-to-failure distribution, Technometrics 35(2):161–174, May 1993.

17. S. T. Tseng, M. S. Hamada, and C. H. Chiao, Using degradation data from a fractional factorial experiment to improve fluorescent lamp reliability, IIQP Research Report RR-94–05, University of Waterloo, April 1994.

Chapter 20

Developing an Integrated Reliability Test Program

20.1 INTRODUCTION

Chapters 17 to 19 include discussions of the mechanics of designing and conducting accelerated reliability tests, the types of acceleration models to be used, and methods to interpret results from accelerated tests. In this chapter, we discuss ways to incorporate a comprehensive accelerated reliability testing program into the product design and development cycle.

Accelerated testing is one of the most misunderstood topics in the reliability discipline. Most of the confusion results from misconceptions about *why* and *when* the various types of accelerated tests are conducted, rather than about *how* to conduct them. Indeed, the equipment and processes are similar for most accelerated reliability tests, and this in itself contributes to confusion, because those new to reliability engineering have difficulty distinguishing among the various types of tests, such as step stress, environmental stress screening, HALT, STRIFE, HASS, HASA, qualification test, design verification test, periodic requalification, and a host of others.

The *how* and the *why* of accelerated reliability testing are best understood in relationship to the product design and development cycle. Although there is no unanimity in the literature regarding all the various types of tests, most references, e.g., Ref. 1, locate them on a flowchart of the design-development process.

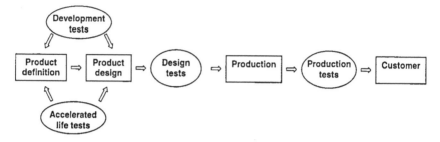

FIGURE 20-1   Flowchart showing where the various types of testing fit into the product design-development-production process.

For this discussion, we will use the somewhat oversimplified flowchart of Figure 20-1.

The on-line processes are shaded in Figure 20-1. On-line processes are those which are included in the budget and schedule for a specific product. As such, they must be justified solely on the basis of their value to a given program, and they are done under the control of the product manager. Off-line processes, which are shown unshaded in Figure 20-1, are not within the scope of a single program, and therefore may be justified on the basis of their value to a range of programs.

Reliability tests are shown as ellipses in Figure 20-1. The four major categories of reliability test are (1) accelerated life tests, (2) development tests, (3) design tests, and (4) production tests. Each is discussed in the following sections. It must be borne in mind that the distinctions among these types of test are not always exact, and that there are different viewpoints about their definitions.

The four reliability test categories are associated with different types of failure, as shown by their relationships to the bathtub curve in Figure 20-2. Accel-

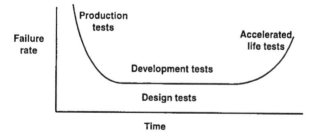

FIGURE 20-2   The bathtub curve showing the main areas of applicability of the various types of reliability tests.

erated life testing is usually associated with detecting wearout failures, and with demonstrating that the lifetime of a given test unit is of sufficient length to meet the reliability goals of the product. Production tests, e.g., environmental stress screening (ESS), are usually associated with ensuring that infant mortality failures due to manufacturing defects are minimized. Design and development tests usually are associated with ensuring that the design of the unit under test is sufficiently robust with respect to expected operating and environmental loads.*

20.2 ACCELERATED LIFE TESTS

The purpose of an accelerated life test is to evaluate the expected lifetimes of components, materials, or subassemblies.† Usually, the following features characterize accelerated life tests.

- They are conducted off-line.
- They are conducted on samples obtained specifically for the purpose of accelerated life testing, since the test is almost always destructive.
- The test units are usually small and homogeneous, so that only a single failure mechanism is expected. Test units are usually individual components, material samples, or simple subassemblies.
- Most failures are unrepairable.
- They are intended to evaluate units under test that are designed and manufactured without flaws; therefore samples may be selected to ensure that they contain no manufacturing defects.

Since accelerated life tests can be expensive and time consuming, they are usually conducted on components, materials, and subassemblies that are used in a variety of products, so that the costs can be spread over the highest possible product volume.

The result of an accelerated life test is an estimate of the lifetime capability of the component, material, or subassembly, in other words, how long it can be expected to last before wearout occurs. Figure 20-3 shows a probability density function (pdf) of a normally distributed failure mechanism for a typical accelerated life test. The result of this test is the time required for some portion of the sample to fail. Typically, this is a very small portion of the total sample, such

---

* Although these associations are usually correct, there are situations in which they do not apply. For example, sometimes a design or development test may be conducted to ensure that infant mortality failures are minimized, or that the lifetime of the product before wearout is sufficiently long.
† Most of the examples in Chapters 17 and 19 are accelerated life tests.

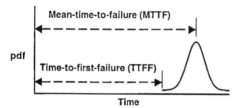

FIGURE **20-3** Probability density function for a normally distributed failure mechanism in accelerated life testing, showing time to first failure (TTFF) and mean time to failure (MTTF).

as time to first failure (TTFF), since the test engineer usually is interested in reporting the expected minimum lifetime of the product.

The mean time to failure (MTTF) for an unrepairable item is the t_{50} point. Usually, only a single failure mechanism is considered relevant in an accelerated life test, and random failures associated with defects in the unit under test, or with handling damage, or with test equipment failure, are not counted as failures. If a failure rate is to be estimated from an accelerated life test, it is usually done for the period prior to the first failure.

Accelerated life testing is perhaps the best-understood category of accelerated reliability test. Accelerated life tests are common in the electronic components industry, and the results of these tests are used by equipment manufacturers to estimate the reliability of their products. It is less common to perform accelerated reliability tests on materials or subassemblies of higher-level products; however, they should be evaluated along with the components. As discussed briefly in Chapter 11, Design of Experiments can be used for efficient and effective evaluation of components, materials, and subassemblies in a single experiment. As a bonus, equipment design features can be evaluated in the same experiment.

One of the criticisms of the probabilistic approach to reliability is that it does not account for interactions among components, materials, and processes. The failure rate for a component is considered to be the same for a given component regardless of the way it is used in the design, or of the process used to assemble it into the final product. Even if the same process is used by two different assemblers, their methods of implementation can cause differences. By combining all of these factors into a designed experiment in an accelerated life test, the impacts of their interactions on reliability can be evaluated.

The advantages of off-line accelerated life tests are: (1) The test conditions and samples are well-understood and controlled; (2) the tests can be designed and conducted to evaluate specific samples, environments, and failure mechanisms, with minimum interference from extraneous events; and (3) since the tests are conducted off-line, there is less temptation to compromise the results by taking shortcuts to meet production schedules.

The disadvantages of off-line accelerated life tests are: (1) They are relatively expensive and time consuming; and (2) extrapolation of the results and conclusions to use conditions is, at best, an inexact science. Because of these limitations, it is usually unrealistic to perform a separate set of accelerated life tests for each program.

The most realistic way to plan and conduct reliability tests is to integrate them with the company's marketing, engineering, and manufacturing plans. In that way, the tests can be conducted independently of product development schedules, results can be used across a range of products, and their costs can be spread accordingly.

20.3   DEVELOPMENT TESTS

Development tests are conducted for the following purposes:

- To find weaknesses in the design of the product
- To ensure that the product goes into service with "mature" reliability
- To quantify, to the extent possible, the design margins with repect to operating and environmental stresses

Development tests are shown as off-line tests in Figure 20-1; however, it is not unusual for them to be conducted on-line. Usually, they are conducted on end item products, or on major subassemblies of the end item. They should be conducted as soon as possible after the equipment or major subassembly has been defined, so that test results can be used to evaluate the design and, if necessary, make changes before committing resources to production. The test units should be as near to the final configuration as is resonably possible.

The goal of a development test is to provide every opportunity for a relevant failure to occur. In that sense, failures in test are desirable. It is not reasonable to say that a unit that did not fail in development testing has "passed" the test if no failures are observed; rather, it is relevant to say that a development test was successful if relevant failures were observed, and changes made to reduce their likelihood when the product is placed in service. If a development test is completely successful, all the potential failures due to poor design will be detected and eliminated prior to the product being placed into service.

It is highly desirable, although not always possible, to quantify the margins available in the design. In that sense, development tests should be conducted at stress levels as high as possible, even if those levels are beyond the specification range of the product. Figure 20-4 shows some of the stress limits that may be considered. The *specification limits* usually are imposed by the user of the product, and represent the maximum operating and environmental stress ranges to

Stress

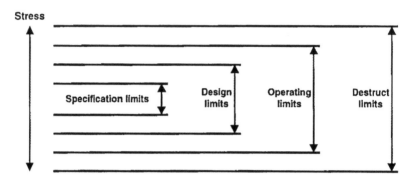

FIGURE 20-4 Stress limits.

which the product will be exposed. The *design limits* are wider than the specification limits. They are the stress levels which the designer uses to ensure that the product will operate successfully in the narrower specification limits. The *operating limits* are wider than the design limits, and represent the stress levels at which the product will survive, even if it does not function properly. The *destruct limits* are those at which the product will fail destructively or irreversibly.

Not all products have distinct specification, design, operating, and destruct limits. For many mechanical products, there is often only a small difference between the specification and the destruct limits. On the other hand, for electronic products, the limits often are widely separated. Ideally, a good development test will quantify all of these levels; however, that is not always realistic.

Development testing is complicated because it is difficult to agree upon the types of tests to be performed, and upon how to interpret the results. It is further complicated by the fact that products with even moderate degrees of complexity are quite inhomogeneous with respect to failure mechanisms, and with respect to acceleration factors for those mechanisms. For example, an electronics box might contain a transformer with insulation which degrades above 150°C, but with other components whose acceleration factors require testing above 200°C.

Upadhyayula and Dasgupta [2] list five steps for an effective accelerated development test:

1. *Physics-of-failure based virtual qualification.* Use some type of structured analytical process to determine the sites, mechanisms, and modes of the failures that can be expected in the design.
2. *Accelerated stress planning and development.* Determine the types, levels, and durations of the loads to be applied in testing, which are most likely to precipitate the failures described in step 1. If more than one load type is to be applied, the sequence of application, or other type

of combination, must be determined. The test units must be defined, as well as the test equipment, fixturing, and setup.

3. *Test vehicle characterization.* Conduct preliminary tests to determine the behavior of the test units in the proposed tests, in order to characterize the types of failures that will occur, and the types of loads that will precipitate them. Make modifications as necessary.

4. *Conduct the tests.* Expose the test units to the defined loads for the defined durations, and in the defined sequences and combinations.

5. *Interpret results.* Observe and analyze all failures to determine their mechanisms. Separate relevant from nonrelevant failures. Apply acceleration models if necessary, and assess whether or not the test results indicate that the design is likely to have acceptable reliability in the expected application.

Stadterman et al. [3], suggested a similar set of steps, and illustrated their application to vibration testing of electronic assemblies. Hu et al. [4] describe the vital role of failure mechanism identification in accelerated testing.

20.3.1   Reliability Demonstration Tests

One type of development test is the reliability demonstration test, in which samples are exposed to defined steady-state environmental conditions for specified times. If the sample survives intact for a sufficiently long time period, the product is considered reliable. The stresses are usually defined on the basis of expected operating and environmental conditions and can be quite complicated. This type of test is usually an example of an *elephant test,* in that it is designed to stress the product severely, and if the product survives, it must be reliable.

The goal of most reliability demonstration tests is for the product to pass the test, and thus demonstrate "acceptable" reliability. The challenge is to apply stresses high enough to convince those involved that a real reliability test is being conducted, but not high enough to cause premature failures. Even after the product has passed this type of test, several questions remain: (1) Were the applied stresses high enough to test the product adequately? (2) If they had been just a little higher, would the product have failed? (3) Since the stress combination was complicated, can we really know the acceleration factors? (4) Were the most likely failure mechanisms tested adequately? (5) Finally, but probably most relevant, what is the real reliability of the product?

20.3.2   Step Stress Tests

Step stress testing is gaining popularity as a design test. Its goal is to produce failures in a short time in order to determine the likely failure mechanisms, so that design improvements can be made before production begins [5, 6]. It can provide information about design margins with respect to operating and environmental stresses, and thus can evaluate the robustness of the design. It is performed on a small sample of the near-final product or a subassembly thereof. In some specialized instances, step stress testing is known by various other names, such as STRIFE (STRess-lIFE), HALT (Highly Accelerated Life Testing), and RET (reliability enhancement testing).

Step stress tests are conducted by exposing the units under tests to relatively low levels of stress, and then increasing those levels in a controlled, stepwise manner until at least one of the following occurs:

- Stress levels are reached that are significantly higher than those expected in service.
- All the test units fail irreversibly or unrepairably.
- Irrelevant failures begin to occur or dominate as new failure mechanisms become evident at higher stress levels. Irrelevant failures are those which are not associated with the design of the test unit, such as equipment failure, handling damage, or defects in the production of the test unit.

Figure 20-5 shows a schematic diagram of a generalized form of a step stress test. The first step in setting up such a test is to determine all the operating and environmental stresses that may cause failures. Three such stresses, designated S_1, S_2, and S_3, are shown in Figure 20-5. (Any number of stresses could be used, but only three are considered here for ease of presentation.) The stresses could be factors such as temperature, moisture, voltage, vibration, duty cycle, etc. The smaller box in Figure 20-5 represents the upper limits of stress, obtained from the specification limits of the product, with the outer corner of the box representing the worst-case combination of stresses. This is usually the limit to which the product is designed. The larger box represents a combination of stresses beyond

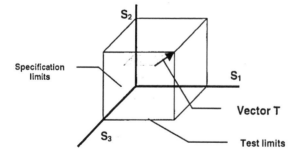

FIGURE 20-5   Diagram showing the general approach to step stress testing.

the specification limits, and it is this combination which is tested in step stress testing. The vector **T**, drawn from the outer corner of the smaller box to the outer corner of the larger box, describes the path through which the stresses are increased in stepwise fashion.

The vector **T** is shown in greater detail in Figure 20-6, where the units of stress are increments in the combination of stresses selected for testing. Traditionally, all steps are of equal time length, although there are different opinions about this. The time lengths may be as short as a few minutes, and are rarely longer than 24 hours. The first step is usually at or below the specification limits. It is desirable to apply all of the stresses, S_1, S_2, and S_3, simultaneously, but they may be applied sequentially. After the first step, the failed items are removed and analyzed. In some cases, a design error, defective part, or manufacturing anomaly is noted, and corrections are made before proceeding further. Most of the time, however, no failures occur at lower stress levels. The remaining items are then stressed at the next level (step 2) for the specified length of time. This process is repeated at successively higher stress levels until one of the criteria listed above for stopping the test is encountered.

Two types of analyses should be conducted on step stress test results: (1) physical analysis of the failed test units and (2) reliability analysis of the failure data. The first, physical analysis, is clearly an engineering function. All failures must be analyzed, and their exact causes at the structural level must be determined. The type of stress, and its level, which actuates each failure mechanism can then be identified.

Based on the results of the physical analysis, it is usually possible to identify causes of failures which may limit the reliability of the product. Changes in the design, manufacturing processes, materials, or components can be made to improve reliability. Less often, it is possible or necessary to impose limits on the

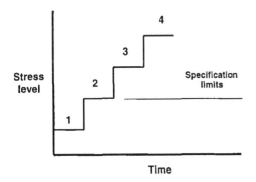

FIGURE 20-6 Diagram of the vector **T** from Figure 20-5, showing stepwise increases.

operation or operating environment of the product to assure reliability. Perhaps the major benefit of this type of analysis is the level of understanding of the product which is developed by design and manufacturing personnel. This understanding often makes it possible to improve the product, and subsequent generations of it, by means which might not be documented formally, but which are quite effective. For this reason, design and production personnel should be involved in the development test planning, conduction, and analysis; and the test should not be relegated completely to reliability personnel or others with no product responsibility.

The second type of analysis, reliability or failure rate analysis, is quite difficult and even impossible if sufficient data are not collected. This type of analysis is directed toward determining the failure rate distribution, e.g., lognormal, Weibull, etc., and its important parameters, for each failure mechanism. It is illustrated by Figures 20-7 and 20-8.

Figure 20-7 shows plots of generalized cumulative distribution functions (cdf) for failures generated by testing at four different stress levels, such that $s_1 < s_2 < s_3 < s_4$ (note that these small s's represent steps, not individual stresses, as represented by the large S's in Figure 20-5). The failure rate curves are steeper for higher stress levels, meaning that higher percentages of test units fail in shorter times. In step stress testing, the samples are held for a time at s_1, during which failures occur according to the cdf for s_1, as shown by the bold portion of the s_1 cdf. At the conclusion of step 1, a percentage of the samples equal to F_1 will have failed. After step 1 is completed, the samples are stressed for the same length of time at s_2, during which failures occur according to the s_2 cdf, as indicated by the bold portion of that cdf. At the end of this step, the cumulative percentage of failures (the sum of steps 1 and 2) is equal to F_2. The process is repeated

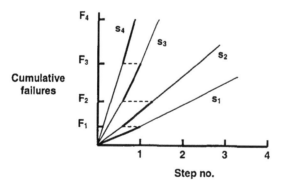

FIGURE 20-7   Schematic representation of the segments of the cumulative distribution curve produced by step stress testing.

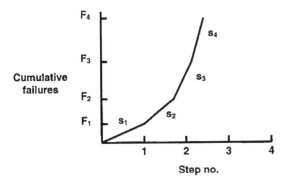

FIGURE 20-8 Cumulative failure distribution composed of segments from step stress testing.

through four steps, after which the total cumulative failure percentage is equal to F_4.

The bold portions of the cdf's from Figure 20-7 can then be "pieced together" to form the cumulative failure distribution for the entire step stress process, as shown in Figure 20-8.

This explanation illustrates the underlying theory of analyzing results from step stress testing. In practice, we work backwards. The data collected from a step stress test should have a cumulative distribution like that of Figure 20-8. By plotting the failure data from the various steps, and by knowing the acceleration factors for the applied stresses, we should be able to construct the failure rate at any stress level, such as those shown in Figure 20-7.

If enough data are available (from step stress and other accelerated tests, such as ESS and accelerated life tests) the failure rate distribution for each failure mechanism can be determined. This allows us to make detailed and accurate predictions about the reliability of the product. This type of conclusive information is difficult to collect and analyze, and it should be considered a bonus, not a goal, of step stress testing. More detailed information on the setup and analysis of step stress tests is contained in Refs. 7–14.

An example of a step stress test is shown in Table 20-1, for a power supply. In this example, step 2 represents the specification limits. Five samples were tested, and the test time was 24 hours at each step. When the samples were turned on at step 1, all of them failed. The failures were due to a resistor with a value specified too low, and to some defective capacitors. After the resistors were replaced with ones of higher value, and the defective capacitors were replaced, all samples survived through step 4. Total test time was less than 2 weeks, including analysis and repairs.

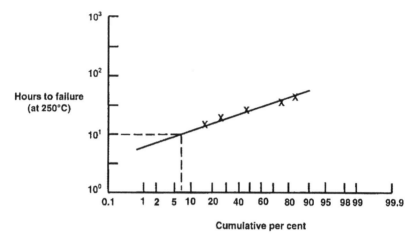

FIGURE 20-9 Estimated cumulative failure distribution at 250°C from step stress testing of high-temperature sensing devices.

As a second example, consider a test to determine the reliability of high-temperature sensing devices after 10 hours of operation at 250°C. The failure mechanism of concern is a thermally activated one with a known activation energy of 1.0 eV. In this case, only one stress type, steady-state operating temperature, needs to be considered. Ten samples were tested. The first step consisted of operating the samples at 250°C for 10 hours, and subsequent steps were conducted at successively higher 10°C temperature intervals for 2 hours each. Acceleration factors for each temperature were determined using the Arrhenius equation.

TABLE 20-1 Stress Levels for a Step Stress Test of a Power Supply

			Stress parameters			
Step	ΔT, °C/min	ΔT, °C	Line voltage, % of rated	No. cycles	Power cycling	Output load, %
1	3	0 to +45	115	20	4 (25%)	80
2	5	−10 to +55	100	20	9 (50%)	100
3	7	−20 to +65	90	20	12 (75%)	120
4	10	−30 to +71	80	20	14 (100%)	140

TABLE 20-2 Results of Step Stress Testing High-Temperature Sensing Devices

Step no.	Test temp., °C	Test time, hours		Acceleration factor	Equivalent test time at 150°C		No. of failures	Cum. failures	Cum. % failures
		Inc.	Cum.		Inc.	Cum.			
1	250	10	10	1.0	10	10	0	0	0
2	260	2	12	1.6	3.2	13.2	2	2	16.3
3	270	2	14	2.5	5.0	18.2	1	3	26.0
4	280	2	16	3.6	7.2	25.4	2	5	45.2
5	290	2	18	5.0	10.0	35.4	3	8	74.0
6	300	2	20	7.7	15.4	50.8	1	9	83.7

Table 20-2 shows the test conditions, equivalent accelerated test conditions, and failures observed for the test. A plot of percent cumulative failures vs. equivalent time at 250°C is shown in Figure 20-7. By extrapolation, it is seen that approximately 6% failures could be expected after 10 hours, and the expected reliability is approximately 0. 94.

From the above discussion and examples, it is obvious that great care must be taken to avoid misleading conclusions in setting up and analyzing data from step stress tests. McLinn [15] suggested the following ground rules for step stress testing.

1. Determine one or more appropriate stresses that approximate the operating and environmental conditions in service.
2. Start at a stress level that is high enough to begin precipitating failures quickly. Generally, this is at or above the maximum service stress level.
3. Estimate the maximum destruct level prior to the start of the test.
4. Divide the difference between the starting point and the destruct level into three levels.
5. Identify maximum and minimum times for the length of each step (the steps do not need to be equal in length).
6. Select stress levels as far apart as practical (the intervals need not be equal between all steps).
7. Be aware of possible interactions among the stresses.
8. Assume an acceleration model for each failure mechanism.
9. Select sample sizes as large as practical to get the maximum amount of information. If possible, 10–30 samples should be used.

10. Permit at least three failures in each step.
11. Samples need not be catastrophic, but may reflect degradation of output parameters.
12. Consider hypothetical results prior to the start of the test, in order to anticipate outcomes that could reduce the value of the test.

Reference 15 also describes a hypothetical step stress test, with temperature and relative humidity as the stresses. The steps are shown in Table 20-3, along with three hypothetical outcomes. The author says that, if outcome 1 had been achieved, there would have been too few failures to provide the desired information. The test could have been improved by increasing the stress levels, increasing the times for the steps, or by adding steps at higher stress levels. If outcome 2 had been achieved, the number of failures is about right, and the test would have been a good one. If outcome 3 had been achieved, too many failures would have been caused, and there would have been opportunities to reduce the cost and time of the test.

Before leaving the subject of step stress testing, it must be emphasized again that testing a product beyond its specification limits involves some risk. In general, electronic products can survive at stress levels significantly higher than their specification or design limits. Mechanical products, on the other hand, generally do not have this capability. For mechanical products, other methods besides applying higher stresses must be considered. A typical approach for rotating machinery, for example, is to operate it with insufficient lubrication, or with degraded lubricant, or with an increased duty cycle.

Often, a "run-in" test is used as a development test for mechanical products. Run-in tests involve operating within, or only slightly above, the specification levels, but doing so for a time long enough to demonstrate significant product life. A good example of this type of test is the endurance tests applied to many

TABLE **20-3** Step Stress Test for Temperature and Relative Humidity with Three Hypothetical Outcomes

Step	Temperature, °C	Relative humidity, %	Test time	Hypothetical outcomes		
				1	2	3
1	65	70	1 week	0/15	0/15	3/15
2	75	80	1 week	0/15	2/15	8/12
3	86	65	1 week	1/15	3/15	3/4
4	95	95	1 week	2/14	3/10	1/1

Source: Ref. 15. © 1998, John Wiley & Sons, used by permission.

automotive products. A run-in test might fall under the definition of a reliability demonstration test, rather than that of a step stress test.

20.4 DESIGN TESTS

Design tests are conducted within the budget and schedule of the product, and thus are intended to verify, rather than improve, the performance and reliability of the product. The test units are representative of the design that will be put into production and shipped to the customer; therefore they are expected to pass the test. Unless something unexpected occurs, there is no intent to make changes to the product design as a result of the test. The two types of design test discussed in this section are design verification tests and qualification tests.

20.4.1 Design Verification Tests

Design verification tests are conducted on test units that are identical to those that will be manufactured and shipped to the customer. The purpose is to verify that the product will perform its intended purpose at the extreme limits of all its specified operating and environmental conditions. The extremes are sometimes called "corner conditions." The design verification test is conducted by exposing the test unit to the extreme environmental conditions, either sequentially or in combination, while at the same time operating it at the extreme functional conditions of the specification range.

20.4.2 Qualification Tests

Qualification tests are similar to design verification tests; with the exception that, usually, they are somewhat formal in nature and are often defined in the contract between the producer and the customer. They may not include tests at all the corner conditions, as does the design verification test, but they do include representative conditions.

If the product is designed and produced for a single customer according to a contract, the customer may specify the types of tests to be performed, the test units to be used, and the pass-fail criteria. In some cases, the customer may witness the qualification test. Also, there may be some contractual requirements for maintaining records from the qualification test. In some cases, the qualification test unit may be kept in its as-tested configuration for a specified length of time.

If the product is designed and produced for sale "off the shelf," the producer defines the qualification tests, test units, and pass-fail criteria. These specifications may be defined by the best criteria available, and often may be defined according to standard industry practices.

TABLE 20-4 Comparison of Step Stress, Design Verification, and Qualification Tests

	Step stress	Design verification	Qualification
Purpose	Find and eliminate design weaknesses, ensure reliability, quantify margins	Verify that the design will function in all corner conditions of the specification range	Pass the specified test requirements
Pass-fail?	No	Yes	Yes
Test units	Represents final product, but may be different. May be a component or subassembly.	Identical to final product	Identical to final product
Applied stresses	Range from specification limits to destruct limits	Within specification limits	Within specification limits, and specified by the customer or marketing department
Success criteria	Find and eliminate design weaknesses	Successful operation at extremes of environmental and operating conditions	Pass test requirements
Degree of formality	Informal	Formal, may be contractual	Formal, often contractual
Part of program schedule and budget?	Maybe	Yes	Yes

Often, there is some confusion among step stress, design verification, and qualification tests. Some comparisons are shown in Table 20-4.

20.5   PRODUCTION TESTS

Production tests are conducted on-line, and the test units usually are shippable product, or subassemblies thereof. Usually, the test units are removed from the production line for testing. If the test is a destructive one, the test units are discarded after the test; if it is nondestructive, the test units are shipped to the customer. The purpose of a production test is to detect, remove, and correct defects introduced during the production process. Since the failures are of the type that would have been observed early in the service life in the product, they are called infant mortality failures. When viewed this sense, production tests may be considered to be a form of inspection process.

Usually, production test involve the application of some type of accelerated stress, and the level of the stress may be above the specification range. Care should be taken, however, to ensure that the stress level is not high enough to cause irreversible damage to the product, or to remove significant life. Typical production tests involve the application of stresses resulting from increased operating or environmental duty cycles, in an attempt to simulate the early service life of the product and precipitate early service failures.

Among the common types of production test are environmental stress screening, ongoing reliability test, ongoing accelerated life test, and periodic requalification. They may be used separately or in combination.

20.5.1   Environmental Stress Screening

There is a wide spectrum of understanding about environmental stress screening (ESS), and thus there is also a wide range of effectiveness achieved by its users. Some manufacturers view ESS as just another activity to satisfy contractual requirements, or as an insurance policy to protect them from future blame, or as just a very expensive inspection process. They typically employ "standard" ESS procedure to all products at the same stage of manufacturing, without regard to the types of flaws which may exist in the product. For them, ESS adds cost but not value to the product.

At the other end of the spectrum are those who develop and apply ESS processes effectively to improve product cost, quality, and reliability. They base their ESS processes on a thorough understanding of the product, the stresses to which it will be exposed in service, and its potential failure mechanisms. Results from ESS are used to provide further insights into product improvement, and to upgrade the ESS process itself.

The cost of misunderstanding the purpose of ESS is illustrated by Table 20-5, which shows a range of ESS profiles, imposed by the customers, on a single manufacturer of aerospace electronics products. Each of the 18 profiles was required by a different program, but the products for all of the programs were practically identical. Since many of the programs required dedicated ESS chambers and other test equipment, for which the utilization factor was quite low, this type of thinking caused significant waste in capital equipment cost, record keeping, and other factors. If the equipment manufacturer had been allowed to define the ESS profiles, with concurrence by the customers, cost and time could have been reduced significantly. Another advantage is that the equipment manufacturer, who knows the most about the likely types and mechanisms of failure, would be able to define the ESS profiles to detect manufacturing flaws most effectively.[1]

ESS is just what its name says it is: a screen. A screen is another term for an inspection step in the manufacturing process sequence; that is, its purpose is to examine the product and expose weaknesses which may result in field failures if corrective action is not taken. There is no value added directly to the product at this point and, as with all inspection processes, a realistic goal is to improve the manufacturing processes to the point where ESS can be reduced or eliminated to reduce cost.

ESS is intended to detect latent defects introduced into the product after the last prior inspection step. It is not intended to find design flaws, or to screen materials or components whose acceptability has been established in earlier inspections, or by use of SPC data, or by other means. The goal of ESS is to apply stresses of sufficient magnitude and duration to cause failures in products with manufacturing, component, or material flaws, but to avoid damaging or reducing the lifetime of "good" products.

Figure 20-10 shows the relationship between the probability density function (pdf) and the bathtub curve for a typical product with latent defects. The pdf shows a number of small distributions representing subpopulations of latent defects which would cause infant mortality failures in service, with the main population of the product failing at a much later time. The infant mortality region of the failure rate curve results from these latent defect distributions. The goal of ESS is to apply the proper stresses, at the proper levels, so that all of the infant mortality failures are removed prior to placing the product in service, but to avoid using up a significant portion of the product's useful life.

---

[1] Fortunately, since the information in Table 20-5 was assembled in 1993, the aerospace industry has made significant progress in understanding the purpose of ESS, and the inefficiencies illustrated by Table 20-5 have been greatly reduced.

TABLE 20-5 ESS Profiles, Imposed by the Customers on a Single Aerospace Electronics Equipment Manufacturer for a Range of Nearly Identical Products

Part no.	Time, hours	No. of cycles	Total, hours	Temperature cycle conditions			Power	Vibration
				Lower limit, °C	Upper limit, °C	Rate, °C/min		
1	8	2	16	−40	+71	5	Hot	Random
2	5	10	50	−54	+55	5	Cycled	Random
3	24	2	48	+22	+65	1	—	
4	24	2	48	+22	+65	2	—	
5	8	4	32	−40	+70	5	—	
6	4	12	48	−40	+70	5	Cycled	Sine-1g
7	8	4	32	−40	+71	5	On all	
8	4	8	32	−40	+80	Max	On all	
9	4	8	32	−40	+71	5	On all	
10	8	4	32	−40	+71	Max	On all	
11	4	12	48	−40	+70	5	Cycled	Sine-1g
12	7	7	49	−40	+70	5	Cycled	Sine-2.2g
13	8	12	96	−54	+71	5	Cycled	Sine-1g
14	1	12	48	−40	+71	5	Cycled	Sine-1g
15	5	10	50	−54	+71	5	Cycled	Sine-2.2g
16	4	12	48	−40	+71	5	Cycled	Sine-2.2g
17	8	12	96	−51	+71	7	Cycled	Random
18	2.6	5	26	−40	+70	5	Cycled	Random

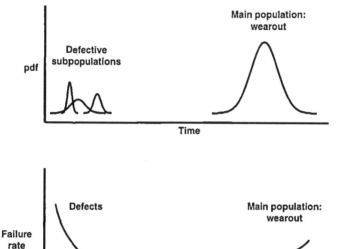

FIGURE 20-10   An illustration of the purpose of ESS. The top plot shows a probability density function (pdf) of a product with latent defects due to defective subpopulations, and the bottom plot shows the bathtub curve for the same product. The purpose of ESS is to apply stresses that will precipitate failures due to defective subpopulations.

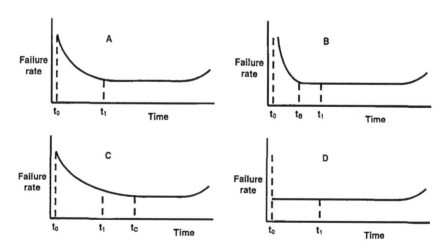

FIGURE 20-11   Examples in which (A) the ESS time t_c is about right, (B) ESS is too long, (C) ESS is too short, and (D) ESS is not necessary.

Figure 20-11 illustrates several different scenarios for a typical ESS process that extends to a time equal to t_1. In part (a), t_1 is long enough to precipitate the infant mortality failures, and it does not extend into the constant failure rate region of the bathtub curve. This is an effective ESS process, because it will detect latent defects, and thus reduce the initial failure rate of the product in service without causing unecessary failures in testing. In part (b), the ESS time t_1 is too long, and it will reduce the useful life of the product and precipitate additional failures without improving the failure rate after the optimum time, t_B, is reached. In part (c), the ESS time t_1 is too short and it leaves latent defects in the product. It should be extended to a time equal to t_C. In part (d), there is no infant mortality region, and ESS is unnecessary

It may be noted from this discussion that ESS is effective only if the failure rate curve initially shows a decreasing slope. Therefore, it is imperative to collect sufficient data, and to analyze all failures, so that separate failure rate curves can be plotted for each failure mechanism. Based on this information, the most effective types, levels, and durations of the ESS stresses can be determined.

The literature lists numerous examples of the benefits of an effective ESS process. Tang et al. [16] increased the lifetime of laser diodes by several orders of magnitude by introducing a controlled atmosphere burn-in to the manufacturing process. Chik and Devenyi [17] showed that early life failure modes of laser diodes could be eliminated by the use of an ESS process that included burn-in at 165°C and a current density of 10 kA/cm^2 for 1–2 hours. Furthermore, they were able to use degradation data from the process to predict the lifetimes of the unfailed devices. LoVasco and Lo [18] found that the combination of temperature cycling and high-temperature burn-in was much more effective than either stress applied separately in detecting defects in an optical communication system. Parker and Harrison [19] reduced the field failure rate of telecommunications systems by a factor of 2 when they applied a comprehensive program of ESS, combined with failure mode analysis and corrective action.

The steps in setting up an effective ESS process are conceptually quite simple, but they must be based on a good understanding of the product design, its reliability requirements, its operating environment, and its functional requirements, as well as the manufacturing processes and the principles of accelerated testing and analysis. They are listed here.

1. *Identify potential infant mortality failure mechanisms, failure sites, and failure stresses.* The most effective ESS processes are product- or process-specific. Failure mechanisms and failure sites depend on materials, components, and manufacturing technologies. The entire production process sequence should be examined to determine what possible defects may be introduced into the product, and at what point they may be introduced.

Although step stress testing, discussed in the previous section, is usually used as a design test, it also may be used at various stages in the manufacturing process to determine potential manufacturing failure mechanisms and to aid in setting up the ESS process.

2. *Identify the combinations of types, levels, and durations of stresses that will activate the suspected failure mechanisms.* Since most products have several potential infant mortality failure mechanisms, it is likely that more than one stress condition will be required to expose them. In general, the types of stresses to be considered are temperature, temperature cycling, vibration, high current, high voltage, spatial temperature gradients, mechanical shock, thermal shock, humidity, and chemical attack. Table 20-6 lists some potential failures of electronic products which, experience has shown, can be precipitated by three typical screening stresses. Different types of ESS stresses may be applied simultaneously or sequentially.

If ESS stresses are applied at accelerated levels, it is critical to keep them low enough to avoid causing failure mechanisms not possible during the service life of the product. Results from step stress testing, either from design verification or from process verification, can be a valuable source of information in this regard.

Over the years, some manufacturers have developed "standard" ESS conditions, based on previous experience and knowledge of the product. Some of them are listed in Refs. 20–27. They can be used as baselines, but should be evaluated in every case to see if they need to be modified to fit specific requirements.

Figure 20-12 shows a "typical" ESS profile for a moderately complex electronics box, i.e., one that contains approximately 5–10 printed wiring assemblies. It shows temperature cycling, vibration, and power cycling in combination. Usually, somewhere between 4 and 12 such cycles would be applied in an ESS process. Several observations may be made about this cycle, illustrating several general "rules" for ESS.

- The initial temperature excursion is from the starting temperature to the high temperature, to avoid condensation and freezing of moisture trapped in the test chamber prior to the start of the test.
- The dwell times are of the order of about 30 minutes, or just long enough to allow the temperature to stabilize at the extremes.
- The power is turned on during the temperature rise periods to aid in heating, and during the dwell times. It is turned off during the high-to-low temperature excursion to allow more rapid cooling.
- Vibration is conducted only for about ten minutes after the temperature

TABLE 20-6 Potential Failures That Can Be Precipitated by Temperature Cycling, Vibration, and the Combination of the Two

Temperature cycling	Vibration	Temperature cycling and vibration
Parameter drift	Large particle contamination	Defective solder joints
Printed wiring board opens and shorts	Chafed or pinched wires	Loose wires or fasteners
Cracked solder joints	Contact between adjacent components	Defective components
Wrong component	Contact between adjacent boards	Broken components
Hermetic seal failures	Poorly seated printed wiring assemblies	PWB etch defects
Defective wire terminations	Poor solder joints	Mechanical handling damage
Software timing errors	Mechanical cracks or other defects	Mold cracks
Cracks in molding compounds	High-mass or high-center-of-gravity components improperly secured	Interface cracks
Some chemical contamination	Loose wire bonds	Die passivation cracks
Die cracks	Structural weaknesses	
Die attachment failures		
Loose wire attachments		

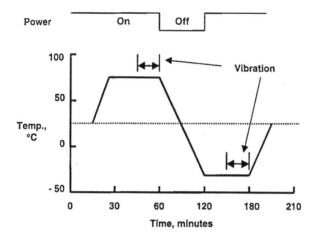

FIGURE 20-12   A cycle of a typical ESS profile for an electronics product.

has stabilized at the extreme temperatures. If a number of temperature cycles are used, vibration testing may be limited to only a portion of the temperature cycles.

* The test units should be monitored throughout the ESS cycle, so that if intermittent failures are precipitated, they are detected; otherwise, the test units may have been weakened and will fail early in service.

3. *Determine where in the process sequence to apply various ESS processes.* The more potential failure mechanisms there are, the more complicated the ESS process becomes. For this reason, ESS should be conducted as early in the manufacturing process flow as possible to detect a given defect. Table 20-7 shows typical points in the process flow where various ESS stresses are applied to an electronic product. If no ESS were conducted until the final product were completed, it would be impossible to detect some latent defects, and extremely expensive to repair others.

TABLE 20-7   Typical Points in the Manufacturing Process Flow Where ESS May be Applied to Electronic Products

Component or piece part	Subassembly	System
Temperature cycling	Temperature cycling	Power
Steady-state temperature	Power	Random vibration
Temperature-humidity-bias	Vibration	

Since ESS is an inspection step, and therefore does not add value directly to the product, it is appropriate to try to eliminate it. The steps to reducing and, ultimately, eliminating ESS are listed below.

1. *Collect failure rate data during the ESS process.* Sufficient data of sufficient resolution must be collected between time t_0 and t_1 (from Figure 20-11) to plot a failure rate curve. For example, if ESS involves eight temperature cycles, the product could be tested functionallly at one- or two-cycle intervals. Since the failure rate is calculated by dividing failures by operating hours (or cycles, etc.), data from *all* ESS attempts, even those which produce no failures, must be recorded. The ESS process, environmental and test equipment, and management system must be set up to provide this information.

2. *Prepare a probabilistic failure rate plot.* An overall probabilistic plot of failure rate vs. time can be prepared from these data. If the failure rate is constant or increasing, one of two possibilities exists: (1) The process is not set up to detect and precipitate relevant flaws, and must be revised; or (2) the product does not have an infant mortality region, and ESS can be eliminated immediately. If the plot shows a decreasing failure rate, there is opportunity for ESS reduction. A probabilistic curve is too general to be used for corrective action, but it can be used as a metric to chart progress in product reliability improvement.

3. *Analyze failures and separate them into homogeneous groups, according to failure mechanism and environment.* There may be several distinct failure mechanisms and failure-causing environments operant during ESS. Each of these mechanisms must be analyzed independently of the others. Analyses of all failures must be conducted to determine the failure mechanisms and the accelerated test conditions that caused them.

4. *Prepare deterministic reliability plots for each failure mechanism.* These are plots of failure rate vs. time (or cycles, etc.) for all failure mechanisms in the environmental conditions that caused them. These plots must also include data from all ESS attempts. Unlike the probabilistic plots of step 2 above, these plots can be used to implement corrective action, since they describe physical phenomena.

5. *Improve the product.* Using the information generated in ESS and failure analysis, changes in the product design, manufacturing processes, or materials can be made to eliminate failures due to specific mechanisms. Failures can be prioritized using Pareto or other methods, and prospective causes and solutions can be evaluated using Design of Experiments. Sometimes corrective action for one failure mechanism will improve the product with respect to other causes of failures. Also,

changes in the product may result in the introduction of new failure mechanisms, and the ESS process may have to be modified to expose potential failures due to them.

6. *Collect and analyze ESS data for improved product.* If the steps taken to improve the product are successful, the slope of the failure rate curve in the infant mortality region will decrease, both for the deterministic reliability curve for the individual failure mechanism, and for the overall probabilistic curve. Sometimes, the slope will not decrease, but the time required to reach a constant failure rate will be shorter. In this case, ESS must continue under the accelerated conditions concerned, but the time may be reduced.

7. *Modify ESS conditions to reflect the new failure rates.* In some cases, implemented changes will make it possible to eliminate a particular ESS test environment. In others, it will be possible to reduce the time. Sometimes, new conditions will have to be implemented to account for potential new failure mechanisms due to product changes.

8. *Reduce or eliminate ESS as warranted.* An effective ESS reduction plan will result in a continuously decreasing failure rate in the infant mortality region. The initial failure rate does not have to be constant in order to make ESS reduction an economically attractive option. If the process described here is followed, there will be enough data for an economic analysis of the ESS process.

The cost of ESS always must be balanced against its value in reducing manufacturing defects and field failures. The economic analysis of ESS effectiveness is based on conditions specific to the product, manufacturing costs, and the cost of failures in manufacturing and in service. A discussion of this analysis is beyond the scope of this book, but several papers which discuss ways to evaluate the economic tradeoffs are listed as Refs. 28–34.

20.5.2 Ongoing Reliability Testing

Ongoing Reliability Testing (ORT) involves removing a sample from each production lot, and placing it on test at slightly elevated stresses for a period of, typically, 1–2 weeks. Usually, the stress levels are low enough, and the times are short enough, that the test units can be shipped if they pass the test. The purpose of this test is to see if there are any lot-related production defects that must be corrected. The test time is sufficiently short to allow identification and correction of problems before the given lot goes into service by the customer. ORT is usually applied to high-volume products, such as disk drives.

20.5.3 Ongoing Accelerated Life Testing

Ongoing accelerated life testing involves the removal of a specified sample of production units from each lot, and placing it on test in simulated operating conditions until failure occurs. The purpose of this test is to provide assurance that the production lot is defect-free, and to provide notification if a problem occurs any time during the service life of the product. It is not particularly good at triggering corrective action, because some defects may not occur for several months or years. Ongoing accelerated life tests often are used in products that are subject to regulatory oversight, such as pacemakers and other medical electronic products.

20.5.4 Periodic Requalification

In periodic requalification, the qualification test, or a variation thereof, is repeated at specified intervals during the production cycle. This type of test is common in the semiconductor industry, in which production volumes are large, and the designs and manufacturing processes are being continuously improved. Usually, the periodic requalification test process is well defined, and results may be published to aid in the marketing effort to insure customers that the design and processes are under control.

20.6 SUMMARY

This chapter describes the various types of tests that may be used in a comprehensive accelerated reliability test program. Although the chapter contains many technical details, it is apparent that setting up such a program is as much a management challenge as a technical one. Most manufacturing operations generate data which can be a gold mine of valuable information for improving operations and reducing costs. The benefits of this information are rarely obtained, however, because sufficient time and effort are not devoted to organizing and analyzing it. If management has the insight to devote resources to setting up and operating (1) an effective, comprehensive, data collection and analysis system, (2) a capable failure analysis system, and (3) an effective data feedback and response system, coupled with the ability to improve the product, it will have gone a long way toward developing a reliability test program that provides a significant competitive edge.

REFERENCES

1. H. A. Malec, Accelerated testing—Design, production and field returns, Quality and Reliability Engineering International, 14:449–451, 1998.
2. K. Upadhyayula and A. Dasgupta, Physics-of-failure guidelines for accelerated qualification of electronic systems, Quality and Reliability Engineering International, 14: 433–447, 1998.
3. T. J. Stadterman, W. Connon, and D. Barker, Accelerated vibration life testing of electronic assemblies, Proceedings of the Institute for Environmental Sciences, pp. 233–238, 1997.

4. J. M. Hu, D. Barker, A. Dasgupta, and A. Arora, Role of failure mechanism identification in accelerated testing, Journal of the Institute for Environmental Sciences, pp. 39–46, July/August 1993.

5. G. E. Fornall, Life testing advances prove high reliability C^3I systems, Signal, pp. 62–69, April 1991.

6. R. J. Allen and W. J. Roesch, Stringent lifetesting ensures superior GaAs IC reliability, Solid State Technology, pp. 103–108, September 1990.

7. L. R. Pollock, A wide parametric, Bayesian methodology for system-level, step-stress, accelerated life testing, Ph.D. dissertation, Florida Institute of Technology, 1989.

8. M. Shaked and N. D. Singpurwalla, Inference for step-stress accelerated life tests, Journal of Statistical Planning and Inference 7:295–306, 1983.

9. J. S. Bora, Step stress accelerated life testing of diodes, Microelectronics Reliability, vol. 19, Pergamon Press, Ltd., 1979, pp. 279–280.

10. W. Nelson, Accelerated life testing—Step stress models and data analysis, IEEE Transactions on Reliability R-29(2):103–108, June 1980.

11. R. Miller and W. Nelson, Optimum simple step-stress plans for accelerated life testing, IEEE Transactions on Reliability R-32(1):59–65, April 1983.

12. W. Nelson, Graphical analysis of accelerated life test data with a mix of failure modes, IEEE Transactions on Reliability, R-24(4):230–237, October 1975.

13. D. S. Bai, M. S. Kim, and S. H. Lee, Optimum simple step-stress accelerated life tests with censoring, IEEE Transactions on Reliability R-38(5):528–532, December 1989.

14. G. Iuculano and A. Zanini, Evaluation of failure models through step-stress tests, IEEE Transactions on Reliability R-35(4):409–413, October 1986.

15. J. A. McLinn, Ways to improve the analysis of multi-level accelerated life testing, Quality and Reliability Engineering International 14:393–401, 1998.

16. W. C. Tang, E. H. Altendorf, H. J. Rosen, D. J. Webb, and P. Vettiger, Lifetime extension of uncoated AlGaAs single quantum well lasers by high power burn-in in inert atmospheres, Electronics Letters 30(2):143–145, January 20, 1994.

17. K. D. Chik and T. F. Devenyi, The effects of screening on the reliability of GaAlAs/ GaAs semiconductor lasers, IEEE Transactions on Electron Devices 35(7):966–969, July 1988.

18. F. LoVasco and K. Lo, Relative effectiveness of thermal cycling versus burn-in: A Case Study, Proceedings of the 42nd Electronic Components and Technology Conference, pp. 185–189, 1992.

19. T. P. Parker and G. L. Harrison, Quality improvement using environmental stress screening, AT&T Technical Journal, pp. 10–23, July/August 1992.

20. MIL-STD-883.

21. Military Volume I: Military Quality Assurance, Intel Corporation, Santa Clara, CA, pp. 1–3, 1990.

22. Components Quality and Reliability Handbook, Intel Corporation, Santa Clara, CA, pp. 9–18, 1989.

23. Reliability Manual and Data Summary, LSI Logic, Milpitas, CA, pp. 12–18, 1990.

24. Military Products: Products Spectrum Nomenclature and Cross Reference, Texas Instruments, Dallas, TX, 1988.

25. Discrete and Materials Technologies Group, Reliability Audit Report, First Quarter, Motorola, Inc., Phoenix, AZ, 1991.

26. High Performance CMOS Databook Supplement, Santa Clara, CA, 1989.

27. BiCMOS/CMOS Databook, Cypress Semiconductor, San Jose, CA, 1990.

28. W. B. Smith, Jr., Integrated product and process design to achieve high reliability in both early and useful life of the product, Proceedings of the Reliability and Maintainability Symposium, IEEE, pp. 66–71, 1987.

29. W. B. Smith and N. Khory, Does the burn-in of integrated circuits continue to be a meaningful course to pursue? Proceedings of the 38th Electronic Components Conference, IEEE, pp. 507–510, 1988.

30. D. Pantic, Benefits of integrated-circuit burn-in to obtain high reliability parts, IEEE Transactions on Reliability R-35(1):3–6, 1986.

31. M. Shaw, Recognizing the optimum burn-in period, Quality and Reliability Engineering International 3:259–263, 1987.

32. H. H. Huston, M. H. Wood, and V. M. DePalma, Burn-in effectiveness—theory and measurement, Proceedings of the International Reliability Physics Symposium, IEEE, pp. 271–276, 1991.

33. A. Suydo and S. Sy, Development of a burn-in time reduction algorithm using the principles of acceleration factors, Proceedings of the International Reliability Physics Symposium, IEEE, pp. 264–270, 1991.

34. D. C. Trindade, Can burn-in screen wearout mechanisms? Reliability modeling of defective sub populations—A case study, Proceedings of the International Reliability Physics Symposium, IEEE, pp. 260–263, 1991.

Chapter 21

Reliability Improvement with Design of Experiments

21.1 INTRODUCTION

In a survey among high-technology manufacturing companies to rate the effectiveness of various disciplines in producing reliable products, Design of Experiments (DoE) ranked sixth out of the 26 disciplines surveyed [1], and none of the respondents listed DoE as a requirement. Probabilistic prediction methods such as those of MIL-HDBK-217 were required by over half the respondents, but ranked last in effectiveness. From results like these, it is easy to see that there is a significant potential for industry to improve competitiveness by implementing DoE into its operations.

Until recently, the literature contained almost no references to the use of DoE in reliability work. Even now the number is still relatively small, and there are no scientific data to explain this. It is apparent to those who are experienced in the use of DoE, that the major limitation is lack of vision regarding the opportunities to use it in a variety of non-traditional situations. One of the most common characterizations of DoE is that it is very good for process optimization, and little else. In this chapter, we consider some examples of how DoE can be applied to reliability assurance, and describe its systematic application to the entire product development process, from product definition through shipment to the customer. The information presented here is based on all the previous chapters of this text;

but reference is made specifically to Chapter 11, where the topic is first introduced.

The flow of activity in a traditional product cycle is shown in Figure 21-1. It shows product definition, design, process development, manufacturing, and use by the customer occurring sequentially. With the recent attention given to concurrent engineering, some communication back and forth between adjacent steps might have been added, but there has been little progress toward a comprehensive, concurrent approach to the entire process. It is tempting to show a follow-on figure with this comprehensive approach neatly outlined, but this is difficult if not impossible.

To be effective, a comprehensive approach to concurrent engineering throughout the entire development cycle must include timely availability of information from across the entire cycle. For example, even in the product definition stage, it is necessary to know something about the capability of the manufacturing process, or the expected reliability of the product. In Figure 21-1, the reliability functions are shown outside the mainstream of the product development flow. It is rare to find an operation in which reliability assurance is truly integrated into this flow. Typically, design and manufacturing decisions are made, and their effects on expected reliability are evaluated off-line, either subjectively or with a probabilistic "cookbook" approach such as MIL-HDBK-217. Sometimes these evaluations are delayed until after the product is shipped, and customer complaints or field failures are used as the evaluation criteria. A structured approach, such as Design of Experiments, can be used to integrate reliability analyses and assurance methods into the product development cycle, and examples of that approach are presented in this chapter. Also included is a discussion of the further application of DoE to the product development process, introduced in earlier chapters.

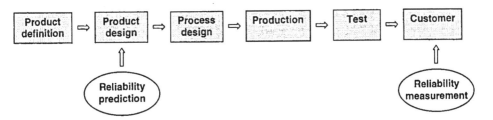

FIGURE **21-1**   Traditional use of reliability engineering in the product development cycle.

21.2 APPLICATION OF DOE TO RELIABILITY TESTING

In the simplest form of application of DoE to reliability determination [2–6], the environmental or operating stresses are considered as the factors, and the samples tested are all made with the same design and manufacturing parameters. In other words, the inner array factors are the applied stresses, and there are no outer array factors. This approach is best illustrated by the helium-neon laser example [7].

21.2.1 Helium-Neon Laser Example

The object of this experiment was to improve the reliability of a helium-neon laser by maximizing the stability of the design, so that it would be less susceptible to the stresses imposed upon it during operation. Thirteen factors were selected as most likely to affect the stability of the design, and three of them were considered proprietary by the manufacturer of the laser. Since the experiment was conducted jointly by the manufacturer of the laser and the customer, the levels of these three factors were identified only as 0.5x, 1.0x, and 1.5x. A series of experiments was conducted, but the initial one was considered most effective in identifying the design and process parameters with the greatest effect on laser stability. The experimental array is shown in Table 21-1; it evaluated ten factors at two levels each, and three more factors at three levels each. The responses were various measures of laser diode stability, and also the reliability of the printer in which it was used.

It may be noted that the array of Table 21-1 is unbalanced because, for each factor, the different levels are not represented equally. Although this is different from most of the examples in Chapters 3–11, the analysis of results is conducted in exactly the same manner as a balanced array. That analysis is not described in detail here, but the results of this experiment were used to increase the laser life by an order of magnitude, and also to provide better stability. Perhaps more significant was the fact that the responses included not only those of the laser, but of the customer's application: a printer in which it was used. This is an excellent example of concurrent engineering, not only between different functions within the same organization (such as design and production), but between organizations. DoE was used in this instance not only to improve the product, but to improve communications between the supplier and the customer. This type of activity is the essence of a successful long-term relationship between supplier and customer; it results in good marketing for the supplier, and supply chain stability for the customer.

TABLE **21-1A**   Columns A through G of the He-Ne Laser Experiment

Run	A Getter	B Cathode	C Surface	D Cleaning	E Air filter	F Bakeout	G Geometry
1	Bulk	2024	WB	HEDD	900	300	MOD
2	Bulk	6061	E	HEDD	0	300	MOD
3	Bulk	6061	E	HEDD	0	0	MOD
4	Bulk	6061	E	SPEC	0	300	PRES
5	Bulk	6061	E	SPEC	900	300	MOD
6	Comb	6061	E	SPEC	900	0	MOD
7	Comb	2024	WB	SPEC	900	0	PRES
8	Bulk	2024	WB	SPEC	900	0	PRES
9	Bulk	6061	E	HEDD	900	300	PRES
10	Comb	6061	E	HEDD	0	300	MOD
11	Bulk	2024	WB	SPEC	0	300	MOD
12	Comb	6061	WB	SPEC	900	300	MOD
13	Comb	2024	E	HEDD	900	300	MOD
14	Bulk	2024	WB	SPEC	0	0	MOD
15	Comb	6061	WB	HEDD	900	0	PRES
16	Comb	2024	E	HEDD	0	300	PRES
17	Comb	2024	WB	SPEC	0	300	MOD
18	Comb	2024	WB	HEDD	900	0	MOD
19	Bulk	2024	WB	HEDD	0	300	PRES
20	Comb	6061	WB	HEDD	0	0	MOD
21	Bulk	2024	E	HEDD	0	0	PRES
22	Comb	6061	WB	SPEC	0	300	PRES
23	Bulk	2024	E	HEDD	900	0	MOD
24	Bulk	6061	WB	SPEC	0	0	PRES
25	Bulk	6061	E	HEDD	900	0	PRES
26	Comb	6061	E	SPEC	0	0	PRES
27	Bulk	2024	E	SPEC	900	300	PRES
28	Bulk	6061	WB	SPEC	900	0	MOD
29	Comb	6061	E	HEDD	900	300	PRES
30	Comb	2024	E	SPEC	0	0	MOD
31	Comb	2024	WB	SPEC	900	300	PRES
32	Comb	2024	WB	HEDD	0	0	PRES

TABLE **21-1B**   Columns H through M of the He-Ne Laser Experiment

Run	H He:Ne ratio	I Fill pressure	J No. of cycles	K O_2 pressure	L O_2 current	M O_2 time
1	20:1	2.15	4	0. 5	0.5	1.5
2	17:1	2.50	2	0. 5	1.5	1.0
3	20:1	2.15	4	1. 0	1.0	0.5
4	20:1	2.50	2	0. 5	0.5	1.0
5	20:1	2.50	4	1. 5	1.5	1.5
6	17:1	2.50	4	0. 5	0.5	1.0
7	20:1	2.15	4	1. 5	0.5	1.0
8	20:1	2.50	2	1. 0	1.0	1.0
9	17:1	2.50	4	1. 5	0.5	0.5
10	17:1	2.15	4	1. 0	1.0	0.5
11	17:1	2.15	4	1. 5	0.5	1.0
12	20:1	2.15	2	1. 0	1.0	0.5
13	20:1	2.50	2	1. 0	1.0	0.5
14	20:1	2.50	4	1. 0	1.0	0.5
15	20:1	2.50	4	0. 5	1.5	1.0
16	20:1	2.50	4	1. 0	1.0	1.0
17	17:1	2.50	4	1. 0	1.0	1.0
18	17:1	2.15	4	1. 5	1.5	1.0
19	20:1	2.15	2	1. 5	1.5	1.0
20	20:1	2.50	2	1. 5	0.5	1.5
21	17:1	2.50	4	1. 0	1.0	1.5
22	20:1	2.15	4	1. 0	1.0	1.0
23	17:1	2.50	2	1. 0	1.0	1.0
24	17:1	2.15	4	1. 0	1.0	1.5
25	20:1	2.15	2	1. 0	1.0	1.0
26	17:1	2.50	2	1. 5	1.5	0.5
27	17:1	2.15	4	0. 5	1.5	0.5
28	17:1	2.15	2	1. 0	1.0	1.0
29	17:1	2.15	2	1. 0	1.0	1.5
30	20:1	2.15	2	0. 5	1.5	1.5
31	17:1	2.50	2	1. 0	1.0	1.5
32	17:1	2.15	2	0. 5	0.5	0.5

Source: From Ref. 7. © 1994, John Wiley & Sons, used by permission.

21.2.2   Fluorescent Lamp Degradation Experiment

The goal of the fluorescent lamp reliability experiment [8] was to improve the reliability of a product that was already highly reliable. The approach was to evaluate the effects of three steps in the manufacturing process, on the expected lifetime of the fluorescent lamps. Since the expected lifetime of these lamps can be quite long, the experimenters decided not to take the time to leave the samples on test until they failed, but to measure the degradation of their outputs over time prior to failure. The results were then used to calculate a degradation rate, which in turn was used to calculate the expected lifetime according to a linear relationship.

The factors selected for evaluation were:

A—   Amount of electric current in the exhaustive process used to remove CO_2 from the $BaCO_3$ coating on the inner surfaces of the tubes, leaving BaO*

B—   Concentration of the mercury dispenser in the process used to dispense mercury vapor into the tubes

C—   Concentration of argon used in the process to fill the tubes with argon

A full factorial array to evaluate three factors would require eight runs, but the experimenters decided to use a four-run fractional factorial array to reduce costs. Five lamps were selected at random for each run. After a 100 hour burn-in period, and a measurement at 500 hours to verify that the samples were peforming normally, the luminous flux, in lumens, of each of the 20 lamps was measured at times of 1000, 2000, 3000, 4000, 5000, and 6000 hours. The resulting data were plotted versus time, and the degradation rates were obtained from the plots. Degradation rates were then used to calculate the estimated lifetimes of the samples.

The lifetime of a lamp was defined as the time required for its luminous flux to degrade to 60% of its value at a time of 100 hours. The mathematical model for the luminous flux $\Lambda(t)$, at time t is

$$\ln \Lambda(t) = \theta + \lambda t \tag{21-1}$$

where θ is the initial luminous flux and λ is the degradation rate. With the 60% failure criterion, the time to failure is

---

* The reader is referred to Ref. 8 for a more detailed description of the fluorescent lamp production processes.

$$t = \frac{\ln 0.6}{\lambda} + 100 \qquad\qquad (21\text{-}2)$$

Table 21-2 shows the fractional factorial array and the responses for the experiment. The column labeled "θ (estimate)" shows that the initial luminous flux was essentially equal for all of the samples. The degradation rates, in the column labeled "$-\lambda$ (estimate)," were found to be lognormally distributed, and from Table 21-2, it may be seen that the degradation rates of the samples in run 2 were lower than those in the other three runs. Because of this, the estimated lifetimes were longer for the samples in run 2. The lifetimes also were found to be lognormally distributed, and the logs of the lifetimes for each of the four runs are shown in the far right column of Table 21-2.

The results indicate that only factors B, mercury concentration, and C, argon concentration, are important in influencing the lifetime of the lamps. The lifetime can be improved by setting them at their plus levels. Factor A, electric current, was set at the minus level, its pre-experiment setting. When this is done, the estimated lifetime is approximately 23,000 hours. The estimated lifetime at the pre-experiment settings (all minus) is approximately 13,800 hours, so this experiment resulted in a lifetime that is 67% longer than it was before the experiment was performed. This improvement did not come without cost, however, since the new settings raised the cost of the product by about 10%. The tradeoff between longer life and higher cost must be evaluated as a business decision.

21.2.3   Surface Mount Capacitor Example

A more comprehensive approach to the use of DoE in reliability testing is to evaluate design and manufacturing factors simultaneously with operating and environmental factors, using reliability as a response. This approach is outlined in Ref. 9. This type of experiment can get quite large, but an abbreviated example is the surface mount capacitor experiment. In this example, both manufacturing and reliability factors were considered in a single experiment. Two manufacturing factors were evaluated in the inner array; and one operating and one environmental factor were evaluated in the outer array.

The manufacturing factors were dielectric composition and processing temperature, and they were evaluated at two levels each in an L_4 inner array. They were considered the controllable factors.

The uncontrollable factors, evaluated in the outer array, were ambient temperature (an environmental factor) and voltage (an operating factor); and they were evaluated at accelerated conditions. Operating voltage was evaluated at four levels: 200, 250, 300, and 350 V; and ambient temperature was evaluated at two levels: 175 and 190°C. These conditions were arranged in an L_8 outer array. The response was a measure of reliability, which in this case was time to failure, in hours.

TABLE 21-2 Fractional Factorial Array and Responses for the Fluorescent Lamp Experiment

Run	Lamp no.	A Current	B Mercury	C Argon	θ (estimate)	$-\lambda(10^{-5})$ (estimate)	Estimated lifetime	Estimated log lifetime
1	1	−	−	−	7.8440	3.10	14762.98	9.53
	2				7.8279	2.68	16145.74	
	3				7.8490	3.38	13429.86	
	4				7.8375	3.34	12941.79	
	5				7.8697	3.75	12127.27	
2	6	−	+	+	7.8488	1.66	26380.14	10.04
	7				7.8565	1.84	22860.88	
	8				7.8374	1.64	24436.63	
	9				7.8408	2.04	20694.20	
	10				7.8497	2.16	20468.02	
3	11	+	−	+	7.8865	3.28	15028.19	9.67
	12				7.8578	2.99	15914.50	
	13				7.8786	3.46	13708.69	
	14				7.8657	2.73	17285.64	
	15				7.8722	2.66	17578.56	
4	16	+	+	−	7.8691	2.05	20349.38	9.81
	17				7.8456	3.01	15000.83	
	18				7.9378	2.61	17600.54	
	19				7.9218	2.34	19395.51	
	20				7.9265	2.45	18746.54	

Source: From Ref. 8, used by permission.

The factors and levels for the inner and outer arrays are shown in Table 21-3, and the arrays are shown in Table 21-4. Thirty-two unique treatment combinations of inner and outer array factors were evaluated, and ten samples were collected for each combination.

The results for each treatment combination of the surface mount capacitor experiment were found to be lognormally distributed, and the mean of the distribution was used as the response. Table 21-5 is a response table showing mean time to failure (MTTF) in hours for each level of the inner array factors, at the outer array conditions tested.

The results from this experiment can be used to select the optimum manufacturing conditions (the inner array factors), and to estimate the expected MTTF at nonaccelerated operating conditions. For this analysis, these nonaccelerated operating conditions are chosen as 50 V and 50°C.

From Table 21-5, it is apparent that dielectric material A_2 yielded a product with longer MTTF than did A_1, for all outer array conditions. It is also apparent that processing temperature B_2 is superior to B_1. Furthermore, there is no significant interaction (A × B) between them. Since the inner array is a full factorial array, no further analysis is necessary, and it is an easy choice to specify level A_2 for dielectric material and level B_2 for processing temperature.

A confirmation run could be conducted at this point, but run 4 of the inner array (see Table 21-4) already contains both A_2 and B_2, and we can use these results as the confirmation run results. Table 21-6 contains a summary of these results.

The task is now to estimate the MTTF of the optimized product at the expected operating conditions of 50 V and 50°C. We begin by constructing plots of MTTF vs. operating voltage, and of MTTF vs. ambient temperature, using the data from the confirmation run.

The plot of MTTF vs. operating voltage is shown in Figure 21-2. The relationship between operating voltage and MTTF is logarithmic, and data for each ambient temperature are plotted. (Care must be exercised in using these data, and in drawing conclusions from them. Physical analysis of all failures should be conducted to assure that the mechanisms are known, and the results of the failure analysis must support the conclusions drawn from the data.) There is considerable scatter in the data, but for our purposes, it is necessary to determine only the slopes, and not the locations, of the lines. Lines determined by the method of least squares for the two sets of data in Figure 21-2 had nearly equal slopes, which is consistent with earlier work. Therefore, the two lines shown in Figure 21-2 were drawn with the same slope, which is the average of the two slopes determined by least squares.

The general equation for the lines plotted in Figure 21-2 is

$$\text{MTTF} = e^{\gamma_0 - \gamma_1 V} \qquad\qquad (21\text{-}3)$$

TABLE 21-3 Factors and Levels for the Surface Mount Capacitor Reliability Experiment

	Inner-array factor levels			Outer-array factor levels			
Factor	Level 1	Level 2	Factor	Level 1	Level 2	Level 3	Level 4
A: Dielectric composition	A_1	A_2	C: Operating voltage	200	250	300	350
B: Processing temperature	B_1	B_2	D: Operating temperature	175°	190°C		

TABLE 21-4 L₄ Inner Array and L₈ Outer Array for the Surface Mount Capacitor Reliability Experiment

Inner array				Outer array								Average response (lifetime in hours)
				1	1	2	2	3	3	4	4	C
				1	2	1	2	1	2	1	2	D
				1	2	2	1	2	1	1	2	C × D
Run	A	B	A × B									
1	1	1	1	430	950	560	210	310	230	250	230	396
2	1	2	2	1080	1060	890	450	430	320	340	430	625
3	2	1	2	890	1060	680	310	310	310	250	230	505
4	2	2	1	1100	1080	1080	460	620	370	580	430	715
Avg. response				875	1038	803	358	418	308	355	330	561

TABLE 21-5   Response Table for the Surface Mount Capacitor Experiment

Inner array factors		Outer array factors					
		Operating voltage				Operating temp.	
Factor	Level	200 V	250 V	300 V	350 V	175°C	190°C
Dielectric material	A_1	880	528	323	313	536	485
	A_2	1033	633	403	373	689	531
Processing temperature	B_1	833	440	290	240	460	441
	B_2	1080	720	435	445	765	575
A × B	$(A \times B)_1$	890	578	383	373	616	495
	$(A \times B)_2$	1023	583	343	313	609	521

where V is the operating voltage and γ_0 and γ_1 are constants [10, 11]. The parameter of interest to us is γ_1, the slope of the line. From Figure 21-2, the average γ_1 is 0.005.

The plot of MTTF vs. ambient temperature shown in Figure 21-3 was also obtained by plotting the data from Table 21-6. Each line represents a different operating voltage. The temperature dependence of the MTTF of the capacitors is an Arrhenius one, of the form

$$\text{MTTF} = A_0 e^{E_a/kt} \qquad (21\text{-}4)$$

where

$A_0 = $ constant
$E_a = $ activation energy
$k = $ Boltzmann's constant
$T = $ operating temperature

TABLE 21-6   Summary of Confirmation Run Results From the Surface Mount Capacitor Experiment*

Operating temperature	Operating voltage			
	200 V	250 V	300 V	350 V
175°C	1100	1080	620	580
190°C	1080	460	370	430

* These results are from run 4 of the original experiment.

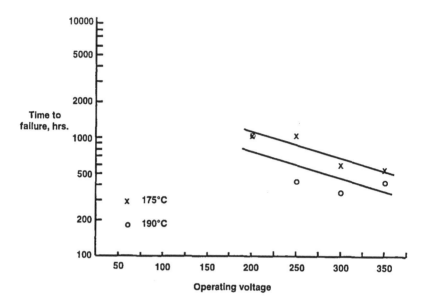

FIGURE 21-2 Time to failure vs. operating voltage for a product manufactured with optimized processing conditions from the surface mount capacitor experiment.

As with operating voltage, we are only interested in the slopes of the lines E_a. These were determined in a manner similar to that used for operating voltage, and the average E_a was found to be 0.46 eV.

The acceleration model for these results is a form of the Eyring equation, which includes terms for both ambient temperature and operating voltage. It is a combination of Eqs. (21–1) and (21–2):

$$\text{MTTF} \propto e^{-0.0005V} e^{0.46/kT} \tag{21-5}$$

The acceleration factor for extrapolation between a test condition of 200 V, 175°C and a use condition of 50 V, 50°C is obtained by

$$\text{AF} = \frac{e^{-0.005(50)}}{e^{-0.005(200)}} \cdot \exp\left[\frac{0.46}{8.617 \times 10^{-5}}\left(\frac{1}{323} - \frac{1}{448}\right)\right] \tag{21-6}$$

$$\approx 2.65 \cdot 83.98 \approx 222.5$$

We can obtain an estimate of the expected MTTF at use conditions of 50 V, 50°C by multiplying the MTTF at test conditions of 200 V, 175°C by 222.5. From Table 21-6, the MTTF for these test conditions using the optimized dielectric material and processing temperature is 1100 hours. Thus our predicted MTTF at use conditions is

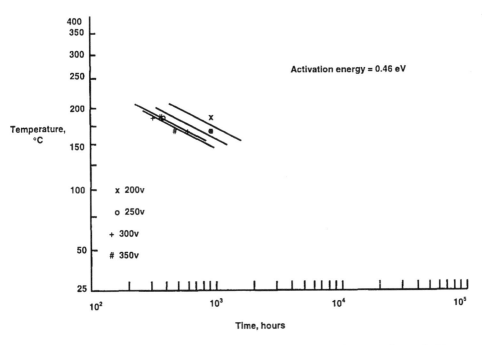

FIGURE 21-3 Time to failure vs. operating temperature for a product manufactured with optimized processing conditions from the surface mount capacitor experiment.

$$1100 \times 222.5 = 244,750 \text{ hours} \tag{21-7}$$

For ease of presentation, a rather small array was chosen for this example. We could have evaluated more factors in larger inner and outer arrays. If we had wished, we also could have included some design factors in the experiment. Also, we used the means of the distributions as responses, but we could have used different distribution parameters, such as μ-σ, etc. Obviously, this example contains assumptions which may be questioned, and there is some scatter in the data. Neither of these features is unusual in accelerated testing. The example illustrates a technique for obtaining both manufacturing and reliability data from a single experiment. This method can be used for concurrent design, manufacturing, and reliability engineering.

21.2.4   The Via Hole Reliability Experiment

An experiment that is similar in nature to the surface mount capacitor experiment is the experiment performed to evaluate the effects of manufacturing and design process variables on the fatigue life of via holes [12]. Via holes are drilled through printed circuit cards, and then plated to provide electrical connections between the conducting layers of the cards. The most common failure mechanism of this type of structure is fatigue failure due to cracks in the metal plating, caused by the difference in coefficients of thermal expansion of the metal and the organic card materials, when the cards are exposed to thermal cycling.

The experimenters identified five factors that might be considered important in the fatigue life of the via holes: ·

A. Via hole plating metallization thickness
B. Organic material thickness
C. Angle of inclination of the via hole wall due to variations in the top and bottom width of the via hole
D. Ductility coefficient of the conductor material
E. Strain concentration factor

Factors A and B are circuit card design factors, C and E are processing factors, and D is a material property factor.

The response for the experiment was fatigue life, which was estimated by using finite element analysis. A full factorial experiment of 32 runs was performed, and analysis of variance was used to determine the sensitivity of fatigue life to the various main effects and interactions of the factors. Results of the analysis indicated that the most signficant factors, in order of importance, were

1. Strain concentration (factor E)
2. Ductility coefficient of the conductor material (factor D)
3. Interaction between factors D and E
4. Via hole plating metallization thickness (factor A)
5. Angle of inclination of the via hole wall (factor C)

The experimenters concluded that the electroplating process should be tightly controlled to improve the strain concentration factor, and that high quality copper should be used to improve the ductility of the conductor material. The metallization should be as thick as practical and economically viable, and the angle of inclination of the wall should be between 50 and 80°.

21.2.5 Environmental Stress Screening Experiment

Pachuki [13] used a Taguchi array to evaluate the effectiveness of a proposed new environmental stress screening (ESS) profile in detecting latent defects in a circuit card assembly.† The test unit was a double-sided CPU card, approximately 8.5 × 11 in, with approximately 350 electronic components. The goal was to

---

† As with many of the examples in this text, the actual experiment was quite complicated, and it has been simplified here for ease of understanding. The reader is referred to the original reference for further details of the experiment as it was actually performed, and of the analysis and interpretation of results.

reduce the amount of time required for ESS from 3.5 hours to 40 min by increasing the temperature change rate from 10–60°C per minute.

A stylized version of the profile to be evaluated is shown in Figure 21-4. (Some of the details of the profile have been eliminated or simplified for ease of presentation.) The previous ESS process (3.5 hours total time) and the proposed new process (40 min total time) each involved four cycles of the profile. The factors evaluated in the experiment, and their levels, were:

Stress	Level 1	Level 2
Random vibration	Applied	Not applied
Temperature cycling	Applied	Not applied
Power cycling	Applied	Not applied
Diagnostic test	Functional	Self-test

The temperature cycling range was chosen on the basis of previous experience, which showed that the selected limits were effective in precipitating defects without damaging non-defective boards. The rapid temperature ramp rate was chosen to decrease ESS time, and was not considered a significant risk, based on the properties of the materials involved. The power cycle was the same as

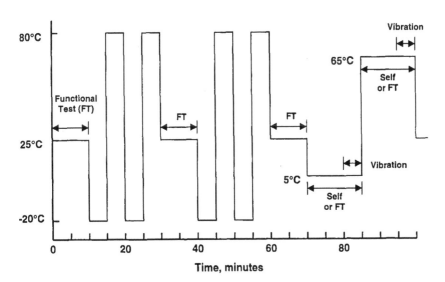

FIGURE 21-4   Stylized ESS profile for the ESS experiment. (From Ref. 13. © 1993, IEST, used by permission.)

TABLE 21-7 Experimental Array for the ESS Experiment

Run	Random vibration	Temperature cycling	Vibration × temperature cycling	Power cycling	Vibration × power cycling	Temperature cycling × power cycling	Diagnostic test
1	1	1	1	1	1	1	1
2	1	1	1	2	2	2	2
3	1	2	2	1	1	2	2
4	1	2	2	2	2	1	1
5	2	1	2	1	2	1	2
6	2	1	2	1	1	2	1
7	2	2	1	1	1	2	1
8	2	2	1	2	1	1	2

Source: Adapted from Ref. 14. © 1993, IEST, used by permission.

that used prior to the experiment. The vibration levels were chosen by performing step stress tests on sample test units to determine the operational and technological limits.

The four factors were evaluated at two levels each in a Taguchi L_8 array, as shown in Table 21-7. This is a fractional factorial array, and allows the evaluation of the two-factor interactions between random vibration and temperature cycling, temperature cycling and power cycling, and random vibration and power cycling. The three-factor interaction among random vibration, temperature cycling, and power cycling is confounded with the diagnostic test factor. 225 cards were evaluated in each of the eight runs, for a total of 1800 cards.

After completion of the ESS process, all the samples were exposed to an additional 8 hours in a system aging test, designed to detect early failures that may have escaped in the ESS process.

Results of the experiment are shown in Table 21-8. From this table, it may be observed that the proposed new ESS profile was effective in detecting workmanship failures, but not very effective in detecting component failures.

To evaluate the effectiveness of the various stresses applied in the proposed new profile, a response table was made, and is shown in Table 21-9. This table shows that random vibration was the only stress that precipitated failures consistently when it was applied, versus when it was not applied, in both ESS and system aging tests. The experimenters verified this with analysis of variance.

Temperature cycling and power cycling were not significant factors as viewed from the test fallout data. The experimenters concluded that the proposed

TABLE 21-8   Results of the ESS Experiment

Run (225 PWBs each)	Component failures		Workmanship failures	
	ESS	System aging	ESS	System aging
1	3	5	2	0
2	6	4	4	0*
3	0	4	0	0*
4	5	1	2	0
5	2	2	0	0
6	1	0	0	0
7	1	0	1	0
8	0	3	0	0
Totals	18	19	9	0

* One failure was observed, but it was analyzed and traced to circuitry that was intentionally not tested in ESS. It is therefore not counted here.

TABLE 21-9 Response Table for the ESS Experiment

Factor	ESS failures		System aging failures	
	Level 1	Level 2	Level 1	Level 2
Random vibration	22	5	5	0
Temperature cycling	18	9	11	8
Power cycling	9	18	11	8
Diagnostic test	15	12	6	13

ESS temperature cycle was insufficient, both in number of cycles and in duration of cycles, and further work is required to develop an effective temperature cycling process. They also concluded that the proposed power cycle was too short. Based on previous experience, however, they concluded that its parameters did not need to be changed.

The experimenters also observed that 90% of the system aging failures occurred in the first 15 min. This indicated that, although the failures did not occur in ESS, they were almost ready to occur. They speculated, but did not verify, that if they moved the vibration portion from the end to the beginning of the profile, its effectiveness would have increased.

This experiment, although not conclusive in itself, pointed the experimenters to some future work that promised significant improvements in ESS time and cost.

21.3 APPLICATION OF DoE TO THE RELIABILITY PROCESS

Application of DoE to many of the steps of the deterministic reliability process, as described earlier in Section 3.6, has already been discussed in individual examples. This section contains a summary of the use of DoE in those applications.

21.3.1 Define Realistic System Requirements

Specific experiments to define the system requirements are difficult to conceive. The most realistic application of DoE here is to develop a list of specific requirements, and to use this list as responses for designed experiments. Examples of such responses are output characteristics or product features such as voltage, weight, size; and reliability measures such as MTBF, MTTF.

21.3.2   Define the Design Usage Environment

Designed experiments to define the usage environment are also rare. Usage environmental conditions are often used as outer array factors, as exemplified in the previous section.

21.3.3   Identify Potential Failure Sites and Failure Mechanisms

Potential failure sites and failure mechanisms are often identified on the basis of information from many sources. One of these sources is step stress testing. A DoE variation of step stress testing is to design an array of operating and environmental factors at several levels; and to test samples according to this array instead of increasing all of the stresses simultaneously. The effect of each factor in precipitating failures can then be quantified. This variation is sometimes more efficient and less expensive than the traditional approach, and has been used successfully by some manufacturers. An example of this type of approach is the ESS example described earlier in this chapter.

21.3.4   Purchase Reliable Materials and Components

21.3.5   Design Reliable Products, Within the Capabilities of the Materials and Manufacturing Processes

21.3.6   Qualify the Manufacturing and Assembly Processes

These three steps are listed together because, with Design of Experiments, it is possible to accomplish them simultaneously. This is one of the most fruitful areas for the application of DoE. Traditionally, control of a supplier's processes has been considered beyond the scope of a manufacturer. If supplier factors were considered at all, each supplier was considered a separate level of a single factor; e.g., supplier A was level 1, and supplier B was level 2. The manufacturer was allowed no control, and often no knowledge, of the details of the supplier's process. With the advent of the partnering concept to provide the necessary cooperation, and DoE to provide the technical capability, we can now combine a supplier's factors with the user's factors in a single experiment to optimize the overall system.

An example of a joint supplier-customer experiment is shown in Table 21-10, which is an experiment to optimize the design and manufacture of a high-speed printed circuit card assembly. (In this example, the assembler is also the

TABLE 21-10   Factors for a Combined Experiment by the Laminator and Assembler of a High-Speed Printed Circuit Card Assembly

Factor name	Levels	Type of factor	Who controls?	Relationship
A: Adhesive material	1. Material A 2. Material B	Manufacturing	Laminator	Supplier
B: Lamination process	1. *T-t* profile A 2. *T-t* profile B	Manufacturing	Laminator	Supplier
C: Line width	1. 0.008 in 2. 0.010 in	Design	Assembler	Customer
D: Line spacing	1. 0.008 in 2. 0.101 in	Design	Assembler	Customer
E: Type of shielding	1. Microstrip 2. Stripline	Design	Assembler	Customer
F: Method of assembly	1. IR solder 2. Vapor phase	Manufacturing	Assembler	Customer

designer of the assembly.) These six main effects can be evaluated in an L_8 array, combined with up to nine interactions in an L_{16} array, or combined with all possible two-factor interactions in an L_{32} array. (There are many other possible experimental designs, including screening experiments at three or more levels.) Responses can be functional effects, such as rise time; quality effects, such as yield; and reliability effects, such as MTTF.

This approach can be used to specify, qualify, and procure purchased materials and components, in designing products, and in developing and optimizing processes. (Many of the earlier examples in this book describe experiments to design products and develop processes separately.) It can include as many suppliers and customers in the vertical product flow as are willing to participate. Not only will it improve the product, but it also will shorten development time, reduce cost, and improve communications among all involved parties.

21.3.7   Control the Manufacturing and Assembly Processes

The most common method of process control is SPC, which is primarily an information-gathering process. It is an excellent method for identifying the existence of problems, and facilitates immediate remedies for obvious ones. If the problem is complex, however, additional tools such as DoE must be used. Some of the examples in earlier chapters of this text deal with process improvement.

21.3.8 Manage the Life Cycle Usage of the Product

Life cycle management is one of the most difficult activities of reliability work, and the application of DoE to it is no easy task. As we have seen from earlier considerations of deterministic reliability methods, it is imperative that the manufacturer and the customer work together to measure and improve the reliability of a product. Since reliability depends at least in part on how the product is used, the customer is in a unique position to provide this type of information for use in life cycle management.

Different levels of use conditions can be evaluated as outer array conditions in long term reliability measurements of products in service. For example, if ambient temperature is considered an important factor in the long term reliability of a product such as diesel engines or automobile tires, then an experiment could be designed in which one level of this outer array factor contains samples which operate primarily in cold climates, and another level could be samples operating in warmer climates. This type of experiment is conceptually no more difficult than the accelerated tests described earlier. The difficult part is the long term commitment to collect and analyze relevant data. For this to be successful, it will be necessary for manufacturers and users of products to commit to a higher level of interest in reliability, and of working together to improve it, than now exists in most industries.

21.4 SUMMARY

In this chapter, we have discussed innovative ways to use Design of Experiments to improve reliability. It is apparent that the limiting factor in both DoE and reliability is the imagination of the users of the tools available to them. If we understand and appreciate what we can do, and proceed with confidence, we can improve our products to a degree not currently considered possible.

REFERENCES

1. M. Lindsley, Reliability task effectiveness survey report, University of Washington unpublished research, 1992.
2. W. Weibull, Statistical design of fatigue experiments, Journal of Applied Mechanics 19(1):109–113, March 1952.
3. M. Zelen, Factorial experiments in life testing, Technometrics 1(3):269–289, August 1959.
4. G. C. Derringer, Consideration in single and multiple stress accelerated life testing, Journal of Quality Technology, 14(3):130–134, July 1982.
5. G. R. Bandurek, H. L. Hughes, and D. Crouch, The use of Taguchi methods in

performance demonstrations, Quality and Reliability Engineering International 6: 121–131, 1990.

6. R. G. Bullington, S. Lovin, D. M. Miller, Improvement of an industrial thermostat using designed experiments, Journal of Quality Technology 25(4):262–271, October 1993.

7. S. J. Keene and P. Vattano, Practical reliability testing considerations, Quality and Reliability Engineering International 10:391–398, 1994.

8. S. T. Tseng, M. S. Hamada, and C. H. Chiao, Using Degradation Data from a Fractional Factorial Experiment to Improve Fluorescent Lamp Reliability, Research Report no. RR-94-05, The Institute for the Improvement in Quality and Productivity, University of Waterloo, Waterloo, Ontario, Canada, 1994.

9. L. Condra and M. Lindsley, Using design of experiments to improve product and process integrity, Proceedings of the National Aerospace Electronics Conference, Dayton, OH, 1991.

10. W. Nelson, Accelerated Testing, John Wiley & Sons, New York, p. 96, 1990.

11. L. Simoni, Voltage Endurance of Electrical Insulation, Tecnoprint, Bologna.

12. A. O. Ogunjimi, S. MacGregor, and M. G. Pecht, and J. W. Evans, The effect of manufacturing and design process variabilities on the fatigue life of the high density interconnect vias, Journal of Electronics Manufacturing 5(2):111–119, June 1995.

13. D. E. Pachuki, Environmental stress screening experiment using the Taguchi method, Proceedings of the Institute of Environmental Sciences and Techology, vol. II, pp. 211–119, 1993.

Chapter 22

Using Design of Experiments to Integrate Reliability into the Organization

22.1 INTRODUCTION

As reliability engineering has developed since its inception as a formal discipline, the following major emphases have evolved [1, 2]:

1940s and 1950s	Component quality and reliability
1960s and 1970s	Module and subsystem quality and reliability
1970s and 1980s	System reliability and field service experience
1990s	(Unexpected) use of the product
2000s	Quality of business processes

The above trends, along with our own experience, tell us that reliability is progressing to higher levels in the business organization; that it is no longer limited to specific tests or analyses for hardware; and that its progress warrants its position in the mainstream of the enterprise. From a somewhat peripheral activity, reliability has become one of the elements required for success, and the purpose of this chapter is to describe, very briefly, how Design of Experiments (DoE) can be used to integrate reliability into the organization.* This is done in three

---

* For a more extensive treatment of this subject, the reader is referred to the author's book, *Value-Added Management with Design of Experiments*, Chapman & Hall, 1995.

345

parts: (1) managing the culture, (2) managing the technology, and (3) managing the data.

22.2 MANAGING THE CULTURE

Throughout this text, a major emphasis has been on viewing all of the various disciplines present in the organization (design of experiments, reliability, accelerated testing, product design, production, etc.) as parts of an integrated whole. Among other things, this implies that no discipline in the modern organization can operate successfully unless it takes into account the various exchanges and interfaces among all of the disciplines. One of the most critical features of the integrated organization is its ability to communicate both within itself and with outside entities.

Figure 22-1 shows a functional view of the product development and implementation process. The boxes indicate the separate subprocesses within the larger process. Usually, the subprocesses are under the control of different individuals or organizations, and the temptation is for those in charge of the subprocesses to optimize each of them with little regard for optimizing the process as a whole. Even when those who are in control of the subprocesses wish to cooperate in optimizing the larger process, they lack the communication tools to do so. As discussed in Chapter 21, and illustrated in Table 21-12, DoE is an excellent tool for such cooperative activity. In Figure 22-1, dashed ovals illustrate that DoE projects can be conducted among any number of the subprocesses to optimize

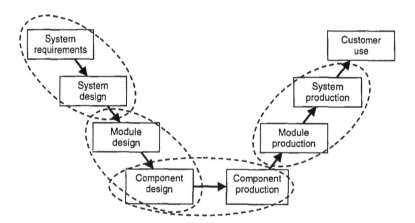

FIGURE 22-1 A schematic diagram of the system design and implementation process. The dashed ovals indicate examples of subprocesses which can be evaluated together in designed experiments.

the entire system. In reality, almost any combination of boxes of Figure 22-1 can be included to contribute factors for evaluation in a designed experiment.

The cultural contribution of DoE to the cooperative development process goes far beyond the optimization of individual subprocesses or product features; and it goes beyond optimizing just the technical aspects of the entire process. Used properly, DoE creates a culture of cooperation, and provides the means to implement and maintain it.

Use of a common language is essential to the culture of cooperation. This can be difficult when the various groups within the organization have different backgrounds and orientation. DoE can facilitate the common language, because of its structured approach to describing a project. Once the members of the organization learn the basic DoE terms, such as factor, level, response, it is easy for them to understand the purpose of any project, and to follow its progress.

Because of its emphasis on constant improvement, DoE can facilitate the emphasis on constant change and continuous improvement necessary for an organization to survive. With a tool such as DoE, change can be planned, structured, monitored, and controlled. The results generated from DoE can be used to improve existing products and processes, and also to plan for future ones. When coupled with reliability improvement, it can be used as an index of the maturity of the organization and of its products. It also contributes a sense of continuous learning and growth for the entire organization, because continuous application of DoE provides a constant stream of new information that can be used to improve the organization, its processes, and its products.

22.3 MANAGING THE TECHNOLOGY

It is common to categorize businesses according to their use of technology. "High-tech" companies of the 1960s and 1970s were invariably aerospace or mainframe computer manufacturers. In the 1980s, software and personal computer companies began to represent the leading edge of technology, and in the 1990s and beyond, high-technology companies include internet, communications, and even toys and games. "High-tech" is no longer limited to a few segments of the economy: everyone is involved in it.

New technologies impact us in ways that we never expected before; and these changes can come from any source. Two examples from the electronics industry illustrate this point. The first is in the area of assembly of electronic components onto printed circuit boards (an area that has contributed many of the examples used earlier in this text). Up to about 1980, the standard package for integrated circuits was a dual-in-line package (DIP), which was standard in the industry. The packages had standard lead dimensions and spacings; standard lead materials and finishes; and standard package dimensions and materials. The configurations and materials of the printed circuit boards also were standard. The electronics assembler's task was straightforward: optimize the materials, processes, and equipment to assemble DIPs onto FR-4 circuit cards with eight layers or less. In the last 20 years, the standard DIP package has practically ceased to exist. It has been replaced by scores of new package types, each with its unique dimensions, materials, and requirements for assembly. Also, due to the need for higher and higher processing speeds, the number of circuit card and other assembly materials to be considered has also grown explosively. The electronics assembler must have a tool such as DoE that allows efficient, structured, cost-effective evaluation of a large number of options in a very short time. It is easy to find similar challenges in almost any industry today, including those that are not considered high tech. In fact, in all areas of life, we are faced with a dazzling array of choices, and often we do not have a structured way to evaluate them.

A second example from the electronics industry is the requirement to eliminate lead from the solders used widely for electronic assembly, driven by environmental requirements. Every business is subject to unexpected requirements, from unexpected sources, to rapidly accommodate new conditions that were not, and could not have been, expected when the processes were developed. Many of the responses to these unexpected requirements are made unconsciously, and many of them are made simply by abdication: by letting events dictate the response. DoE is a tool that can be used to organize the available knowledge, and to evaluate potential responses in a structured, efficient manner.

22.4 MANAGING THE DATA

Even though DoE can be used to produce information in a very efficient and timely manner, there are times when the information must be in place before the decision-making process begins. Also, it is economically practical to develop databases that can be applied to many different problems and projects.

Almost every activity of the modern organization produces data. The data may come from design decisions, design and process parameters, quality and inspection data, reliability test data, etc. They may even come from areas not directly related to the design and production processes, such as business systems and marketing. Too often, the data production and analysis process consists of collecting a small amount of information directed toward making a specific decision, and then never using the information again. Also too often, the data collection process is structured so that the data can be used for only one type of decision. Data collection is almost always expensive and time consuming; therefore it is important to structure the data collection process to provide the maximum amount of information, retrievable and usable in the maximum number of applications, for the minimum amount of cost in time and expense.

In earlier chapters, examples are presented for setting up, conducting, analyzing, and drawing conclusions from experiments designed for various purposes.

This usually involves defining the problem to be solved, selecting the factors that might impact the desired response, designing an experiment to collect relevant information, collecting and analyzing the data, and drawing conclusions. A more proactive approach is to consider not only the problem at hand but to anticipate other variations of the problem that may occur in the future. For example, it may be necessary to design a solder reflow process to assemble printed circuit cards with existing line-space specifications, e.g., 0.006 in lines and 0.006 in spaces. Instead of simply designing an experiment to optimize the process to produce the maximum yield for today's requirements, it may be desirable to consider requirements which may be imposed in the future, such as 0.004 in lines and 0.004 in spaces. In this case, a separate section could be added to the sample artwork to produce additional data for the experiment, and the data could be collected at almost no additional cost. It would then be available for use when needed.

As we have seen earlier, most experimental designs have room for additional factors which can be evaluated with no increase in the size of the experiment. Even if a larger array is required to obtain additional information, the incremental cost to do so is almost always less than that required to set up and conduct a separate experiment at a later date.

A small amount of reflection on the above approach will reveal that it can produce tangible benefits limited only by the imagination of the experimenter. In addition, there are many intangible benefits. Three of them are listed here, but it should not be difficult for the reader to think of others.

1. It encourages experimenters to consider their work in a much wider context, and to communicate with other functions within the organization. In the above example, the process engineer should consider not only how to improve today's process, but also how to anticipate and improve tomorrow's product. This consideration requires communication with product designers and marketers to understand future requirements. Furthermore, results from the more comprehensive experiment can be communicated back to the designers and marketers to help define realistic products based on realistic manufacturing capabilities.

2. It may provide solutions not considered in the more narrow approach. Often, an experiment conducted to optimize an existing process for an existing product consists only of optimizing current capability. If future, more stringent requirements are considered, creative approaches outside the traditional boundaries must be taken. In many cases, these new approaches result in dramatic improvements in the existing process, that would not have been attained otherwise.

3. After an experimenter gets into the mode of generating information in this manner, he or she often finds that it is possible to set up and conduct experiments that provide data, the need for which has not even

been anticipated. Some experimenters have been able to go back to the data or samples from previous work, and to design and conduct "retroactive experiments" in which no additional data or samples were used.

There is no formalism for proactive design of an experiment to generate a comprehensive database comprehensively and proactively. The most important rule is to consider the problem in the widest possible context, and to plan to collect data that will have the widest possible application. This approach is illustrated by some examples.

22.4.1 Microcircuit Example

Today's complex microcircuits contain many different materials, design features, and combinations thereof. Furthermore, the technology has advanced to the point where it is not practical to test individual product samples to determine either their quality or reliability. One approach to this problem is the use of test structures [3–5].

The test structure method involves the analysis of product failures in test and use to determine the most likely mechanisms, as well as where and under what conditions they occur. Based on this knowledge, a series of test structures can be designed and built to evaluate each potential failure mechanism separately. Each test structure is so designed and constructed that only a single mechanism can occur. Table 22-1 shows examples of some such mechanisms [4]. Although all the mechanisms in this table are thermally activated, they have different activation energies, and thus qualify as separate mechanisms.

Accelerated reliability tests are conducted on samples of each structure, using only the operating and environmental conditions that activate its unique

TABLE 22-1 Failure Mechanisms Evaluated by the Microcircuit Test Structure Accelerated Testing for Development of a Reliability Database

Mechanism	Activation energy (eV)	Classification
Sinking gates	2.5	High
Airbridge electromigration	0.4	Low
NiCr resistor drift	1.0	Medium
Metal interdiffusion	2.2	High

Source: Adapted from Ref. 4.

failure mechanism, and a reliability database is developed for each mechanism. Using designed experiments, the effects of various design and processing factors on the final product can be determined. The database thus developed is used for future product and process design. This type of experimentation is conducted on an almost continuous basis.

At the time of publication of Ref. 4, the manufacturer had postulated 32 different possible failure mechanisms to be represented by test structures, and had tested over 7900 test structures in over 7 million device hours of accelerated testing.

22.4.2   Pyrotechnics Example

A defense contractor used DoE to develop a reliability database for the design and production of pyrotechnic tracer ammunition [5]. The approach is shown schematically in Figure 22-2, and a general outline of the experimental plan is shown in Figure 22-3. The inner array, or control factors include design parameters (cavity diameter, depth, and wall thickness), process parameters, ignition parameters, and chemical composition parameters. The outer array, or noise, factors, include functional features (spin rates) and environmental conditions (temperature range). Responses include performance data such as burn times and intensities.

The database for the tracer ammunition experiment was collected in less than 3 man-months, compared to an estimated 800 man-months using traditional methods. The database can be used to determine both product and process design

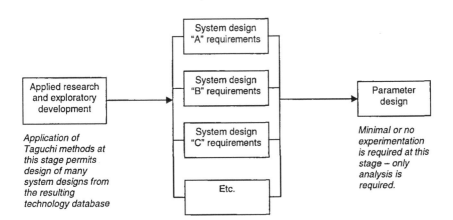

FIGURE 22-2   Schematic diagram describing the use of experimental design methods to produce a database for pyrotechnic tracer ammunition design and manufacturing. (From Ref. 5. © 1987 American Supplier Institute, used by permission.)

Test Condition	Control Factors\nA B C D..........	Noise Matrix\nOuter Array of spin rates and temperature factors selected for each product family	
1\n2\n3\n4\n5\n.\n.\n.\n.	Inner array of controllable factors:\n\nProcess factors\nIgniter factor\nComposition factors\nSize factors\nGeometry factors	Performance data and properties:\n\nBurn times\nColor\nLight intensity	Analysis\n\nMany analyses are conducted as the requirements for different system designs arise

FIGURE 22-3   General outline of the experimental design used to develop the pyrotechnic tracer ammunition database. (From Ref. 5. © 1987 American Supplier Institute, used by permission.)

parameters for a wide range of anticipated new designs. The experimenters estimate the value of the database at over $60 million.

22.4.3   Circuit Card Assembly Example

In setting up and optimizing a process for vapor phase reflow soldering of circuit card assemblies, the usual process factors of reflow profile, flux, time, etc. are used as inner array factors. Instead of using a current product design, or a test sample representing current design rules, a test board was designed with a wide range of features, incorporating design features and components whose use was anticipated over the next several years. The samples were subjected to operating and environmental tests in order to obtain reliability as well as quality data. This was not a one-time project, but is part of a continuous-improvement effort.

The test boards for this type of experiment are used not only to develop a process in a one-time effort, but also to benchmark the process for comparison to suggested design and process changes in the future. They also are used to evaluate the process when production anomalies occur, and to set up the process after a period of down time. The data from this ongoing experimentation are recorded, and the database is used for a variety of applications.

22.5   SUMMARY

The development and use of DoE to integrate reliability into the organization is more a management discipline than a technical one. The examples presented in this chapter involve only technical capabilities that have been available for decades. It is the responsibility of management to provide education in these disciplines, and to ensure their systematic application to comprehensive and proactive design, production, and support activities.

Our society in general is becoming more and more dependent on the accurate collection, analysis, and distribution of data. Management must view their operations as primarily a source of information that can be used to improve their products and services. They must exploit the advantages provided by this viewpoint to continually improve their products and serve their customers. If they do this effectively, they will assure their competitive position in the marketplace.

REFERENCES

1. A. J. M. Huibjen, Reliability improvement in high-volume consumer electronics, presented at The Reliability Challenge Symposium, organized by the Finn Jensen Reliability Consultancy, London, 1996.
2. P. C. Sander and A. C. Brombacher, Product quality and reliability as a function of quality of business processes, Quality and Reliability Engineering International, 15: 409–410, 1999.
3. H. Schafft, D. A. Baglee, and P. E. Kennedy, Building in reliability: Making it work, Proceedings of the International Reliability Physics Symposium, IEEE, 1991. Also in RAC Quarterly, Reliability Analysis Center, Rome, NY, Spring 1999.
4. W. J. Roesch, Finding low activation energy failure mechanisms, Proceedings of the Advanced Microelectronics Technology Qualification, Reliability, and Logistics Workshop, the DoD Tri-Service MIMIC Qualification Committee, the IIT Research Institute, and the Reliability Analysis Center, pp. 308–311, 1991.
5. J. F. Kowalick and G. Hayer, The use of Taguchi methods to establish a broad technology database for system-design applications in the defense industry, Proceedings of the Fifth Symposium on Taguchi Methods, American Supplier Institute, Dearborn, MI, pp. 3–12, 1987.

Glossary

Accelerated test. A test in which the applied stresses are at higher levels than those expected in service, but not high enough to induce failures which would not occur in service. Its purpose is to shorten the time in which the effects of the stresses are observed.

Accumulated damage failure. A type of failure resulting from gradual degradation of a property of an item due to long-term exposure to environmental or operating stresses.

Aliasing pattern. The effects that one contrasts in a fractional factorial experiment estimate. An alias is an effect which cannot be distinguished from another effect.

Allocation. Assigning reliability requirements to individual items within a system to attain the specified system reliability. Sometimes called apportionment.

ANOVA. Acronym for analysis of variance. A method of analysis which quantifies the contribution of each factor to the total variation of the experimental data. Also determines the amount of variation produced by a change in the level of a factor, or the variation due to random error.

Arrhenius equation. A rate equation in which time for an event or process to occur is an exponential function of the inverse of temperature.

Assessed reliability. The reliability of an item as determined on the basis of observed reliability of nominally identical items. Sometimes a confidence level is specified.

Attribute data. Data which cannot be quantified, but only sorted into categories; nonparametric.

Bathtub curve. A curve of failure rate vs. time which has an early life region in which the failure rate decreases with time, a useful life region in which failure rate is nearly constant with time, and a wearout region in which failure rate increases with time.

Bimodal distribution. A probability distribution with two peaks representing two different subpopulations.

Brainstorming. A structured process for discussion of a problem and its potential solutions.

B-type quality characteristic. An experimental result in which the largest value is desired.

Burn-in. An inspection process in which an item is operated under an accelerated stress condition (usually electrical and thermal) in order to precipitate infant mortality failures. See also *environmental stress screening*.

Cell. A unique combination of two or more factor level treatments. The same as a treatment combination.

Censored data. Failure data in which exact times to failure are not known for all samples because they were removed due to irrelevant failures, or due to truncation of the tests before all items failed, or were collected as interval data.

Challenge-response failure. A failure caused by a defect which goes undetected until a specific event challenges it.

Closed-form model. A failure model which can be solved as an equation. Usually, it is based on a description of the failure mechanism at the structural level of the product. See also *constitutive, physics-of-failure,* and *structural* models.

Column downgrading. Modification of a set of orthogonal array columns to allow fewer levels of a factor to be investigated.

Column upgrading. Modification of a set of orthogonal array columns to allow more levels of a factor to be investigated.

Combination design. A method of studying two factors in a single array column.

Condition. A specific combination of factors and levels to be investigated. The same as a treatment combination.

Confidence interval. Range about a given statistical estimate within which the true value is said to be located with some specified degree of confidence.

Confirmation run. Repetition of an experiment using the optimum treatment combination.

Confounding. An experimental situation in which the effects of one factor cannot be distinguished from those of another factor.

Constitutive model. A failure model based on a description of the failure mechanism at the structural level of the product. See also *closed-form, physics-of-failure*, and *structural* models.

Controllable factors. Factors that are intentionally varied in an experiment.

Correction factor. The grand total of all the data for an experiment, squared, and divided by the total number of runs. The sum of the squares due to the grand mean, S_m.

Corrective action. The systematic and documented process of changing designs, materials, and processes to reduce failures.

Critical value. A value from an F-table above which it is unlikely that a factor variance and the error variance could come from the same population.

Cumulative distribution function (cdf). A function which describes the probability that an item will have failed by a given time.

Damage-endurance failure. A failure which results from the accumulation of damage from stresses below the strength of the product.

Degrees of freedom. The number of ways observations are free to vary if the mean is known. For a Taguchi array, the degrees of freedom of a factor is always equal to the number of levels of that factor, minus one.

Design verification test. An accelerated test conducted near the end of the product design period to assess the capability of the product to operate successfully over the anticipated range of operating and environmental conditions.

Deterministic reliability. Reliability methods based on consideration of structural changes within an item which may cause failures.

Effect. The influence exerted by a factor on the results of an experiment. The deviation of a factor level mean from the total mean.

Effective number of replications. The total number of observations divided by the degrees of freedom for all significant factors, plus one.

Environmental conditions. The ambient conditions in which a product must operate. Significant environmental conditions include temperature and variations thereof, humidity, vibration, chemical contamination, etc.

Environmental stress screening (ESS). A more general case for burn-in, in which accelerated stresses are applied to an item in order to expose latent defects.

Error effect. The variation in results of an experiment due to unknown or uncontrolled causes, i.e., causes other than the factors evaluated by the experimental array.

Error variation. The random variation of the entire data base of an experiment.

EVOP. Evolutionary operations. A method of improving a process by "systematic tinkering." Factor levels are changed, but stay within the limits required to produce good product.

Exponential distribution. A special case of the Weibull distribution which describes a constant failure rate.

Extrinsic failure. A failure resulting from a cause external to the failed item.

Eyring model. An accelerated testing model which contains the Arrhenius equation for temperature, and one other stress.

Factor. A variable suspected of having influence on the outcome of an experiment. An independent variable. An input variable.

Failure. The termination of the ability of an item to perform its required function within specified limits.

Failure-free-operating period (FFOP). A time during which an item operates without failure.

Failure mechanism. The structural change within an item that causes failure of the item.

Failure mode. The way in which a failure is manifested, such as fracture, short, open, etc.

Failure mode, effects, and criticality analysis (FMECA). A failure analysis method in which each mode of failure of each component in a system is considered, along with its likelihood of occurrence, and its effects on the system. Potential failures are classified according to their severity.

Failure rate. The total number of failures within a population, divided by the total number of life units expended by that population, during a specified time interval.

Failure terminated test. A reliability test which is terminated after a specified number of failures has occurred.

Fishbone diagram. A graphical representation of the possible causes of an effect. Sometimes called an Ishikawa diagram.

Fractional factorial experiment. An experiment in which not all possible factor level treatment combinations are evaluated.

Full factorial experiment. An experiment in which all possible factor level treatment combinations are run.

F value. The ratio of the variance of a given factor to that of the error variance in an experiment.

Grand mean effect. The variation due to all sources in an experiment.

Hazard rate. A function which describes the probability that an item will fail in the next time interval, given that it has not yet failed. The instantaneous failure rate.

Infant mortality. Early failures due to latent defects resulting from manufacturing errors, defective materials, etc. On the bathtub curve, this is a region of decreasing failure rate.

Inherent reliability. The reliability that an item would exhibit if it contained no design or process defects, and if it were operated and supported under ideal specified conditions.

Inner array. The array of controllable factors.

Interaction. A phenomenon which occurs when the effect of one factor is different at different levels of another factor. Observed when the change in response between levels for one factor is not the same for all levels of another factor.

Interval data. Data collected only at specified time intervals.

Intrinsic failure. A failure resulting from a cause internal to the failed item.

Inverse power law. An accelerated test model used when the life of a system is inversely proportional to the applied stress.

Latent defect. A weakness in a item which is likely to cause failure early in its service life.

Level. The value or condition at which a factor is set for a treatment combination.

Linear graph. A tool for identifying sets of interacting columns in an orthogonal array.

Location parameter. A parameter of a life distribution which locates it in time, e.g., the mean of a normal distribution.

Lognormal distribution. A distribution represented by a bell-shaped curve whose values are symmetric about the mean on a logarithmic scale.

Loss function. A formal method of estimating the total cost of variation from the optimum process or design parameter.

Main effect. The effect of a factor on the results when the factor is changed from one level to another.

Markov analysis. A system failure analysis method in which the reliability of each component is modeled in terms of its probability of going from an unfailed to a failed state, or vice versa, in a given time interval.

Mean. The arithmetic average of a set of numbers.

Mean square deviation. A value used in calculating the signal-to-noise ratio.

Mean-time-between-failures (MTBF). A basic measure of reliability for repairable items. The mean number of life units between failures for a set of items during a specified measurement interval of time in test or use.

Mean-time-to-failure (MTTF). A basic measure of reliability for non-repairable items. Usually applied to components, rather than to systems. The total number of life units divided by the total number of failures for a set of items during a specified measurement interval of time in test or use.

Median. The value of the middle item when the data are arranged in increasing or decreasing order of magnitude.

Miner's rule. A method used to apportion product life among several levels of the applied stresses.

Mode. The most common value occurring in a set of numbers.

Nested factor. A factor which results from the combination of two other variables in an experiment. It is not the interaction of the two other variables.

Noise factors. Factors in an experiment which may contribute to variation, but which are not controlled.

Nominal group theory. A brainstorming technique in which each participant contributes ideas in turn without comment from others, followed by voting. A prior commitment is made to use the results.

Normal distribution. A distribution represented by a bell-shaped curve, whose values are symmetrical about the mean on a linear scale.

N-type quality characteristic. An experimental result in which a nominal value is desired.

Observation. Results of an individual experimental run.

Off-line quality control. Experimental work conducted on non-shippable product.

One-factor-at-a-time experiment. An experiment in which only one factor is varied while the others are held constant.

On-line quality control. Design or process improvements made on shippable product without interrupting production.

Optimum conditions. The combination of factor levels which produces the most desirable results in an experiment.

Orthogonal array. A matrix of numbers which represent the levels of factors to be investigated in an experiment. The column of the matrix represent the factors, and the rows represent the treatment combinations.

Outer array. An array of uncontrolled factors in an experiment, which exists outside the experimental array.

Parameter design. The process of selecting the optimum values for the various factors which influence the quality of a product or process.

Parametric data. Data which can be collected in quantifiable, numeric form; as opposed to attribute data.

Per cent contribution. The amount of influence a factor has on the results of an experiment.

Physics-of-failure model. A failure model based on a description of the failure mechanism at the structural level of the product. See also *closed-form, constitutive*, and *structural* models.

Pooled error. The estimate of random variation of an experiment based on the effects of non-significant factors.

Predicted value. An expected value for an effect, based on experimental data.

Probabilistic reliability. Reliability methods based only on statistical methods, without consideration of the structural changes within an item which cause failure.

Probability density function (pdf). A function which describes the probability that an item will fail at a given time.

Probability plotting. A method of representing a cumulative probability density function by plotting its failure times on paper with scales specialized to various types of distributions.

Pure variation of the source. The variation of a factor with the portion due to error variance removed.

Quality function deployment. A systematic, quantitative process for converting customer requirements into design features.

Random failure. A failure whose occurrence is predictable only in a probabilistic sense.

Range. The difference between the largest and smallest observations.

Ranked data. Data which are sorted according to their values, from lowest to highest.

Redundancy. The improvement of system reliability by providing more than one means of accomplishing one or more of the system functions.

Reliability. The likelihood of performing the desired function, and only the desired function, when the user wants the function to be performed.

Reliability function. A function describing the probability that an item will not have failed by a given time.

Reliability growth. Improvement in the reliability of a product by systematic testing and correction of the causes of failure.

Repairable item. An item which can be restored to service after failure.

Repetition. The number of times an experiment is run.

Replication. The number of times a single run is conducted in an experiment. Can also mean the number of observations for a factor at a given level.

Robust design. A product or process design which is insensitive to uncontrollable factors.

Run. The production of a sample or set of samples made with a single treatment combination. A row of an orthogonal array.

Saturated design. An experimental design in which only main effects are investigated.

Screening. An inspection process to expose and remove defective items. In reliability, defective items are those which might exhibit early life failures.

Screening experiment. An experiment in which as many factors are investigated as possible; usually at the expense of obtaining information about interactions or more than two levels of a factor.

Secondary failure. Failure of an item caused by the failure of another item. Sometimes called a dependent failure.

Shape parameter. A parameter of a distribution which describes its shape, or spread, e.g., the standard deviation of a normal distribution.

Signal-to-noise ratio. Comparison of the variation of a factor with the underlying variation of the process being investigated. A data transformation which allows quantitative evaluation of a factor considering both the mean and variation of its effect.

Significant factor. A factor whose effect is large enough that it is unlikely to have occurred by chance.

Skew. The measure of the asymmetry of a distribution.

Source variance. The variation of the source corrected for the number of degrees of freedom.

Source variation. The variation of a factor in a designed experiment.

Standard deviation. The square root of the variance.

Standard error of the mean. Square root of the error variance divided by the effective number of replications.

Step stress test. A form of design test; an accelerated test in which the levels of a combination of stresses are increased in a stepwise fashion, to cause failures.

Sterilized experiment. An experiment which yields results which cannot be applied to real life situations because the usual noise factors have been suppressed.

Stress-strength failure. A failure caused when the applied stress exceeds the strength of a product.

Structural model. A failure model based on a description of the failure mechanism at the structural level of the product. See also, *closed-form, physics-of-failure*, and *constitutive* models.

Subpopulation. A small group of items within a population which have a common characteristic which distinguishes them from the main population.

Sum of squares. A measure of variation defined in terms of the sum of the squared deviations from the mean.

System design. The selection and specification of equipment, materials, and instructions for a process or product.

Time terminated test. A reliability test which is terminated after a specified time.

Time-to-first-failure (TTFF). A basic measure of useful life. The time at which the first failure is observed for a set of items in test or use. TTFF may also be defined at a point at the beginning of the wearout distribution, such as t_{01} or t_1. TTFF may also be called the failure-free-operating-period (FFOP).

Tolerance design. The specification of allowable limits for a design or process feature.

Tolerance-response failure A failure in which the product operates within specification, but with degraded performance.

Total variation. The sum of the squared observations.

Treatment combination. A unique set of factors in an experiment.

Uncontrollable factor. A factor which for a variety of reasons, cannot be controlled in production.

Unrepairable item. An item which must be discarded after it fails.

Unsaturated design. An experiment in which some columns are available for the evaluation of error terms or interactions of main effects.

Useful life. The period of life in which an item or set of items operates with a constant, acceptable failure rate. On the bathtub curve, useful life falls between infant mortality and wearout.

Variance. A measure of variability. The sum of squared deviation measures from a calculated mean divided by the degrees of freedom associated with the measures.

Wearout. The last phase of the lifetime of an item. Region of increasing failure rate on the bathtub curve.

Weibull distribution. A specialized reliability distribution which can, by changing its parameter values, be made to fit a wide variety of failure data.

Appendix A

Taguchi Orthogonal Arrays and Linear Graphs

The L_{12} and the L_{18} are specially designed arrays which are used primarily to evaluate main effects. Except for an interaction between columns 1 and 2 of the L_{18}, the interactions are distributed evenly among the columns, and cannot be analyzed separately.

The Taguchi orthogonal arrays and linear graphs in this text are taken from G. Taguchi and S. Konishi, *Orthogonal Arrays and Linear Graphs*, American Supplier Institute, Dearborn, MI, 1987. They are used by the generous permission of the publisher. The reader is referred to this publication for a more complete list of orthogonal arrays and linear graphs. © American Supplier Institute, Livonia, Michigan, used by permission.

366

L₄ (2³)

Run no.	1	2	3
1	1	1	1
2	1	2	2
3	2	1	2
4	2	2	1

```
1        3        2
●─────────────────●
```

L₈ (2⁷)

Run no.	1	2	3	4	5	6	7
1	1	1	1	1	1	1	1
2	1	1	1	2	2	2	2
3	1	2	2	1	1	2	2
4	1	2	2	2	2	1	1
5	2	1	2	1	2	1	2
6	2	1	2	2	1	2	1
7	2	2	1	1	2	2	1
8	2	2	1	2	1	1	2

i

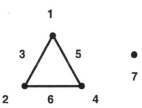

ii

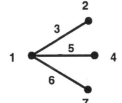

L_{16} (2^{15})

Run	1	2	3	4	5	6	7	8	9	10	11	12	13	14	15
1	1	1	1	1	1	1	1	1	1	1	1	1	1	1	1
2	1	1	1	1	1	1	1	2	2	2	2	2	2	2	2
3	1	1	1	2	2	2	2	1	1	1	1	2	2	2	2
4	1	1	1	2	2	2	2	2	2	2	1	1	1	1
5	1	2	2	1	1	2	2	1	1	2	2	1	1	2	2
6	1	2	2	1	1	2	2	2	2	1	1	2	2	1	1
7	1	2	2	2	2	1	1	1	1	2	2	2	2	1	1
8	1	2	2	2	2	1	1	2	2	1	1	1	1	2	2
9	2	1	2	1	2	1	2	1	2	1	2	1	2	1	2
10	2	1	2	1	2	1	2	2	1	2	1	2	1	2	1
11	2	1	2	2	1	2	1	1	2	1	2	2	1	2	1
12	2	1	2	2	1	2	1	2	1	2	1	1	2	1	2
13	2	2	1	1	2	2	1	1	2	2	1	1	2	2	1
14	2	2	1	1	2	2	1	2	1	1	2	2	1	1	2
15	2	2	1	2	1	1	2	1	2	2	1	2	1	1	2
16	2	2	1	2	1	1	2	2	1	1	2	1	2	2	1

i

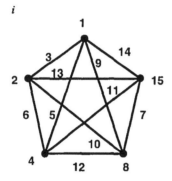

ii

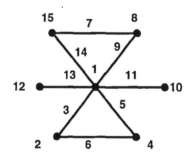

iii

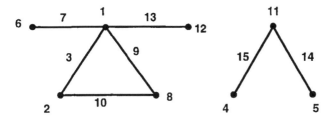

iv

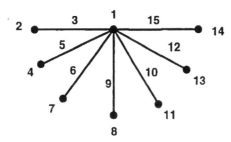

v

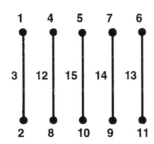

$$L_{32}\,(2^{31})$$

Run	1	2	3	4	5	6	7	8	9	10	11	12	13	14	15
1	1	1	1	1	1	1	1	1	1	1	1	1	1	1	1
2	1	1	1	1	1	1	1	1	1	1	1	1	1	1	1
3	1	1	1	1	1	1	1	2	2	2	2	2	2	2	2
4	1	1	1	1	1	1	1	2	2	2	2	2	2	2	2
5	1	1	1	2	2	2	2	1	1	1	1	2	2	2	2
6	1	1	1	2	2	2	2	1	1	1	1	2	2	2	2
7	1	1	1	2	2	2	2	2	2	2	2	1	1	1	1
8	1	1	1	2	2	2	2	2	2	2	2	1	1	1	1
9	1	2	2	1	1	2	2	1	1	2	2	1	1	2	2
10	1	2	2	1	1	2	2	1	1	2	2	1	1	2	2
11	1	2	2	1	1	2	2	2	2	1	1	2	2	1	1
12	1	2	2	1	1	2	2	2	2	1	1	2	2	1	1
13	1	2	2	2	2	1	1	1	1	2	2	2	2	1	1
14	1	2	2	2	2	1	1	1	1	2	2	2	2	1	1
15	1	2	2	2	2	1	1	2	2	1	1	1	1	2	2
16	1	2	2	2	2	1	1	2	2	1	1	1	1	2	2
17	2	1	2	1	2	1	2	1	2	1	2	1	2	1	2
18	2	1	2	1	2	1	2	1	2	1	2	1	2	1	2
19	2	1	2	1	2	1	2	2	1	2	1	2	1	2	1
20	2	1	2	1	2	1	2	2	1	2	1	2	1	2	1
21	2	1	2	2	1	2	1	1	2	1	2	2	1	2	1
22	2	1	2	2	1	2	1	1	2	1	2	2	1	2	1
23	2	1	2	2	1	2	1	2	1	2	1	1	2	1	2
24	2	1	2	2	1	2	1	2	1	2	1	1	2	1	2
25	2	2	1	1	2	2	1	1	2	2	1	1	2	2	1
26	2	2	1	1	2	2	1	1	2	2	1	1	2	2	1
27	2	2	1	1	2	2	1	2	1	1	2	2	1	1	2
28	2	2	1	1	2	2	1	2	1	1	2	2	1	1	2
29	2	2	1	2	1	1	2	1	2	2	1	2	1	1	2
30	2	2	1	2	1	1	2	1	2	2	1	2	1	1	2
31	2	2	1	2	1	1	2	2	1	1	2	1	2	2	1
32	2	2	1	2	1	1	2	2	1	1	2	1	2	2	1

L_{32} (2^{31}), cont'd.

Run	16	17	18	19	20	21	22	23	24	25	26	27	28	29	30	31
1	1	1	1	1	1	1	1	1	1	1	1	1	1	1	1	1
2	2	2	2	2	2	2	2	2	2	2	2	2	2	2	2	2
3	1	1	1	1	1	1	1	1	2	2	2	2	2	2	2	2
4	2	2	2	2	2	2	2	2	1	1	1	1	1	1	1	1
5	1	1	1	1	2	2	2	2	1	1	1	1	2	2	2	2
6	2	2	2	2	1	1	1	1	2	2	2	2	1	1	1	1
7	1	1	1	1	2	2	2	2	2	2	2	2	1	1	1	1
8	2	2	2	2	1	1	1	1	1	1	1	1	2	2	2	2
9	1	1	2	2	1	1	2	2	1	1	2	2	1	1	2	2
10	2	2	1	1	2	2	1	1	2	2	1	1	2	2	1	1
11	1	1	2	2	1	1	2	2	2	2	1	1	2	2	1	1
12	2	2	1	1	2	2	1	1	1	1	2	2	1	1	2	2
13	1	1	2	2	2	2	1	1	1	1	2	2	2	2	1	1
14	2	2	1	1	1	1	2	2	2	2	1	1	1	1	2	2
15	1	1	2	2	2	2	1	1	2	2	1	1	1	1	2	2
16	2	2	1	1	1	1	2	2	1	1	2	2	2	2	1	1
17	1	2	1	2	1	2	1	2	1	2	1	2	1	2	1	2
18	2	1	2	1	2	1	2	1	2	1	2	1	2	1	2	1
19	1	2	1	2	1	2	1	2	2	1	2	1	2	1	2	1
20	2	1	2	1	2	1	2	1	1	2	1	2	1	2	1	2
21	1	2	1	2	2	1	2	1	1	2	1	2	2	1	2	1
22	2	1	2	1	1	2	1	2	2	1	2	1	1	2	1	2
23	1	2	1	2	2	1	2	1	2	1	2	1	1	2	1	2
24	2	1	2	1	1	2	1	2	1	2	1	2	2	1	2	1
25	1	2	2	1	1	2	2	1	1	2	2	1	1	2	2	1
26	2	1	1	2	2	1	1	2	2	1	1	2	2	1	1	2
27	1	2	2	1	1	2	2	1	2	1	1	2	2	1	1	2
28	2	1	1	2	2	1	1	2	1	2	2	1	1	2	2	1
29	1	2	2	1	2	1	1	2	1	2	2	1	2	1	1	2
30	2	1	1	2	1	2	2	1	2	1	1	2	1	2	2	1
31	1	2	2	1	2	1	1	2	2	1	1	2	1	2	2	1
32	2	1	1	2	1	2	2	1	1	2	2	1	2	1	1	2

L_{12} (2^{11})

Run	1	2	3	4	5	6	7	8	9	10	11
1	1	1	1	1	1	1	1	1	1	1	1
2	1	1	1	1	1	2	2	2	2	2	2
3	1	1	2	2	2	1	1	1	2	2	2
4	1	2	1	2	2	1	2	2	1	1	2
5	1	2	2	1	2	2	1	2	1	2	1
6	1	2	2	2	1	2	2	1	2	1	1
7	2	1	2	2	1	1	2	2	1	2	1
8	2	1	2	1	2	2	2	1	1	1	2
9	2	1	1	2	2	2	1	2	2	1	1
10	2	2	2	1	1	1	1	2	2	1	2
11	2	2	1	2	1	2	1	1	1	2	2
12	2	2	1	1	2	1	2	1	2	2	1

L_{18} ($2^1 3^7$)

Run	1	2	3	4	5	6	7	8
1	1	1	1	1	1	1	1	1
2	1	1	2	2	2	2	2	2
3	1	1	3	3	3	3	3	3
4	1	2	1	1	2	2	3	3
5	1	2	2	2	3	3	1	1
6	1	2	3	3	1	1	2	2
7	1	3	1	2	1	3	2	3
8	1	3	2	3	2	1	3	1
9	1	3	3	1	3	2	1	2
10	2	1	1	3	3	2	2	1
11	2	1	2	1	1	3	3	2
12	2	1	3	2	2	1	1	3
13	2	2	1	2	3	1	3	2
14	2	2	2	3	1	2	1	3
15	2	2	3	1	2	3	2	1
16	2	3	1	3	2	3	1	2
17	2	3	2	1	3	1	2	3
18	2	3	3	2	1	2	3	1

$L_9 (3^4)$

Run no.	1	2	3	4
1	1	1	1	1
2	1	2	2	2
3	1	3	3	3
4	2	1	2	3
5	2	2	3	1
6	2	3	1	2
7	3	1	3	2
8	3	2	1	3
9	3	3	2	1

$L_{16} (4^5)$

Run no.	1	2	3	4	5
1	1	1	1	1	1
2	1	2	2	2	2
3	1	3	3	3	3
4	1	4	4	4	4
5	2	1	2	3	4
6	2	2	1	4	3
7	2	3	4	1	2
8	2	4	3	2	1
9	3	1	3	4	2
10	3	2	4	3	1
11	3	3	1	2	4
12	3	4	2	1	3
13	4	1	4	2	3
14	4	2	3	1	4
15	4	3	2	4	1
16	4	4	1	3	2

$$L_{27}(3^{13})$$

Run	1	2	3	4	5	6	7	8	9	10	11	12	13
1	1	1	1	1	1	1	1	1	1	1	1	1	1
2	1	1	1	1	2	2	2	2	2	2	2	2	2
3	1	1	1	1	3	3	3	3	3	3	3	3	3
4	1	2	2	2	1	1	1	2	2	2	3	3	3
5	1	2	2	2	2	2	2	3	3	3	1	1	1
6	1	2	2	2	3	3	3	1	1	1	2	2	2
7	1	3	3	3	1	1	1	3	3	3	2	2	2
8	1	3	3	3	2	2	2	1	1	1	3	3	3
9	1	3	3	3	3	3	3	2	2	2	1	1	1
10	2	1	2	3	1	2	3	1	2	3	1	2	3
11	2	1	2	3	2	3	1	2	3	1	2	3	1
12	2	1	2	3	3	1	2	3	1	2	3	1	2
13	2	2	3	1	1	2	3	2	3	1	3	1	2
14	2	2	3	1	2	3	1	3	1	2	1	2	3
15	2	2	3	1	3	1	2	1	2	3	2	3	1
16	2	3	1	2	1	2	3	3	1	2	2	3	1
17	2	3	1	2	2	3	1	1	2	3	3	1	2
18	2	3	1	2	3	1	2	2	3	1	1	2	3
19	3	1	3	2	1	3	2	1	3	2	1	3	2
20	3	1	3	2	2	1	3	2	1	3	2	1	3
21	3	1	3	2	3	2	1	3	2	1	3	2	1
22	3	2	1	3	1	3	2	2	1	3	3	2	1
23	3	2	1	3	2	1	3	3	2	1	1	3	2
24	3	2	1	3	3	2	1	1	3	2	2	1	3
25	3	3	2	1	1	3	2	3	2	1	2	1	3
26	3	3	2	1	2	1	3	1	3	2	3	2	1
27	3	3	2	1	3	2	1	2	1	3	1	3	2

i

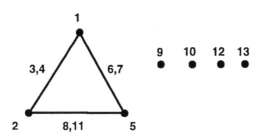

ii

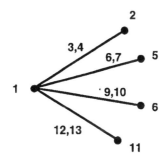

iii

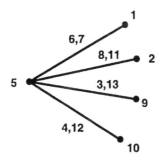

Appendix B

Values of the F-Distribution

All tables in this appendix are from A. Hald, *Statistical and Formulas*, John Wiley & Sons, Inc., New York, 1952. They are used by permission of the publisher.

F- Table for 99%

f_1: degrees of freedom in the numerator
f_2: degrees of freedom in the denominator

f_2 \ f_1	1	2	3	4	5	6	7	8	9	10	12	15	20	24	30	40	60	120	∞
1	4052.2	4999.5	5403.3	5624.6	5763.7	5859.0	5928.3	5981.6	6022.5	6055.8	6106.3	6157.3	6208.7	6234.6	6260.7	6286.8	6313.0	6339.4	6356.0
2	98.503	99.000	99.166	99.249	99.299	99.332	99.356	99.374	99.388	99.399	99.416	99.432	99.449	99.458	99.466	99.474	99.483	99.491	99.501
3	34.116	30.817	29.457	28.710	28.237	27.911	27.672	27.489	27.345	27.229	27.052	26.872	26.690	26.598	26.505	26.411	26.316	26.221	26.125
4	21.198	18.000	16.694	15.977	15.522	15.207	14.976	14.799	14.659	14.546	14.374	14.198	14.020	13.929	13.838	13.745	13.652	13.558	13.463
5	16.258	13.274	12.060	11.392	10.967	10.672	10.456	10.289	10.158	10.051	9.8883	9.7222	9.5527	9.4665	9.3793	9.2912	9.2020	9.1118	9.0204
6	13.745	10.925	9.7795	9.1483	8.7459	8.4661	8.2600	8.1016	7.9761	7.8741	7.7183	7.5590	7.3958	7.3127	7.2285	7.1432	7.0568	6.9690	6.8801
7	12.246	9.5466	8.4513	7.8467	7.4604	7.1914	6.9928	6.8401	6.7188	6.6201	6.4691	6.3143	6.1554	6.0743	5.9921	5.9084	5.8236	5.7372	5.6495
8	11.259	8.6491	7.5910	7.0060	6.6318	6.3707	6.1776	6.0289	5.9106	5.8143	5.6668	5.5151	5.3591	5.2793	5.1981	5.1156	5.0316	4.9460	4.8588
9	10.561	8.0215	6.9919	6.4221	6.0569	5.8018	5.6129	5.4671	5.3511	5.2565	5.1114	4.9621	4.8080	4.7290	4.6486	4.5667	4.4831	4.3978	4.3105
10	10.044	7.5594	6.5523	5.9943	5.6363	5.3858	5.2001	5.0567	4.9424	4.8492	4.7059	4.5582	4.4054	4.3269	4.2469	4.1653	4.0819	3.9965	3.9090
11	9.6460	7.2057	6.2167	5.6683	5.3160	5.0692	4.8861	4.7445	4.6315	4.5393	4.3974	4.2509	4.0990	4.0209	3.9411	3.8596	3.7761	3.6904	3.6025
12	9.3302	6.9266	5.9526	5.4119	5.0643	4.8206	4.6395	4.4994	4.3875	4.2961	4.1553	4.0096	3.8584	3.7805	3.7008	3.6192	3.5355	3.4494	3.3608
13	9.0738	6.7010	5.7394	5.2053	4.8616	4.6204	4.4410	4.3021	4.1911	4.1003	3.9603	3.8154	3.6646	3.5868	3.5070	3.4253	3.3413	3.2548	3.1654
14	8.8616	6.5149	5.5639	5.0354	4.6950	4.4558	4.2779	4.1399	4.0297	3.9394	3.8001	3.6557	3.5052	3.4274	3.3476	3.2656	3.1813	3.0942	3.0040

F- Table for 99%

f_1: degrees of freedom in the numerator
f_2: degrees of freedom in the denominator

f_2 \ f_1	1	2	3	4	5	6	7	8	9	10	12	15	20	24	30	40	60	120	∞
15	8.6631	6.3589	5.4170	4.8932	4.5556	4.3183	4.1415	4.0045	3.8948	3.8049	3.6662	3.5222	3.3719	3.2940	3.2141	3.1319	3.0471	2.9595	2.8684
16	8.5310	6.2262	5.2922	4.7726	4.4374	4.2016	4.0259	3.8896	3.7804	3.6909	3.5527	3.4089	3.2588	3.1808	3.1007	3.0182	2.9330	2.8447	2.7528
17	8.3997	6.1121	5.1850	4.6690	4.3359	4.1015	3.9267	3.7910	3.6822	3.5931	3.4552	3.3117	3.1615	3.0835	3.0032	2.9205	2.8348	2.7459	2.6530
18	8.2854	6.0129	5.0919	4.5790	4.2479	4.0146	3.8406	3.7054	3.5971	3.5082	3.3706	3.2273	3.0771	2.9990	2.9185	2.8354	2.7493	2.6597	2.5660
19	8.1850	5.9259	5.0103	4.5003	4.1708	3.9386	3.7653	3.6305	3.5225	3.4338	3.2965	3.1533	3.0031	2.9249	2.8442	2.7608	2.6742	2.5839	2.4893
20	8.0960	5.8489	4.9382	4.4307	4.1027	3.8714	3.6987	3.5644	3.4567	3.3682	3.2311	3.0880	2.9377	2.8594	2.7785	2.6947	2.6077	2.5168	2.4212
21	8.0166	5.7804	4.8740	4.3688	4.0421	3.8117	3.6396	3.5056	3.3981	3.3098	3.1729	3.0299	2.8796	2.8011	2.7200	2.6359	2.5484	2.4568	2.3603
22	7.9454	5.7190	4.8166	4.3134	3.9880	3.7583	3.5867	3.4530	3.3458	3.2576	3.1209	2.9780	2.8274	2.7488	2.6675	2.5831	2.4951	2.4029	2.3055
23	7.8811	5.6637	4.7649	4.2635	3.9392	3.7102	3.5390	3.4057	3.2986	3.2106	3.0740	2.9311	2.7805	2.7017	2.6202	2.5355	2.4471	2.3542	2.2559
24	7.8229	5.6136	4.7181	4.2184	3.8951	3.6667	3.4959	3.3629	3.2560	3.1681	3.0316	2.8887	2.7380	2.6591	2.5773	2.4923	2.4035	2.3099	2.2107
25	7.7698	5.5680	4.6755	4.1774	3.8550	3.6272	3.4568	3.3239	3.2172	3.1294	2.9931	2.8502	2.6993	2.6203	2.5383	2.4530	2.3637	2.2695	2.1694
26	7.7213	5.5263	4.6366	4.1400	3.8183	3.5911	3.4210	3.2884	3.1818	3.0941	2.9579	2.8150	2.6640	2.5848	2.5026	2.4170	2.3273	2.2325	2.1315
27	7.6767	5.4881	4.6009	4.1056	3.7848	3.5580	3.3882	3.2558	3.1494	3.0618	2.9256	2.7827	2.6316	2.5522	2.4699	2.3840	2.2938	2.1984	2.0965
28	7.6356	5.4529	4.5681	4.0740	3.7539	3.5276	3.3581	3.2259	3.1195	3.0320	2.8959	2.7530	2.6017	2.5223	2.4397	2.3535	2.2629	2.1670	2.0642
29	7.5976	5.4205	4.5378	4.0449	3.7254	3.4995	3.3302	3.1982	3.0920	3.0045	2.8685	2.7256	2.5742	2.4946	2.4118	2.3253	2.2344	2.1378	2.0342
30	7.5625	5.3904	4.5097	4.0179	3.6990	3.4735	3.3045	3.1726	3.0665	2.9791	2.8431	2.7002	2.5487	2.4689	2.3860	2.2992	2.2079	2.1107	2.0062
40	7.3141	5.1785	4.3126	3.8283	3.5138	3.2910	3.1238	2.9930	2.8876	2.8005	2.6648	2.5216	2.3689	2.2880	2.2034	2.1142	2.0194	1.9172	1.8047
60	7.0771	4.9774	4.1259	3.6491	3.3389	3.1187	2.9530	2.8233	2.7185	2.6318	2.4961	2.3523	2.1978	2.1154	2.0285	1.9360	1.8363	1.7263	1.6006
120	6.8510	4.7865	3.9493	3.4796	3.1735	2.9559	2.7918	2.6629	2.5586	2.4721	2.3363	2.1915	2.0346	1.9500	1.8600	1.7628	1.6557	1.5330	1.3805
∞	6.6349	4.6052	3.7816	3.3192	3.0173	2.8020	2.6393	2.5113	2.4073	2.3209	2.1848	2.0385	1.8783	1.7908	1.6964	1.5923	1.4730	1.3246	1.0000

F- Table for 95%

f_1: degrees of freedom in the numerator
f_2: degrees of freedom in the denominator

f_2 \ f_1	1	2	3	4	5	6	7	8	9	10	12	15	20	24	30	40	60	120	∞
1	161.45	199.50	215.71	224.58	230.16	233.99	236.77	238.88	240.54	241.88	243.91	245.95	248.01	249.05	250.09	251.14	252.20	253.25	254.32
2	18.513	19.000	19.164	19.247	19.296	19.330	19.353	19.371	19.385	19.396	19.413	19.429	19.446	19.454	19.462	19.471	19.479	19.487	19.496
3	10.128	9.5521	9.2766	9.1172	9.0135	8.9406	8.8868	8.8452	8.8123	8.7855	8.7446	8.7029	8.6602	8.6385	8.6166	8.5944	8.5720	8.5494	8.5265
4	7.7086	6.9443	6.5914	6.3883	6.2560	6.1631	6.0942	6.0410	5.9988	5.9644	5.9117	5.8578	5.8025	5.7744	5.7459	5.7170	5.6878	5.6581	5.6281
5	6.6079	5.7861	5.4095	5.1922	5.0503	4.9503	4.8759	4.8183	4.7725	4.7351	4.6777	4.6188	4.5581	4.5272	4.4957	4.4638	4.4314	4.3984	4.3650
6	5.9874	5.1433	4.7571	4.5337	4.3874	4.2839	4.2066	4.1468	4.0990	4.0600	3.9999	3.9381	3.8742	3.8415	3.8082	3.7743	3.7398	3.7047	3.6688
7	5.5914	4.7374	4.3468	4.1203	3.9715	3.8660	3.7870	3.7257	3.6767	3.6365	3.5747	3.5108	3.4445	3.4105	3.3758	3.3404	3.3043	3.2674	3.2298
8	5.3177	4.4590	4.0662	3.8378	3.6875	3.5806	3.5005	3.4381	3.3881	3.3472	3.2840	3.2184	3.1503	3.1152	3.0794	3.0428	3.0053	2.9669	2.9276
9	5.1174	4.2565	3.8626	3.6331	3.4817	3.3738	3.2927	3.2296	3.1789	3.1373	3.0729	3.0061	2.9365	2.9005	2.8637	2.8259	2.7872	2.7475	2.7067
10	4.9646	4.1028	3.7083	3.4780	3.3258	3.2172	3.1355	3.0717	3.0204	2.9782	2.9130	2.8450	2.7740	2.7372	2.6996	2.6609	2.6211	2.5801	2.5379
11	4.8443	3.9823	3.5874	3.3567	3.2039	3.0946	3.0123	2.9480	2.8962	2.8536	2.7876	2.7186	2.6464	2.6090	2.5705	2.5309	2.4901	2.4480	2.4045
12	4.7472	3.8853	3.4903	3.2592	3.1059	2.9961	2.9134	2.8486	2.7964	2.7534	2.6866	2.6169	2.5436	2.5055	2.4663	2.4259	2.3842	2.3410	2.2962
13	4.6672	3.8056	3.4105	3.1791	3.0254	2.9153	2.8321	2.7669	2.7144	2.6710	2.6037	2.5331	2.4589	2.4202	2.3803	2.3392	2.2966	2.2524	2.2064
14	4.6001	3.7389	3.3439	3.1122	2.9582	2.8477	2.7642	2.6987	2.6458	2.6021	2.5342	2.4630	2.3879	2.3487	2.3082	2.2664	2.2230	2.1778	2.1307

F- Table for 95%

f_1: degrees of freedom in the numerator
f_2: degrees of freedom in the denominator

f_2 \ f_1	1	2	3	4	5	6	7	8	9	10	12	15	20	24	30	40	60	120	∞
15	4.5431	3.6823	3.2874	3.0556	2.9013	2.7905	2.7066	2.6408	2.5876	2.5437	2.4753	2.4035	2.3275	2.2878	2.2468	2.2043	2.1601	2.1141	2.0658
16	4.4940	3.6337	3.2389	3.0069	2.8524	2.7413	2.6572	2.5911	2.5377	2.4935	2.4247	2.3522	2.2756	2.2354	2.1938	2.1507	2.1058	2.0589	2.0096
17	4.4513	3.5915	3.1968	2.9647	2.8100	2.6987	2.6143	2.5480	2.4943	2.4499	2.3807	2.3077	2.2304	2.1898	2.1477	2.1040	2.0584	2.0107	1.9604
18	4.4139	3.5546	3.1599	2.9277	2.7729	2.6613	2.5767	2.5102	2.4563	2.4117	2.3421	2.2686	2.1906	2.1497	2.1071	2.0629	2.0166	1.9681	1.9168
19	4.3808	3.5219	3.1274	2.8951	2.7401	2.6283	2.5435	2.4768	2.4227	2.3779	2.3080	2.2341	2.1555	2.1141	2.0712	2.0264	1.9796	1.9302	1.8780
20	4.3513	3.4928	3.0984	2.8661	2.7109	2.5990	2.5140	2.4471	2.3928	2.3479	2.2776	2.2033	2.1242	2.0825	2.0391	1.9938	1.9464	1.8963	1.8432
21	4.3248	3.4668	3.0725	2.8401	2.6848	2.5727	2.4876	2.4205	2.3661	2.3210	2.2504	2.1757	2.0960	2.0540	2.0102	1.9645	1.9165	1.8657	1.8117
22	4.3009	3.4434	3.0491	2.8167	2.6613	2.5491	2.4638	2.3965	2.3419	2.2967	2.2258	2.1508	2.0707	2.0283	1.9842	1.9380	1.8895	1.8380	1.7831
23	4.2793	3.4221	3.0280	2.7955	2.6400	2.5277	2.4422	2.3748	2.3201	2.2747	2.2036	2.1282	2.0476	2.0050	1.9605	1.9139	1.8649	1.8128	1.7570
24	4.2597	3.4028	3.0088	2.7763	2.6207	2.5082	2.4226	2.3551	2.3002	2.2547	2.1834	2.1077	2.0267	1.9838	1.9390	1.8920	1.8424	1.7897	1.7331
25	4.2417	3.3852	2.9912	2.7587	2.6030	2.4904	2.4047	2.3371	2.2821	2.2365	2.1649	2.0889	2.0075	1.9643	1.9192	1.8718	1.8217	1.7684	1.7110
26	4.2252	3.3690	2.9751	2.7426	2.5868	2.4741	2.3883	2.3205	2.2655	2.2197	2.1479	2.0716	1.9898	1.9464	1.9010	1.8533	1.8027	1.7488	1.6906
27	4.2100	3.3541	2.9604	2.7278	2.5719	2.4591	2.3732	2.3053	2.2501	2.2043	2.1323	2.0558	1.9736	1.9299	1.8842	1.8361	1.7851	1.7307	1.6717
28	4.1960	3.3404	2.9467	2.7141	2.5581	2.4453	2.3593	2.2913	2.2360	2.1900	2.1179	2.0411	1.9586	1.9147	1.8687	1.8203	1.7689	1.7138	1.6541
29	4.1830	3.3277	2.9340	2.7014	2.5454	2.4324	2.3463	2.2782	2.2229	2.1768	2.1045	2.0275	1.9446	1.9005	1.8543	1.8055	1.7537	1.6981	1.6377
30	4.1709	3.3158	2.9223	2.6896	2.5336	2.4205	2.3343	2.2662	2.2107	2.1646	2.0921	2.0148	1.9317	1.8874	1.8409	1.7918	1.7396	1.6835	1.6223
40	4.0848	3.2317	2.8387	2.6060	2.4495	2.3359	2.2490	2.1802	2.1240	2.0772	2.0035	1.9245	1.8389	1.7929	1.7444	1.6928	1.6373	1.5766	1.5089
60	4.0012	3.1504	2.7581	2.5252	2.3683	2.2540	2.1665	2.0970	2.0401	1.9926	1.9174	1.8364	1.7480	1.7001	1.6491	1.5943	1.5343	1.4673	1.3893
120	3.9201	3.0718	2.6802	2.4472	2.2900	2.1750	2.0867	2.0164	1.9588	1.9105	1.8337	1.7505	1.6587	1.6084	1.5543	1.4952	1.4290	1.3519	1.2539
∞	3.8415	2.9957	2.6049	2.3719	2.2141	2.0986	2.0096	1.9384	1.8799	1.8307	1.7522	1.6664	1.5705	1.5173	1.4591	1.3940	1.3180	1.2214	1.0000

F- Table for 90%

f_1: degrees of freedom in the numerator
f_2: degrees of freedom in the denominator

f_2 \ f_1	1	2	3	4	5	6	7	8	9	10	12	15	20	24	30	40	60	120	∞
1	39.864	49.500	53.593	55.833	57.241	58.204	58.906	59.439	59.858	60.195	60.705	61.220	61.740	62.002	62.265	62.529	62.794	63.061	63.328
2	8.5263	9.0000	9.1618	9.2434	9.2926	9.3255	9.3491	9.3668	9.3805	9.3916	9.4081	9.4247	9.4413	9.4496	9.4579	9.4663	9.4746	9.4829	9.4913
3	5.5383	5.4624	5.3908	5.3427	5.3092	5.2847	5.2662	5.2517	5.2400	5.2304	5.2156	5.2003	5.1845	5.1764	5.1681	5.1597	5.1512	5.1425	5.1337
4	4.5448	4.3246	4.1906	4.1073	4.0506	4.0098	3.9790	3.9549	3.9357	3.9199	3.8955	3.8689	3.8443	3.8310	3.8174	3.8036	3.7896	3.7753	3.7607
5	4.0604	3.7797	3.6195	3.5202	3.4530	3.4045	3.3679	3.3393	3.3163	3.2974	3.2682	3.2380	3.2067	3.1905	3.1741	3.1573	3.1402	3.1228	3.1050
6	3.7760	3.4633	3.2888	3.1808	3.1075	3.0546	3.0145	2.9830	2.9577	2.9369	2.9047	2.8712	2.8363	2.8183	2.8000	2.7812	2.7620	2.7423	2.7222
7	3.5894	3.2574	3.0741	2.9605	2.8833	2.8274	2.7849	2.7516	2.7247	2.7025	2.6681	2.6322	2.5947	2.5753	2.5555	2.5351	2.5142	2.4928	2.4708
8	3.4579	3.1131	2.9238	2.8064	2.7265	2.6683	2.6241	2.5893	2.5612	2.5380	2.5020	2.4642	2.4246	2.4041	2.3830	2.3614	2.3391	2.3162	2.2926
9	3.3603	3.0065	2.8129	2.6927	2.6106	2.5509	2.5053	2.4694	2.4403	2.4163	2.3789	2.3396	2.2983	2.2768	2.2547	2.2320	2.2085	2.1843	2.1592
10	3.2850	2.9245	2.7277	2.6053	2.5216	2.4606	2.4140	2.3772	2.3473	2.3226	2.2841	2.2435	2.2007	2.1784	2.1554	2.1317	2.1072	2.0818	2.0554
11	3.2252	2.8595	2.6602	2.5362	2.4512	2.3891	2.3416	2.3040	2.2735	2.2482	2.2087	2.1671	2.1230	2.1000	2.0762	2.0616	2.0261	1.9997	1.9721
12	3.1765	2.8068	2.6055	2.4801	2.3940	2.3310	2.2828	2.2446	2.2135	2.1878	2.1474	2.1049	2.0597	2.0360	2.0115	1.9861	1.9597	1.9323	1.9036
13	3.1362	2.7632	2.5603	2.4337	2.3467	2.2830	2.2341	2.1953	2.1638	2.1376	2.0966	2.0532	2.0070	1.9827	1.9576	1.9315	1.9043	1.8759	1.8462
14	3.1022	2.7265	2.5222	2.3947	2.3069	2.2426	2.1931	2.1539	2.1220	2.0954	2.0537	2.0095	1.9625	1.9377	1.9119	1.8852	1.8572	1.8280	1.7973

F- Table for 90%

f_1: degrees of freedom in the numerator
f_2: degrees of freedom in the denominator

f_2 \ f_1	1	2	3	4	5	6	7	8	9	10	12	15	20	24	30	40	60	120	∞
15	3.0732	2.6952	2.4898	2.3614	2.2730	2.2081	2.1582	2.1185	2.0862	2.0593	2.0171	1.9722	1.9243	1.8990	1.8728	1.8454	1.8168	1.7867	1.7551
16	3.0481	2.6682	2.4618	2.3327	2.2438	2.1783	2.1280	2.0880	2.0553	2.0281	1.9854	1.9399	1.8913	1.8656	1.8388	1.8108	1.7816	1.7507	1.7182
17	3.0262	2.6446	2.4374	2.3077	2.2183	2.1524	2.1017	2.0613	2.0284	2.0009	1.9577	1.9117	1.8624	1.8362	1.8090	1.7805	1.7506	1.7191	1.6856
18	3.0070	2.6239	2.4160	2.2858	2.1958	2.1296	2.0785	2.0379	2.0047	1.9770	1.9333	1.8868	1.8368	1.8103	1.7827	1.7537	1.7232	1.6910	1.6567
19	2.9899	2.6056	2.3970	2.2663	2.1760	2.1094	2.0580	2.0171	1.9836	1.9557	1.9117	1.8647	1.8142	1.7873	1.7592	1.7298	1.6988	1.6659	1.6308
20	2.9747	2.5893	2.3801	2.2489	2.1582	2.0913	2.0397	1.9985	1.9649	1.9367	1.8924	1.8449	1.7938	1.7667	1.7382	1.7083	1.6768	1.6433	1.6074
21	2.9609	2.5746	2.3649	2.2333	2.1423	2.0751	2.0232	1.9819	1.9480	1.9197	1.8750	1.8272	1.7756	1.7481	1.7193	1.6890	1.6569	1.6228	1.5862
22	2.9486	2.5613	2.3512	2.2193	2.1279	2.0605	2.0084	1.9668	1.9327	1.9043	1.8593	1.8111	1.7590	1.7312	1.7021	1.6714	1.6389	1.6042	1.5668
23	2.9374	2.5493	2.3387	2.2065	2.1149	2.0472	1.9949	1.9531	1.9189	1.8903	1.8450	1.7964	1.7439	1.7159	1.6864	1.6554	1.6224	1.5871	1.5490
24	2.9271	2.5383	2.3274	2.1949	2.1030	2.0351	1.9826	1.9407	1.9063	1.8775	1.8319	1.7831	1.7302	1.7019	1.6721	1.6407	1.6073	1.5715	1.5327
25	2.9177	2.5283	2.3170	2.1843	2.0922	2.0241	1.9714	1.9292	1.8947	1.8658	1.8200	1.7708	1.7175	1.6890	1.6589	1.6272	1.5934	1.5570	1.5176
26	2.9091	2.5191	2.3075	2.1745	2.0822	2.0139	1.9610	1.9188	1.8841	1.8550	1.8090	1.7596	1.7059	1.6771	1.6468	1.6147	1.5805	1.5437	1.5036
27	2.9012	2.5106	2.2987	2.1655	2.0730	2.0045	1.9515	1.9091	1.8743	1.8451	1.7989	1.7492	1.6951	1.6662	1.6356	1.6032	1.5686	1.5313	1.4906
28	2.8839	2.5028	2.2906	2.1571	2.0645	1.9959	1.9427	1.9001	1.8652	1.8359	1.7895	1.7395	1.6852	1.6560	1.6252	1.5925	1.5575	1.5198	1.4784
29	2.8871	2.4955	2.2831	2.1494	2.0566	1.9878	1.9345	1.8918	1.8560	1.8274	1.7808	1.7306	1.6759	1.6465	1.6155	1.5825	1.5472	1.5090	1.4670
30	2.8807	2.4887	2.2761	2.1422	2.0492	1.9803	1.9269	1.8841	1.8498	1.8195	1.7727	1.7223	1.6673	1.6377	1.6065	1.5732	1.5376	1.4989	1.4564
40	2.8354	2.4404	2.2261	2.0909	1.9968	1.9269	1.8725	1.8289	1.7929	1.7627	1.7146	1.6624	1.6052	1.5741	1.5411	1.5056	1.4672	1.4248	1.3769
60	2.7914	2.3932	2.1774	2.0410	1.9457	1.8747	1.8194	1.7748	1.7380	1.7070	1.6574	1.6034	1.5435	1.5107	1.4755	1.4373	1.3952	1.3476	1.2915
120	2.7478	2.3473	2.1300	1.9923	1.8959	1.8238	1.7675	1.7220	1.6843	1.6524	1.6012	1.5450	1.4821	1.4472	1.4094	1.3676	1.3203	1.2646	1.1926
∞	2.7055	2.3026	2.0838	1.9449	1.8473	1.7741	1.7167	1.6702	1.6315	1.5987	1.5458	1.4871	1.4206	1.3832	1.3419	1.2951	1.2400	1.1686	1.0000

Appendix C

Median, 5%, and 95% Rank Tables

Median ranks

Sample size, j

i	1	2	3	4	5	6	7	8	9	10
1	50.000	29.289	20.630	15.910	12.945	10.910	9.428	8.300	7.412	6.697
2		70.711	50.000	38.573	31.381	26.445	22.849	20.113	17.962	16.226
3			79.370	61.427	50.000	42.141	36.412	32.052	28.624	25.857
4				84.090	68.619	57.859	50.000	44.015	39.308	35.510
5					87.055	73.555	63.588	55.984	50.000	45.169
6						89.090	77.151	67.948	60.691	54.831
7							90.572	79.887	71.376	64.490
8								91.700	82.038	74.142
9									92.587	83.774
10										93.303

Sample size, j

i	11	12	13	14	15	16	17	18	19	20
1	6.107	5.613	5.192	4.830	4.516	4.240	3.995	3.778	3.582	3.406
2	14.796	13.598	12.579	11.702	10.940	10.270	9.678	9.151	8.677	8.251
3	23.578	21.669	20.045	18.647	17.432	16.365	15.422	14.581	13.827	13.147
4	32.380	29.758	27.528	25.608	23.939	22.474	21.178	20.024	18.988	18.055
5	41.189	37.853	35.016	32.575	30.452	28.589	26.940	25.471	24.154	22.967
6	50.000	45.951	42.508	39.544	36.967	34.705	32.704	30.921	29.322	27.880
7	58.811	54.049	50.000	46.515	43.483	40.823	38.469	36.371	34.491	32.795
8	67.620	62.147	57.492	53.485	50.000	46.941	44.234	41.823	39.660	37.710
9	76.421	70.242	64.984	60.456	56.517	53.059	50.000	47.274	44.830	42.626
10	85.204	78.331	72.472	67.425	63.033	59.177	55.766	52.726	50.000	47.542
11	93.893	86.402	79.955	74.392	69.548	65.295	61.531	58.177	55.170	52.458
12		94.387	87.421	81.353	76.061	71.411	67.296	63.629	60.340	57.374
13			94.808	88.298	82.568	77.525	73.060	69.079	65.509	62.289
14				95.169	89.060	83.635	78.821	74.529	70.678	67.205
15					95.484	89.730	84.578	79.976	75.846	72.119
16						95.760	90.322	85.419	81.011	77.033
17							96.005	90.849	86.173	81.945
18								96.222	91.322	86.853
19									96.418	91.749
20										96.594

Median ranks

Sample size, j

i	21	22	23	24	25	26	27	28	29	30
1	3.247	3.101	2.969	2.847	2.734	2.631	2.534	2.445	2.362	2.284
2	7.864	7.512	7.191	6.895	6.623	6.372	6.139	5.922	5.720	5.532
3	12.531	11.970	11.458	10.987	10.553	10.153	9.781	9.436	9.114	8.814
4	17.209	16.439	15.734	15.088	14.492	13.942	13.432	12.958	12.517	12.104
5	21.890	20.911	20.015	19.192	18.435	17.735	17.086	16.483	15.922	15.397
6	26.574	25.384	24.297	23.299	22.379	21.529	20.742	20.010	19.328	18.691
7	31.258	29.859	28.580	27.406	26.324	25.325	24.398	23.537	22.735	21.986
8	35.943	34.334	32.863	31.513	30.269	29.120	28.055	27.065	26.143	25.281
9	40.629	38.810	37.147	35.621	34.215	32.916	31.712	30.593	29.550	28.576
10	45.314	43.286	41.431	39.729	38.161	36.712	35.370	34.121	32.958	31.872
11	50.000	47.762	45.716	43.837	42.107	40.509	39.027	37.650	36.367	35.168
12	54.686	52.238	50.000	47.946	46.054	44.305	42.685	41.178	39.775	38.464
13	59.371	56.714	54.284	52.054	50.000	48.102	46.342	44.707	43.183	41.760
14	64.057	61.190	58.568	56.162	53.946	51.898	50.000	48.236	46.592	45.056
15	68.742	65.665	62.853	60.271	57.892	55.695	53.658	51.764	50.000	48.352
16	73.426	70.141	67.137	64.379	61.839	59.491	57.315	55.293	53.408	51.648
17	78.109	74.616	71.420	68.487	65.785	63.287	60.973	58.821	56.817	54.944
18	82.791	79.089	75.703	72.594	69.730	67.084	64.630	62.350	60.225	58.240
19	87.469	83.561	79.985	76.701	73.676	70.880	68.288	65.878	63.633	61.536
20	92.136	88.030	84.266	80.808	77.621	74.675	71.945	69.407	67.041	64.852
21	96.753	92.488	88.542	84.912	81.565	78.471	75.602	72.935	70.450	68.128
22		96.898	92.809	89.013	85.507	82.265	79.258	76.463	73.857	71.424
23			97.031	93.105	89.447	86.058	82.914	79.990	77.265	74.719
24				97.153	93.377	89.847	86.568	83.517	80.672	78.014
25					97.265	93.628	90.219	87.042	84.078	81.309
26						97.369	93.861	90.564	87.483	84.603
27							97.465	94.078	90.865	87.896
28								97.555	94.280	91.186
29									97.638	94.468
30										97.716

5% Ranks

Sample size, j

i	1	2	3	4	5	6	7	8	9	10
1	5.000	2.532	1.695	1.274	1.021	0.851	0.730	0.639	0.568	0.512
2		22.361	13.535	9.761	7.644	6.285	5.337	4.639	4.102	3.677
3			36.840	24.860	18.925	15.316	12.876	11.111	9.775	8.726
4				47.237	34.259	27.134	22.532	19.290	16.875	15.003
5					54.928	41.820	34.126	28.924	25.137	22.244
6						60.696	47.930	40.031	34.494	30.354
7							65.184	52.932	45.036	39.338
8								68.766	57.086	49.310
9									71.687	60.584
10										74.113

Sample size, j

i	11	12	13	14	15	16	17	18	19	20
1	0.465	0.426	0.394	0.366	0.341	0.320	0.301	0.285	0.270	0.256
2	3.332	3.046	2.805	2.600	2.423	2.268	2.132	2.011	1.903	1.806
3	7.882	7.187	6.605	6.110	5.685	5.315	4.990	4.702	4.446	4.217
4	13.507	12.285	11.267	10.405	9.666	9.025	8.464	7.969	7.529	7.135
5	19.958	18.102	16.566	15.272	14.166	13.211	12.377	11.643	10.991	10.408
6	27.125	24.530	22.395	20.607	19.086	17.777	16.636	15.634	14.747	13.955
7	34.981	31.524	28.705	26.358	24.373	22.669	21.191	19.895	18.750	17.731
8	43.563	39.086	35.480	32.503	29.999	27.860	26.011	24.396	22.972	21.707
9	52.991	47.267	42.738	39.041	35.956	33.337	31.083	29.120	27.395	25.865
10	63.564	56.189	50.535	45.999	42.256	39.101	36.401	34.060	32.009	30.195
11	76.160	66.132	58.990	53.434	48.925	45.165	41.970	39.215	36.811	34.693
12		77.908	68.366	61.461	56.022	51.560	47.808	44.595	41.806	39.358
13			79.418	70.327	63.656	58.343	53.945	50.217	47.003	44.197
14				80.736	72.060	65.617	60.436	56.112	52.420	49.218
15					81.896	73.604	67.381	62.332	58.088	54.442
16						82.925	74.988	68.974	64.057	59.897
17							83.843	76.234	70.420	65.634
18								84.668	77.363	71.738
19									85.413	78.389
20										86.089

5% ranks

Sample size, j

i	21	22	23	24	25	26	27	28	29	30
1	0.244	0.233	0.223	0.213	0.205	0.197	0.190	0.183	0.177	0.171
2	1.719	1.640	1.567	1.501	1.440	1.384	1.332	1.284	1.239	1.198
3	4.010	3.822	3.651	3.495	3.352	3.220	3.098	2.985	2.879	2.781
4	6.781	6.460	6.167	5.901	5.656	5.431	5.223	5.031	4.852	4.685
5	9.884	9.411	8.981	8.588	8.229	7.899	7.594	7.311	7.049	6.806
6	13.425	12.603	12.021	11.491	11.006	10.560	10.148	9.768	9.415	9.087
7	16.818	15.994	15.248	14.569	13.947	13.377	12.852	12.367	11.917	11.499
8	20.575	19.556	18.634	17.796	17.030	16.328	15.682	15.085	14.532	14.018
9	24.499	23.272	22.164	21.157	20.238	19.396	18.622	17.908	17.246	16.633
10	28.580	27.131	25.824	24.639	23.559	22.570	21.662	20.824	20.050	19.331
11	32.811	31.126	29.609	28.236	26.985	25.842	24.793	23.827	22.934	22.106
12	37.190	35.254	33.515	31.942	30.513	29.508	28.012	26.911	25.894	24.953
13	41.720	39.516	37.539	35.756	34.139	32.664	31.314	30.072	28.927	27.867
14	46.406	43.913	41.684	39.678	37.862	36.209	34.697	33.309	32.030	30.846
15	51.261	48.454	45.954	43.711	41.684	39.842	38.161	36.620	35.200	33.889
16	56.302	53.151	50.356	47.858	45.607	43.566	41.707	40.004	38.439	36.995
17	61.559	58.020	54.902	52.127	49.636	47.384	45.336	43.464	41.746	40.163
18	67.079	63.091	59.610	56.531	53.779	51.300	49.052	47.002	45.123	43.394
19	72.945	68.409	64.507	61.086	58.048	55.323	52.861	50.621	48.573	46.691
20	79.327	74.053	69.636	65.819	62.459	59.465	56.770	54.327	52.099	50.056
21	86.705	80.188	75.075	70.773	67.039	63.740	60.790	58.127	55.706	53.493
22		87.269	80.980	76.020	71.828	68.176	64.936	62.033	59.403	57.007
23			87.788	81.711	76.896	72.810	69.237	66.060	63.200	60.605
24				88.265	82.388	77.711	73.726	70.231	67.113	64.299
25					88.707	83.017	78.470	74.583	71.168	68.103
26						89.117	83.603	79.179	75.386	72.038
27							89.498	84.149	79.844	76.140
28								89.853	84.661	80.467
29									90.185	85.140
30										90.497

95% Ranks

Sample size, j

i	1	2	3	4	5	6	7	8	9	10
1	95.000	77.639	63.160	52.713	45.072	39.304	34.816	31.234	28.313	25.887
2		97.468	86.465	75.139	65.741	58.180	52.070	47.068	42.914	39.416
3			98.305	90.239	81.075	72.866	65.874	59.969	54.964	50.690
4				98.726	92.356	84.684	77.468	71.076	65.506	60.662
5					98.979	93.715	87.124	80.710	74.863	69.646
6						99.149	94.662	88.889	83.125	77.756
7							99.270	95.361	90.225	84.997
8								99.361	95.898	91.274
9									99.432	96.323
10										99.488

Sample size, j

i	11	12	13	14	15	16	17	18	19	20
1	23.840	22.092	20.582	19.264	18.104	17.075	16.157	15.332	14.587	13.911
2	36.436	33.868	31.634	29.673	27.940	26.396	25.012	23.766	22.637	21.611
3	47.009	43.811	41.010	38.539	36.344	34.383	32.619	31.026	29.580	28.262
4	56.437	52.733	49.465	46.566	43.978	41.657	39.564	37.668	35.943	34.366
5	65.019	60.914	57.262	54.000	51.075	48.440	46.055	43.888	41.912	40.103
6	72.875	68.476	64.520	60.928	57.744	54.835	52.192	49.783	47.580	45.558
7	80.042	75.470	71.295	67.497	64.043	60.899	58.029	55.404	52.997	50.782
8	86.492	81.898	77.604	73.641	70.001	66.663	63.599	60.784	58.194	55.803
9	92.118	87.715	83.434	79.393	75.627	72.140	68.917	65.940	63.188	60.641
10	96.668	92.813	88.733	84.728	80.913	77.331	73.989	70.880	67.991	65.307
11	99.535	96.954	93.395	89.595	85.834	82.223	78.809	75.604	72.605	69.805
12		99.573	97.195	93.890	90.334	86.789	83.364	80.105	77.028	74.135
13			99.606	97.400	94.315	90.975	87.623	84.366	81.250	78.293
14				99.634	97.577	94.685	91.535	88.357	85.253	82.269
15					99.659	97.732	95.010	92.030	89.009	86.045
16						99.680	97.868	95.297	92.471	89.592
17							99.699	97.989	95.553	92.865
18								99.715	98.097	95.783
19									99.730	98.193
20										99.744

95% ranks

Sample size, j

i	21	22	23	24	25	26	27	28	29	30
1	13.295	12.731	12.212	11.735	11.293	10.883	10.502	10.147	9.814	9.503
2	20.673	19.812	19.020	18.289	17.612	16.983	16.397	15.851	15.339	14.860
3	27.055	25.947	24.925	23.980	23.104	22.289	21.530	20.821	20.156	19.533
4	32.921	31.591	30.364	29.227	28.172	27.190	26.274	25.417	24.614	23.860
5	38.441	36.909	35.193	34.181	32.961	31.824	30.763	29.769	28.837	27.962
6	43.698	41.980	40.390	38.914	37.541	36.260	35.062	33.940	32.887	31.897
7	48.739	46.849	45.097	43.469	41.952	40.535	39.210	37.967	36.800	35.701
8	53.954	51.546	49.643	47.873	46.221	44.677	43.230	41.873	40.597	39.395
9	58.280	56.087	54.046	52.142	50.364	48.700	47.139	45.673	44.294	42.993
10	62.810	60.484	58.315	56.289	54.393	52.616	50.948	49.379	47.901	46.507
11	67.189	64.746	62.461	60.321	58.316	56.434	54.664	52.998	51.427	49.944
12	71.420	68.874	66.485	64.244	62.138	60.158	58.293	56.536	54.877	53.309
13	75.501	72.869	70.391	68.058	65.861	63.791	61.839	59.996	58.254	56.605
14	79.425	76.728	74.176	71.764	69.487	67.336	65.303	63.380	61.561	59.837
15	83.182	80.444	77.836	75.361	73.015	70.792	68.686	66.691	64.799	63.005
16	86.755	84.006	81.366	78.843	76.441	74.158	71.988	69.927	67.970	66.111
17	90.116	87.397	84.752	82.204	79.762	77.430	75.207	73.089	71.073	69.154
18	93.219	90.589	87.978	85.431	82.970	80.604	78.338	76.173	74.106	72.133
19	95.990	93.540	91.019	88.509	86.052	83.672	81.378	79.176	77.066	75.047
20	98.281	96.178	93.832	91.411	88.994	86.623	84.318	82.092	79.950	77.894
21	99.756	98.360	96.348	94.099	91.771	89.440	87.148	84.915	82.753	80.669
22		99.767	98.433	96.505	94.344	92.101	89.851	87.633	85.468	83.367
23			99.777	98.499	96.648	94.569	92.406	90.232	88.083	85.981
24				99.786	98.560	96.780	94.777	92.689	90.584	88.501
25					99.795	98.616	96.902	94.969	92.951	90.913
26						99.803	98.668	97.015	95.148	93.194
27							99.810	98.716	97.120	95.314
28								99.817	98.761	97.218
29									99.823	98.802
30										99.829

Appendix D

Critical Values of the Chi-Squared Distribution

df = degrees of freedom

$\gamma = \Pr\{X^2_{\gamma;df} \geq \text{table value}\}$

df	.995	.99	.98	.975	.95	.90	.80	.75	.70	.50
1	.04393	.03157	.03628	.03982	.00393	.0158	.0642	.102	.148	.455
2	.0100	.0201	.0404	.0506	.103	.211	.446	.575	.713	1.386
3	.0717	.115	.185	.216	.352	.584	1.005	1.213	1.424	2.366
4	.207	.297	.429	.484	.711	1.064	1.649	1.923	2.195	3.357
5	.412	.554	.752	.831	1.145	1.610	2.343	2.675	3.000	4.351
6	.676	.872	1.134	1.237	1.635	2.204	3.070	3.455	3.828	5.348
7	.989	1.239	1.564	1.690	2.167	2.833	3.822	4.255	4.671	6.346
8	1.344	1.646	2.032	2.180	2.733	3.490	4.594	5.071	5.527	7.344
9	1.735	2.088	2.532	2.700	3.325	4.168	5.380	5.899	6.393	8.343
10	2.156	2.558	3.059	3.247	3.940	4.865	6.179	6.737	7.267	9.342
11	2.603	3.053	3.609	3.816	4.575	5.578	6.989	7.584	8.148	10.341
12	3.074	3.571	4.178	4.404	5.226	6.304	7.807	8.438	9.034	11.340
13	3.565	4.107	4.765	5.009	5.892	7.042	8.634	9.299	9.926	12.340
14	4.075	4.660	5.368	5.629	6.571	7.790	9.467	10.165	10.821	13.339
15	4.601	5.229	5.985	6.262	7.261	8.547	10.307	11.036	11.721	14.339
16	5.142	5.812	6.614	6.908	7.962	9.312	11.152	11.912	12.624	15.338
17	5.697	6.408	7.255	7.564	8.672	10.085	12.002	12.792	13.531	16.338
18	6.265	7.015	7.906	8.231	9.390	10.865	12.857	13.675	14.440	17.338
19	6.844	7.633	8.567	8.907	10.117	11.651	13.716	14.562	15.352	18.338
20	7.434	8.260	9.237	9.591	10.851	12.443	14.578	15.452	16.266	19.337
21	8.034	8.897	9.915	10.283	11.591	13.240	15.445	16.344	17.182	20.337
22	8.643	9.542	10.600	10.982	12.338	14.041	16.314	17.240	18.101	21.337
23	9.260	10.196	11.293	11.688	13.091	14.848	17.187	18.137	19.021	22.337
24	9.886	10.856	11.992	12.401	13.848	15.659	18.062	19.037	19.943	23.337
25	10.520	11.524	12.697	13.120	14.611	16.473	18.940	19.939	20.867	24.337
26	11.160	12.198	13.409	13.844	15.379	17.292	19.820	20.843	21.792	25.336
27	11.808	12.879	14.125	14.573	16.151	18.114	20.703	21.749	22.719	26.336
28	12.461	13.565	14.847	15.308	16.928	18.939	21.588	22.657	23.647	27.336
29	13.121	14.256	15.574	16.047	17.708	19.768	22.475	23.567	24.577	28.336
30	13.787	14.953	16.306	16.791	18.493	20.599	23.364	24.478	25.508	29.336

df = degrees of freedom

$$\gamma = \Pr\{X^2_{\gamma,df} \geq \text{table value}\}$$

γ / df	.30	.25	.20	.10	.05	.025	.02	.01	.005	.001
1	1.074	1.323	1.642	2.706	3.841	5.024	5.412	6.635	7.879	10.827
2	2.408	2.773	3.219	4.605	5.991	7.378	7.824	9.210	10.597	13.815
3	3.665	4.108	4.642	6.251	7.815	9.348	9.837	11.345	12.838	16.268
4	4.878	5.385	5.989	7.779	9.488	11.143	11.668	13.277	14.860	18.465
5	6.064	6.626	7.289	9.236	11.070	12.832	13.388	15.086	16.750	20.517
6	7.231	7.841	8.558	10.645	12.592	14.449	15.033	16.812	18.548	22.457
7	8.383	9.037	9.803	12.017	14.067	16.013	16.622	18.475	20.278	24.322
8	9.524	10.219	11.030	13.362	15.507	17.535	18.168	20.090	21.955	26.125
9	10.656	11.389	12.242	14.684	16.910	19.023	19.679	21.666	23.589	27.877
10	11.781	12.549	13.442	15.987	18.307	20.483	21.161	23.209	25.188	29.588
11	12.899	13.701	14.631	17.275	19.675	21.920	22.618	24.725	26.757	31.264
12	14.011	14.845	15.812	18.549	21.026	23.337	24.054	26.217	28.300	32.909
13	15.119	15.984	16.985	19.812	22.362	24.736	25.472	27.688	29.819	34.528
14	16.222	17.117	18.151	21.064	23.685	26.119	26.873	29.141	31.319	36.123
15	17.322	18.245	19.311	22.307	24.996	27.488	28.259	30.578	32.801	37.697
16	18.418	19.369	20.465	23.542	26.296	28.845	29.633	32.000	34.267	39.252
17	19.511	20.489	21.615	24.769	27.587	30.191	30.995	33.409	35.718	40.790
18	20.601	21.605	22.760	25.989	28.869	31.526	32.346	34.805	37.156	42.312
19	21.689	22.718	23.900	27.204	30.144	32.852	33.687	36.191	38.582	43.820
20	22.775	23.828	25.038	28.412	31.410	34.170	35.020	37.566	39.997	45.315
21	23.858	24.935	26.171	29.615	32.671	35.479	36.343	38.932	41.401	46.797
22	24.939	26.039	27.301	30.813	33.924	36.781	37.659	40.289	42.796	48.268
23	26.018	27.141	28.429	32.007	35.172	38.076	38.968	41.638	44.181	49.728
24	27.096	28.241	29.553	33.196	36.415	39.364	40.270	42.980	45.558	51.179
25	28.172	29.339	30.675	34.382	37.652	40.646	41.566	44.314	46.928	52.620
26	29.246	30.434	31.795	35.563	38.885	41.923	42.856	45.642	48.290	54.052
27	30.319	31.528	32.912	36.741	40.113	43.194	44.140	46.963	49.645	55.476
28	31.391	32.620	34.027	37.916	41.337	44.461	45.419	48.278	50.993	56.893
29	32.461	33.711	35.139	39.087	42.557	45.722	46.693	49.588	52.336	58.302
30	33.530	34.800	36.250	40.256	43.773	46.979	47.962	50.892	53.672	59.703

Index

393

Milton Keynes UK
Ingram Content Group UK Ltd.
UKHW021827071024
449327UK00021B/1453

9 780367 397395